Logic in Asia: Studia Logica Library

Logic in Asia: Studia Logica Library

This book series promotes the advance of scientific research within the field of logic in Asian countries. It strengthens the collaboration between researchers based in Asia with researchers across the international scientific community and offers a platform for presenting the results of their collaborations. One of the most prominent features of contemporary logic is its interdisciplinary character, combining mathematics, philosophy, modern computer science, and even the cognitive and social sciences. The aim of this book series is to provide a forum for current logic research, reflecting this trend in the field's development.

The series accepts books on any topic concerning logic in the broadest sense, i.e., books on contemporary formal logic, its applications and its relations to other disciplines. It accepts monographs and thematically coherent volumes addressing important developments in logic and presenting significant contributions to logical research. In addition, research works on the history of logical ideas, especially on the traditions in China and India, are welcome contributions.

The scope of the book series includes but is not limited to the following:

- Monographs written by researchers in Asian countries.
- Proceedings of conferences held in Asia, or edited by Asian researchers.
- Anthologies edited by researchers in Asia.
- Research works by scholars from other regions of the world, which fit the goal of "Logic in Asia".

The series discourages the submission of manuscripts that contain reprints of previously published material and/or manuscripts that are less than 165 pages/ 90,000 words in length.

Please also visit our webpage: http://tsinghualogic.net/logic-in-asia/background/

Relation with Studia Logica Library

This series is part of the Studia Logica Library, and is also connected to the journal Studia Logica. This connection does not imply any dependence on the Editorial Office of Studia Logica in terms of editorial operations, though the series maintains cooperative ties to the journal.

This book series is also a sister series to Trends in Logic and Outstanding Contributions to Logic.

For inquiries and to submit proposals, authors can contact the editors-in-chief Fenrong Liu at fenrong@tsinghua.edu.cn or Hiroakira Ono at ono@jaist.ac.jp.

More information about this series at http://www.springer.com/series/13080

Shier Ju · Alessandra Palmigiano · Minghui Ma
Editors

Nonclassical Logics and Their Applications

Post-proceedings of the 8th International Workshop on Logic and Cognition

Springer

Editors
Shier Ju
Institute of Logic and Cognition
Sun Yat-sen University
Guangzhou, Guangdong, China

Minghui Ma
Sun Yat-Sen University
Guangzhou, Guangdong, China

Alessandra Palmigiano
School of Business and Economics
Vrije Universiteit Amsterdam
Amsterdam, The Netherlands

Department of Mathematics and Applied Mathematics
University of Johannesburg
Johannesburg, South Africa

ISSN 2364-4613 ISSN 2364-4621 (electronic)
Logic in Asia: Studia Logica Library
ISBN 978-981-15-1341-1 ISBN 978-981-15-1342-8 (eBook)
https://doi.org/10.1007/978-981-15-1342-8

This Springer imprint is published by the registered company Springer Nature Singapore Pte Ltd.
The registered company address is: 152 Beach Road, #21-01/04 Gateway East, Singapore 189721, Singapore

Preface

The 8th instalment of the International Conference on Logic and Cognition (WOLC 2016) took place on 5–9 December 2016 at the Institute of Logic and Cognition of Sun Yat-Sen University in Guangzhou. This instalment focused on non-classical logics and their applications. The conference provided a very interactive environment in which experts from disciplines ranging from formal linguistics and theoretical computer science to analytic philosophy, game theory, social choice, and management science could not only discuss how to advance formal theories so as to meet the main challenges in their own disciplines, but could also find a common ground based on the tools, insights and techniques developed in the study of non-classical logics.

The discussions and research directions with which the conference participants engaged reverberate in their research and the research of colleagues in their network. This volume collects a—necessarily non-exhaustive—sample of ways in which advancements in (the mathematical theory of) non-classical logics are obtained in response to challenges arising in other scientific fields.

Amsterdam, The Netherlands — Alessandra Palmigiano
Guangzhou, China — Shier Ju
Guangzhou, China — Minghui Ma

Contents

Hyperstates of Involutive MTL-Algebras that Satisfy $(2x)^2 = 2(x^2)$

Tommaso Flaminio and Sara Ugolini

Abstract States of MV-algebras have been the object of intensive study and attempts of generalizations. The aim of this contribution is to provide a preliminary investigation for states of prelinear semihoops and hyperstates of algebras in the variety generated by perfect and involutive MTL-algebras (IBP_0-algebras for short). Grounding on a recent result showing that IBP_0-algebras can be constructed from a Boolean algebra, a prelinear semihoop and a suitably defined operator between them, our first investigation on states of prelinear semihoops will support and justify the notion of hyperstate for IBP_0-algebras and will actually show that each such map can be represented by a probability measure on its Boolean skeleton, and a state on a suitably defined abelian ℓ-group.

Keywords IBP_0-algebras · Abelian ℓ-groups · Prelinear semihoop · States of prelinear semihoop · Hyperstates

1 Motivation

States of MV-algebras have been introduced by Daniele Mundici (1995) as averaging processes for truth-values of Łukasiewicz formulas. These are mappings of any MV-algebra in the real unit interval [0, 1] satisfying a normalization condition and a generalized version of the usual additivity law (see Flaminio and Kroupa 2015;

T. Flaminio
IIIA - CSIC, Campus de la Universidad Autònoma de Barcelona S/n, 08193 Bellaterra, Spain
e-mail: tommaso@iiia.csic.es

S. Ugolini (✉)
Department of Computer Science, University of Pisa, Pisa, Italy
e-mail: sara.ugolini@di.unipi.it

S. Ju et al. (eds.), *Nonclassical Logics and Their Applications*, Logic in Asia: Studia Logica Library,
https://doi.org/10.1007/978-981-15-1342-8_1

Mundici 2011 for more details). The states of MV-algebras are strongly connected to states of abelian ℓ-groups (Goodearl 1986) through Mundici's categorical equivalence between the category of MV-algebras with homomorphisms and the category of abelian ℓ-groups with strong order unit (unital ℓ-groups) and unit-preserving ℓ-group homomorphisms (Mundici 1986). Indeed, if $\mathbf{A}$ is an MV-algebra and $\mathbf{G_A}$ is its corresponding unital ℓ-group, then the states of $\mathbf{A}$ and the states of $\mathbf{G_A}$ are in 1-1 correspondence.

MV-algebraic states have been widely studied in the last years (cf. Flaminio and Kroupa 2015 and Mundici 2011 for a brief survey), and many attempts have been made to define states to alternative algebraic structures. In particular, the task of defining states on *perfect* MV-algebras has been the object of several proposals (Di Nola et al. 2000; Diaconescu et al. 2014a, b) since the very notion of state given in Mundici (1995) trivializes when applied to these structures. Indeed, every perfect MV-algebra has only one state: the function s mapping its radical, i.e. the intersection of its maximal filters, $Rad(\mathbf{A})$ in 1 and its co-radical $coRad(\mathbf{A})$ in 0 (see Di Nola et al. 2000 and Sect. 2 below for further details).

In this contribution, mimicking the insights provided by Mundici's categorical equivalence to the study of state theory, we shall define a notion of hyperstate (i.e., hyperreal-valued state) on a wider class of algebras called *IBP*$_0$*-algebras* which properly contains perfect MV-algebras and for which we recently provided a categorical equivalence with respect to a category whose objects are *prelinear-semihoop triples*, that is, systems $(\mathbf{B}, \mathbf{H}, \vee_e)$ where $\mathbf{B}$ is a Boolean algebra, $\mathbf{H}$ is a prelinear semihoop and $\vee_e : B \times H \to H$ is a suitably defined map, intuitively representing the natural join between elements of B and H. If $(\mathbf{B}, \mathbf{H}, \vee_e)$ and $(\mathbf{B}', \mathbf{H}', \vee_e')$ are two triples, a morphism between them is a pair (f, g) where $f : \mathbf{B} \to \mathbf{B}'$ is a Boolean homomorphism, $g : \mathbf{H} \to \mathbf{H}'$ is a prelinear semihoop homomorphism, and for every $(b, c) \in B \times H$, $g(b \vee_e c) = f(b) \vee_e' g(c)$.

The definition of hyperstate that we will present and study in the following sections is grounded on the fact that Boolean algebras already possess a well-established notion of *state*: probability functions. As for prelinear semihoops, we will show in the next section that each of them has a homomorphic image (as abelian ℓ-monoid) in an abelian ℓ-group. Thus, taking into account the categorical equivalence between IBP$_0$-algebras and prelinear semihoop triples, our main result will prove that any hyperstate on an IBP$_0$-algebra splits into a probability measure of the Boolean skeleton and a state of the largest prelinear semihoop contained in it. As a consequence, we will prove that if the IBP$_0$-algebra actually belongs to the variety generated by perfect MV-algebras, then its hyperstates are given by a probability measure on its Boolean skeleton and a state of a suitably defined abelian ℓ-group.

The present paper is structured in the following way: Sect. 2 is devoted to recalling basic notions and results about abelian ℓ-groups, MV-algebras, perfect MV-algebras and their states, while in Sect. 3 we will prove a first result which partially extends the usual Grothendieck group construction to lattice-ordered monoids. That result and its corollary will be used in Sect. 4 to introduce a suitable notion of state of a prelinear semihoop which, in turn, allows to introduce a notion of hyperstate of IBP$_0$-algebras in Sect. 5. In the same Sect. 5 we will prove that every hyperstate of an

IBP_0-algebra splits into a probability measure on its Boolean skeleton and a state of the largest prelinear semihoop contained in it. We end this paper with Sect. 6 which is devoted to concluding remarks and future work on this subject.

2 Abelian ℓ-Groups, MV-Algebras and Their States

An *abelian ℓ-group with strong unit* (or *unital ℓ-group* for short) is a pair $(\mathbf{G}, u)$ where $\mathbf{G}$ is an abelian ℓ-group (see Goodearl 1986) and $u \in G$ satisfies the following requirement: for every $x \in G$ there is a natural number n such that $x \leq u + \cdots + u$ where, in the previous expression, $u + \cdots + u$ is the n-times sum of u in $\mathbf{G}$, and $\leq$ denotes the lattice order of $\mathbf{G}$.

A *state of an ℓ-group* $\mathbf{G}$ is a group homomorphism σ to the additive group $\mathbf{R}$ of reals which further satisfies: for all $x \geq 0$ in G, $\sigma(x) \geq 0$ in $\mathbb{R}$. If $(\mathbf{G}, u)$ is a unital ℓ-group, a state of $(\mathbf{G}, u)$ is any state of $\mathbf{G}$ such that $\sigma(u) = 1$ (see Goodearl 1986 for further details).

MV-algebras can be introduced as those structures $\mathbf{A} = (A, \oplus, \neg, 0, 1)$ of type $(2, 1, 0, 0)$ for which there exists a unital ℓ-group $(\mathbf{G}_A, u)$ such that $A = \{x \in G_A \mid 0 \leq x \leq u\}$, $x \oplus y = (x + y) \wedge u$, $\neg x = u - x$ and $1 = u$. Furthermore, for every MV-algebra $\mathbf{A}$, the unital ℓ-group $(\mathbf{G}_A, u)$ is unique. Indeed, the previous construction induces a categorical equivalence, established by Mundici's functor Γ, between the categories of unital ℓ-groups with unit-preserving ℓ-group homomorphisms and that of MV-algebras with MV-homomorphisms (Mundici 1986). In particular, it is worth noticing that for every morphism h in the category of unital ℓ-groups, $\Gamma(h)$ is an MV-homomorphism that is defined by restriction.

The latter construction suggests that we can speak about *states* of an MV-algebra $\mathbf{A}$ restricting any state of $(\mathbf{G}_A, u)$ both in its domain, which thus becomes A, and its codomain that, since 1 is a strong unit for the ℓ-group $\mathbf{R}$, restricts to the real unit interval $[0, 1]$. Indeed, by a state of $\mathbf{A}$ we mean any map $s : A \to [0, 1]$ such that $s(1) = 1$ and $s(x \oplus y) = s(x) + s(y)$ for all $x, y \in A$ such that $x + y$ (the group sum) coincides with $x \oplus y$ (the MV-sum) (Mundici 1995).

Although states of MV-algebras resemble finitely additive probability measures on Boolean algebras, they are intimately related with Borel (and hence σ-additive) regular measures. Indeed, by the Kroupa-Panti Theorem (Kroupa 2006 and Panti 2009), for every MV-algebra $\mathbf{A}$ the set of its states $\mathcal{S}(\mathbf{A})$ is in 1-1 correspondence with the set of Borel regular measures on the compact and Hausdorff space $Max(\mathbf{A})$ of maximal MV-filters of $\mathbf{A}$. Precisely, for every state s of $\mathbf{A}$ there exists a unique Borel regular measure μ of $Max(\mathbf{A})$ such that s is the Lebesgue integral w.r.t. μ (see also Flaminio and Kroupa 2015; Mundici 2011 for more details).

Every MV-algebra admits at least one state. However, there are relevant examples of MV-algebras whose unique state is trivial, i.e., it only takes Boolean values, 0 and 1. This is the case, for instance, of *perfect* MV-algebras (Di Nola and Lettieri 1994). Mimicking the way we used to introduce MV-algebras in general, perfect MV-algebras are, up to isomorphisms, those MV-algebras of the form $\Gamma(\mathbf{Z} \times \mathbf{G}, (1, 0))$ where Γ is Mundici's functor, $\mathbf{Z}$ is the ℓ-group of integers, $\mathbf{G}$ is any ℓ-group, $\times$

denotes the lexicographic product between ℓ-groups (which is an ℓ-group iff the first component is totally ordered Glass and Holland 1989, Example 3), and $(1, 0) \in \mathbb{Z} \times G$ is indeed a strong unit for $\mathbf{Z} \times \mathbf{G}$. Again, this construction lifts to a categorical equivalences shown by Di Nola and Lettieri between perfect MV-algebras and ℓ-groups (Di Nola and Lettieri 1994).

An immediate consequence of the previous definition shows that every perfect MV-algebra has for domain the disjoint union $G^+ \cup G^-$ for a unique ℓ-group $\mathbf{G}$, where G^+ denotes the positive cone of $\mathbf{G}$, $G^- = \{-x \mid x \in G^+\}$ and $x > y$ for every $x \in G^-$ and $y \in G^+$. Furthermore, in every perfect MV-algebra $\mathbf{A}$, displayed as above, $x \oplus x = 1$ for every $x \in G^-$ and $y \odot y = 0$ for all $y \in G^+$. Therefore, every state s of $\mathbf{A}$ maps G^- in 1 and G^+ in 0, whence s is also unique.

In order to overcome this limitation and noticing that ℓ-groups have more than one trivial state, in Diaconescu et al. (2014b) the authors introduced the notion of *lexicographic states* for a wide class of MV-algebras which includes perfect algebras. For any algebra $\mathbf{A}$ in the variety generated by perfect MV-algebras, a lexicographic state is any map of A to the MV-algebra $\mathscr{L}(\mathbb{R}) = \Gamma(\mathbf{R} \times \mathbf{R}, (1, 0))$ satisfying $s(1) = 1$, $s(x \oplus y) = s(x) + s(y)$ whenever $x \odot y = 0$ and such that the restriction of s to its maximal semisimple quotient, is a state in its usual sense. Diaconescu et al. (Diaconescu et al. (2014b), Corollary 6.7) shows that the class of lexicographic states of a perfect MV-algebra $\Gamma(\mathbf{Z} \times \mathbf{G}, (1, 0))$ is in one-one correspondence with the class of states of $\mathbf{G}$.

As we shall see in Sect. 4 the previous definition is a particular case of a more general construction that we will exhibit along this paper.

In what follows we shall provide preliminary results which will help us, in Sect. 4, to provide a reasonable notion of *state* for a class of algebras, called prelinear semihoops, which play a crucial role in this paper. In particular, following the same lines that we recalled in this section, our axiomatization for states of a prelinear semihoop will follow directly by Goodearl's definition of state of an ℓ-group.

3 From ℓ-Monoids to ℓ-Groups and Hoops

In this section we will prove and recall some basic results we shall need in the rest of this paper. As to begin with, let us recall that a *lattice-ordered monoid* (*ℓ-monoid* for short) is a structure $\mathbf{M} = (M, +, \wedge, \vee, 0)$ such that $(M, +, 0)$ is a commutative monoid, $(M, \wedge, \vee)$ is a lattice, and the following distribution laws hold for all $x, y, z \in M$:

(D1) $x + (y \wedge z) = (x + y) \wedge (x + z)$,
(D2) $x + (y \vee z) = (x + y) \vee (x + z)$.

The following result, that we need to reprove completely, is an extension of the usual Grothendieck group construction to the case of lattice-ordered, and in general not cancellative, monoids. Assuming cancellativity, an analogous construction has been used, for instance, in Cignoli et al. (Cignoli et al. (2000), Sect. 2.4) and Di Nola and Lettieri (1994).

Theorem 3.1 *Let* $\mathbf{M} = (M, +, \wedge, \vee, 0)$ *be a lattice-ordered monoid. Then, there is an abelian* ℓ*-group* $\mathbf{K}(\mathbf{M})$ *and a* ℓ*-monoid homomorphism* $h : \mathbf{M} \to \mathbf{K}(\mathbf{M})$ *which is injective iff* $\mathbf{M}$ *is cancellative.*

Proof Starting from a commutative monoid $\mathbf{M}$ it is possible to define the Grothendieck group of $\mathbf{M}$, namely $\mathbf{K}(\mathbf{M})$ (see Weibel 2013, Chap. II), by means the following construction. Consider the equivalence relation on the cartesian product $M \times M$ given by $(x, y) \sim (x', y')$ if there exists $z \in M$ such that $z + x + y' = z + x' + y$. Let $K(M) = M \times M/\sim$, and for every $[x, y], [x', y'] \in K(M)$, let $[x, y]\hat{+}[x', y'] = [x + x', y + y']$. Let $[0, 0]$ be the identity and let the inverse of $[x, y]$ be $-[x, y] = [y, x]$. Then $\mathbf{K}(\mathbf{M}) = (K(M), \hat{+}, -, [0, 0])$ is an abelian group, and it satisfies the universal property: there exist a monoid homomorphism h such that for any other monoid homomorphism $k : \mathbf{M} \to \mathbf{G}$ into an abelian group $\mathbf{G}$, there exists a unique group homomorphism $l : \mathbf{K}(\mathbf{M}) \to \mathbf{G}$ such that $k = l \circ h$. In particular, $h(x) = [x + x, x]$ for every $x \in M$. Moreover, the homomorphism h is injective iff $\mathbf{M}$ is cancellative (cf. for instance Birkhoff 1967).

Now, we are going to prove that if $\mathbf{M}$ is lattice-ordered, it is possible to define on $\mathbf{K}(\mathbf{M})$ an ℓ-group structure such that the claim holds. First, denoting with $\leq_M$ the lattice order of $\mathbf{M}$, let

$$[x_1, y_1] \leq [x_2, y_2] \text{ if } \exists z : z + x_1 + y_2 \leq_M z + y_1 + x_2.$$

It is easy to see that it is a partial order on $\mathbf{K}(\mathbf{M})$, for instance let us prove transitivity. If $[x_1, y_1] \leq [x_2, y_2]$ and $[x_2, y_2] \leq [x_3, y_3]$, then by definition $\exists z : z + x_1 + y_2 \leq_M z + y_1 + x_2$ and $\exists z' : z' + x_2 + y_3 \leq_M z' + y_2 + x_3$. By monotonicity, $z + x_1 + y_2 + z' + x_2 + y_3 \leq_M z + y_1 + x_2 + z' + y_2 + x_3$, and putting $z'' = z + z' + x_2 + y_2 \in M$, we have that $z'' + x_1 + y_3 \leq_M z'' + y_1 + x_3$, thus $[x_1, y_1] \leq [x_3, y_3]$ and $\leq$ is transitive.

Let us now define lattice operations with respect to the order $\leq$:

$$[x_1, y_1] \sqcup [x_2, y_2] = [x_1 + x_2, (x_1 + y_2) \wedge (x_2 + y_1)],$$

$$[x_1, y_1] \sqcap [x_2, y_2] = [(x_1 + y_2) \wedge (x_2 + y_1), y_1 + y_2].$$

We prove that $\sqcup$ is the join, the proof that $\sqcap$ is the meet being similar. First we prove that it is an upper bound, that is, $[x_1, y_1], [x_2, y_2] \leq [x_1 + x_2, (x_1 + y_2) \wedge (x_2 + y_1)]$. Indeed: $x_1 + (x_1 + y_2) \wedge (x_2 + y_1) \leq x_1 + (x_2 + y_1)$ and similarly $x_2 + (x_1 + y_2) \wedge (x_2 + y_1) \leq x_2 + (x_1 + y_2)$. Now we shall prove that it is the least upper bound, that is, for any $[x_3, y_3] \in K(M)$, if $[x_1, y_1], [x_2, y_2] \leq [x_3, y_3]$, then $[x_1 + x_2, (x_1 + y_2) \wedge (x_2 + y_1)] \leq [x_3, y_3]$. By hypothesis, $\exists z : z + x_1 + y_3 \leq_M z + y_1 + x_3$ and $\exists z' : z + x_2 + y_3 \leq_M z' + y_2 + x_3$. Thus we get

$$z + z' + x_1 + x_2 + y_3 \leq_M z + z' + x_3 + y_1 + x_2, \text{ and}$$

$$z + z' + x_1 + x_2 + y_3 \leq_M z + z' + x_3 + y_2 + x_1$$

Hence $(z + z' + x_1 + x_2 + y_3) \wedge (z + z' + x_1 + x_2 + y_3) \leq_M (z + z' + x_3 + y_1 + x_2) \wedge (z + z' + x_3 + y_2 + x_1)$ and since $\wedge$ distributes over $+$, we obtain that $z + z' + (x_1 + x_2) + y_3 \leq_M z + z' + x_3 + (y_1 + x_2) \wedge (y_2 + x_1)$, which means exactly that $[x_1 + x_2, (x_1 + y_2) \wedge (x_2 + y_1)] \leq [x_3, y_3]$.

In order to prove that $\mathbf{K}(\mathbf{M})$ with $\sqcup, \sqcap$ is an ℓ-group, we need to show that $\hat{+}$ distributes over $\sqcup$, that is:

$$[x_1, y_1] \hat{+} ([x_2, y_2] \sqcup [x_3, y_3]) = ([x_1, y_1] \hat{+} [x_2, y_2]) \sqcup ([x_1, y_1] \hat{+} [x_3, y_3])$$

Now, $[x_1, y_1] \hat{+} ([x_2, y_2] \sqcup [x_3, y_3]) = [x_1, y_1] \hat{+} [x_2 + x_3, (y_2 + x_3) \wedge (y_3 + x_2)] = [x_1 + x_2 + x_3, y_1 + (y_2 + x_3) \wedge (y_3 + x_2)]$, while $([x_1, y_1] \hat{+} [x_2, y_2]) \sqcup ([x_1, y_1] \hat{+} [x_3, y_3]) = [x_1 + x_2, y_1 + y_2] \sqcup [x_1 + x_3, y_1 + y_3] = [x_1 + x_2 + x_1 + x_3, (y_1 + y_2 + x_1 + x_3) \wedge (y_1 + y_3 + x_1 + x_2)]$. It is easy to see that $[x_1 + x_2 + x_1 + x_3, (y_1 + y_2 + x_1 + x_3) \wedge (y_1 + y_3 + x_1 + x_2)] = [x_1 + x_2 + x_3, y_1 + (y_2 + x_3) \wedge (y_3 + x_2)]$, since $x_1 + x_2 + x_3 + x_1 + (y_1 + y_2 + x_1 + x_3) \wedge (y_1 + y_3 + x_2) = x_1 + x_2 + x_3 + (x_1 + y_1 + y_2 + x_3) \wedge (y_1 + y_3 + x_1 + x_2)$, which settles the claim.

In order to conclude the proof we need to prove that the monoid homomorphism $h : \mathbf{M} \to \mathbf{K}(\mathbf{M})$, $h(x) = [x + x, x]$, is also a lattice homomorphism. Let us prove that it respects the meet operation, the proof for the join being similar. We need to show that $h(x \wedge y) = h(x) \sqcap h(y)$, that is to say,

$$[(x \wedge y) + (x \wedge y), x \wedge y] = [x + x, x] \sqcap [y + y, y].$$

Now, $[x + x, x] \sqcap [y + y, y] = [(x + x + y) \wedge (y + y + x), x + y]$. Since $(x + y) + (x \wedge y) = (x + y + x) \wedge (x + y + y)$, we have that $(x \wedge y) + (x \wedge y) + x + y = (x + y + x) \wedge (x + y + y) + (x \wedge y)$, which proves the claim.

Definition 3.2 (Esteva et al. 2003) An algebra $\mathbf{Hal} = (H, \cdot, \to, \wedge, 1)$ of type $(2, 2, 2, 0)$ is a *semihoop* if it satisfies the following conditions:

(i) $(H, \wedge, 1)$ is an inf-semilattice with upper bound;
(ii) $(H, \cdot, 1)$ is a commutative monoid isotonic with respect to the inf-semilattice order;
(iii) For every $c_1, c_2 \in H$, $c_1 \leq c_2$ iff $c_1 \to c_2 = 1$;
(iv) For every $c_1, c_2, c_3 \in H$, $(c_1 \cdot c_2) \to c_3 = c_1 \to (c_2 \to c_3)$.

Furthermore, a semihoop $\mathbf{H}$ is said to be *prelinear* if it satisfies:

(Pre) $(x \to y) \to z \leq ((y \to x) \to z) \to z$.

A prelinear semihoop is said to be a *basic hoop* if it satisfies:

(Div) $x \cdot (x \to y) = y \cdot (y \to x)$.

A basic hoop is said to be *cancellative* if the following holds:

(Canc) $(x \to (x \cdot y)) \to y = 1$.

The class of prelinear semihoops forms a variety that we will denote by $\mathbb{PSH}$, while $\mathbb{BH}$ and $\mathbb{CH}$ denote the variety of basic hoops and cancellative hoops respectively.

Remark 3.3 In a semihoop $\mathbf{H}$ we can always define a *pseudo-join* as follows:

$$x \vee y = ((x \to y) \to y) \wedge ((y \to x) \to x),$$

and $(H, \wedge, \vee, 1)$ is a lattice iff $\vee$ is associative. Furthermore, in a prelinear semihoop $\mathbf{H}$, the pseudo-join $\vee$ is the join operation on $\mathbf{H}$, thus $(H, \wedge, \vee, 1)$ is a lattice (Esteva et al. 2003, Lemma 3.9). Henceforth we will include the $\vee$ in the signature of prelinear semihoops.

The following result is a direct consequence of Theorem 3.1 plus the observation that the reduct $\hat{\mathbf{H}} = (H, \cdot, \wedge, \vee, 1)$ of a prelinear semihoop $\mathbf{H}$ is an ℓ-monoid (written in multiplicative form). In the following result and in the rest of this paper, if $h : \mathbf{M} \to \mathbf{K}(\mathbf{M})$ is an ℓ-monoid homomorphism, we shall denote by $\mathbf{J}_{h[M]}$ the ℓ-subgroup of $\mathbf{K}(\mathbf{M})$ generated by $h[M] = \{h(a) \mid a \in M\}$.

Corollary 3.4 *For every prelinear semihoop* $\mathbf{H}$ *there is an abelian* ℓ*-group* $\mathbf{K}(\hat{\mathbf{H}})$ *and a* ℓ*-monoid homomorphism* $h : \mathbf{H} \to \mathbf{K}(\hat{\mathbf{H}})$ *which is injective iff* $\mathbf{H}$ *is cancellative. Furthermore, for every* $x \in \mathbf{K}(\hat{\mathbf{H}})$*, there exists a* $y \in \mathbf{J}_{h[H]}$ *such that* $y \leq x$*. Consequently, for every* $x \in \mathbf{K}(\hat{\mathbf{H}})$*, there exists a* $y' \in \mathbf{J}_{h[H]}$ *such that* $y' \geq x$*.*

Proof The first part directly follows from Theorem 3.1. As for the second part, let $[a, b]$ be a generic element of $\mathbf{K}(\hat{\mathbf{H}})$ and let, for every $x \in \hat{H}$, $h(x) = [x \cdot x, x]$. Thus, $a \geq a \wedge b$ and hence $a \cdot (a \wedge b) \geq (a \wedge b) \cdot (a \wedge b) \geq (a \wedge b) \cdot (a \wedge b) \cdot b$, since in every prelinear semihoop $z \geq z \cdot k$. Thus, by definition of $\leq$ in $\mathbf{K}(\hat{\mathbf{H}})$, $[a, b] \geq [(a \wedge b) \cdot (a \wedge b), a \wedge b] = h(a \wedge b) \in \mathbf{J}_{h[H]}$. Obviously, since every element of $\mathbf{J}_{h[H]}$ can be equivalently displayed as $-[c, d]$ for some $[c, d] \in \mathbf{J}_{h[H]}$ and since $-$ reverses the order, there is a $y \in \mathbf{J}_{h[H]}$ such that $[c, d] \geq y'$, whence $-[c, d] \leq -y \in \mathbf{J}_{h[H]}$. □

4 States of Prelinear Semihoops

In this section we will introduce states of prelinear semihoops and we will show some basic properties. The following definition naturally arises by following the same lines that inspired the axiomatization of state of an MV-algebra (recall Sect. 2), from Corollary 3.4 and recalling that a state of an ℓ-group $\mathbf{G}$ is a map $\sigma : G \to \mathbb{R}$ which is a group homomorphism, and such that, if $x \geq 0$, then $\sigma(x) \geq 0$.

Definition 4.1 A *state* of a prelinear semihoop $\mathbf{H} = (H, \cdot, \to, \wedge, \vee, 1)$ is a map $w : H \to \mathbb{R}^-$ satisfying the following conditions:

(v1) $w(1) = 0$,
(v2) $w(x \cdot y) = w(x) + w(y)$,
(v3) if $x \leq y$, then $w(x) \leq w(y)$.

Given any prelinear semihoop $\mathbf{H}$ we denote by $\mathcal{W}(\mathbf{H})$ the set of its states. Notice that $\mathcal{W}(\mathbf{H})$ is not empty. Indeed, letting $x \ominus y = \min\{0, x - y\}$ on $\mathbb{R}^-$, $\mathbf{R}^- = (\mathbb{R}^-, +, \ominus, \leq, 0)$ is a prelinear semihoop, and any hoop-homomorphism of $\mathbf{H}$ to $\mathbf{R}^-$ is a state.

Proposition 4.2 *For every prelinear semihoop $\mathbf{H}$ and for every $w \in \mathcal{W}(\mathbf{H})$, the following hold:*

(1) $w(x \wedge y) + w(x \vee y) = w(x) + w(y)$,
(2) if $\mathbf{H}$ is a basic hoop, then $w(x) + w(x \to y) = w(y) + w(y \to x)$,
(3) if $\mathbf{H}$ is divisible, then (v3) is redundant.

Proof (1). The variety $\mathbb{PSH}$ is generated by its linearly ordered members, and in every totally ordered prelinear semihoop $x \cdot y = (x \wedge y) \cdot (x \vee y)$. Thus, the latter equation holds in every $\mathbf{H} \in \mathbb{PSH}$. Therefore, for every $w \in \mathcal{W}(\mathbf{H})$, $w(x) + w(y) = w(x \cdot y) = w((x \wedge y) \cdot (x \vee y)) = w(x \wedge y) + w(x \vee y)$ where the last equality follows from (v2).

(2). Immediate from *(Div)* and *(v2)*.

(3). Let $\mathbf{H}$ be divisible and assume that $x \leq y$. Then, $x = x \wedge y = y \cdot (y \to x)$ and hence $w(x) = w(y \cdot (y \to x)) = w(y) + w(y \to x) \leq w(y)$ where the last inequality holds since, by definition, $w(y), w(y \to x) \in \mathbb{R}^-$. □

Remark 4.3 In any prelinear semihoop $(H, \cdot, \to, \wedge, \vee, 1)$, the reduct $(H, \wedge, \vee)$ is a distributive lattice and indeed Proposition 4.2 (1) above shows that states of prelinear semihoops are *valuations* on their lattice reduct as defined by Birkhoff (1967).

Furthermore, if $\mathbf{H}$ is a basic hoop, then Proposition 4.2 (2) shows that w satisfies Bosbach equation $w(x) + w(x \to y) = w(y) + w(y \to x)$. Thus, every state of a basic hoop can be seen as a *Bosbach state* in the sense of He et al. (2017).

For the next result, recall how the ℓ-group $\mathbf{K}(\hat{\mathbf{H}})$ and the ℓ-monoid homomorphism h are defined in Theorem 3.1 and Corollary 3.4.

Proposition 4.4 *For every prelinear semihoop $\mathbf{H}$ and every $w \in \mathcal{W}(\mathbf{H})$, there is a state σ of the abelian ℓ-group $\mathbf{K}(\hat{\mathbf{H}})$ such that $w = \sigma \circ h$. Conversely, if σ is a state of $\mathbf{K}(\hat{\mathbf{H}})$, then the composition map $w = \sigma \circ h$ is a state of $\mathbf{H}$.*

Proof Let h and $\mathbf{K}(\hat{\mathbf{H}})$ as in Theorem 3.1 and Corollary 3.4 and let $\mathbf{J}_{h[H]}$ be the ℓ-subgroup of $\mathbf{K}(\hat{\mathbf{H}})$ generated by $h[H] = \{h(x) \mid x \in H\}$. For every element $[x, y] \in J_{h[H]}$, let $\hat{\sigma}([x, y]) = w(y) - w(x) \in \mathbb{R}$. □

Claim 1 The map $\hat{\sigma} : J_{h[H]} \to \mathbb{R}$ is a state of $\mathbf{J}_{h[H]}$.

Proof (of the Claim). The neutral element of $\mathbf{K}(\hat{\mathbf{H}})$, which obviously coincide with the neutral element of $\mathbf{J}_{h[H]}$, is $[1, 1]$. Thus, since $w(1) = 0, \hat{\sigma}([1, 1]) = 0$. Moreover, for every positive element $[x, 1]$ of $\mathbf{J}_{h[H]}$, $\hat{\sigma}([x, 1]) = -w(x) \in \mathbb{R}^+$, whence σ is positive.

It is left to show that $\hat{\sigma}$ is a group homomorphism. Let $[x_1, y_1], [x_2, y_2] \in J_{h[H]}$. Then, $\hat{\sigma}([x_1, y_1] + [x_2, y_2]) = \hat{\sigma}([x_1 \cdot x_2, y_1 \cdot y_2]) = w(y_1) + w(y_2) - w(x_1) - w(x_2) = w(y_1) - w(x_1) + w(y_2) - w(x_2) = \hat{\sigma}([x_1, y_1]) + \hat{\sigma}(x_2, y_2)$. □

Turning back to the proof of Proposition 4.4, let $\sigma : \mathbf{K}(\hat{\mathbf{H}}) \to \mathbb{R}$ be a state of $\mathbf{K}(\hat{\mathbf{H}})$ obtained by extending $\hat{\sigma}$ from the ℓ-subgroup $\mathbf{J}_{h[H]}$ of $\mathbf{K}(\hat{\mathbf{H}})$. The existence of σ is hence guaranteed by Goodearl (1986, Proposition 4.2) plus the observation that every element of $\mathbf{K}(\hat{\mathbf{H}})$ is bounded above by an element of $\mathbf{J}_{h[H]}$ (second part of Corollary 3.4).

Now, since every element x of H is represented in $K(\hat{H})$ as $[1, x]$, $w(x) = w(x) - w(1) = \hat{\sigma}([1, x]) = \sigma([1, x])$.

Conversely, let $\sigma : \mathbf{K}(\hat{\mathbf{H}}) \to \mathbb{R}$ be a state. Then $w(1) = \sigma(h(1)) = \sigma(0) = 0$. Moreover, for every $x, y \in H$, $w(x \cdot y) = \sigma(h(x \cdot y)) = \sigma(h(x) + h(y)) = \sigma(h(x)) + \sigma(h(y)) = w(x) + w(y)$. Finally, the monotonicity of w comes from the monotonicity of σ and h. □

5 States of IBP_0-Algebras and Their Representation

MTL-algebras are bounded, commutative, integral residuated lattices $\mathbf{A} = (A, \cdot, \to, \wedge, \vee, 0, 1)$ further satisfying $(x \to y) \vee (y \to x) = 1$. MTL-algebras form a variety that we will denote with $\mathbb{MTL}$. In every MTL-algebra $\mathbf{A}$ we can define further operations and abbreviations in the following manner: $\neg x := x \to 0$, $x \oplus y := \neg x \to y$, $2x := x \oplus x$, $x^2 := x \cdot x$. *GMTL-algebras* are unbounded MTL-algebras and they form a variety which is term-equivalent to the variety $\mathbb{PSH}$ (Noguera et al. 2005).

Definition 5.1 An IBP_0-algebra is an MTL-algebra further satisfying the following equations:

(DL) $(2x)^2 = 2(x^2)$,
(Inv) $\neg\neg x = x$.

Remark 5.2 Every perfect MV-algebra (Di Nola and Lettieri 1994) is an IBP_0-algebra. Indeed, the class of IBP_0-algebras is a variety $\mathbb{IBP}_0$ that properly contains the subvariety of $\mathbb{MV}$ generated by perfect MV-algebras. More precisely, the variety generated by perfect MV-algebras is definable, within $\mathbb{IBP}_0$, by the divisibility equation $x \wedge y = x \cdot (x \to y)$.

For every IBP_0-algebra $\mathbf{A}$, let us define

$$\mathscr{B}(\mathbf{A}) = \{a \in A \mid a \vee \neg a = 1\} \text{ and } \mathscr{H}(\mathbf{A}) = \{x \in A \mid x > \neg x\}.$$

It is known (see Aguzzoli et al. 2019, Proposition 2.5) that $\mathscr{B}(\mathbf{A})$ and $\mathscr{H}(\mathbf{A})$ respectively are the domains of the largest Boolean subalgebra of $\mathbf{A}$ and the domain of the radical of A. In Aguzzoli et al. (2019) we showed a categorical equivalence between the category of IBP_0-algebras with homomorphisms and a category whose objects are triples $(\mathbf{B}, \mathbf{H}, \vee_e)$ where $\mathbf{B}$ is a Boolean algebra, $\mathbf{H}$ is a prelinear semihoop and $\vee_e : B \times H \to H$ is a suitably defined map, intuitively representing the natural join between elements of B and H. If $(\mathbf{B}, \mathbf{H}, \vee_e)$ and $(\mathbf{B}', \mathbf{H}', \vee_e')$ are two triples, a morphism between them is a pair (f, g) where $f : \mathbf{B} \to \mathbf{B}'$ is a Boolean homomorphism, $g : \mathbf{H} \to \mathbf{H}'$ is a prelinear semihoop homomorphism, and for every $(b, c) \in B \times H$, $g(b \vee_e c) = f(b) \vee_e' g(c)$.

The *radical* of an MTL-algebra $\mathbf{A}$, $Rad(\mathbf{A})$, is the intersection of its maximal filters, while the *co-radical* of $\mathbf{A}$ is defined as $coRad(\mathbf{A}) = \{a \in A \mid \neg a \in Rad(\mathbf{A})\}$ (see for instance Cignoli and Torrens 2006).

A direct consequence of the categorical equivalence is the following proposition that we are going to apply later.

Proposition 5.3 *(Aguzzoli et al. 2019) Let* $\mathbf{A}$ *be any IBP*$_0$*-algebra. Then the following holds:*

(i) For every $a \in A$, $a = (b_a \vee \neg c_a) \wedge (\neg b_a \vee c_a)$ *where* $b_a = \neg((\neg a^2)^2)$ *belongs to* $\mathscr{B}(\mathbf{A})$ *and* $c_a = a \vee \neg a$ *belongs to* $\mathscr{H}(\mathbf{A})$.
(ii) For every $b \in \mathscr{B}(\mathbf{A})$ *and every* $c \in \mathscr{H}(\mathbf{A})$, $b \vee c \in \mathscr{H}(\mathbf{A})$.
(iii) $(Rad(\mathbf{A}), \cdot, \to, \wedge, 1)$ *(where operations are obtained by restriction from those of* $\mathbf{A}$*) is a prelinear semihoop isomorphic to* $\mathscr{H}(\mathbf{A})$.

Notation 1 (1). Let $^*[0, 1]$ be a nontrivial ultraproduct of the real unit interval and let ε be an infinitesimal in $^*[0, 1]$. The MV-algebra $\mathscr{L}(\mathbb{R}) = \Gamma(\mathbf{R} \times \mathbf{R}, (1, 0))$ discussed in Sect. 2 is, up to isomorphisms, the MV-subalgebra of $^*[0, 1]_{MV}$ generated by $[0, 1] \cup \{r\varepsilon \mid r \in \mathbb{R}\}$ (see Diaconescu et al. 2014b, Example 6.1). Therefore, every element of $\mathscr{L}(\mathbb{R})$ can be uniquely displayed as $r + \varepsilon s$ for $r \in [0, 1]$ and $s \in \mathbb{R}$. In what follows we will adopt the following notation: for every $x \in \mathscr{L}(\mathbb{R})$, x° and x^* denote those unique elements of $[0, 1]$ and $\mathbb{R}$ respectively, such that $x = x^\circ + \varepsilon x^*$.

(2). In every MTL-algebra $\mathbf{A}$ we abbreviate $\neg a \to b$ as $a \oplus b$ and for every $n, m \in \mathbb{N}$ and every element $x \in A$, $n.a$ stands for $a \oplus \cdots \oplus a$ (n-times) and a^m is $a \cdot \cdots \cdot a$ (m-times).

Definition 5.4 For any IBP$_0$-algebra $\mathbf{A}$, we define a *hyperstate* of $\mathbf{A}$ as a map $s : A \to \mathscr{L}(\mathbb{R})$ such that:

(s1) $s(1) = 1$ and $s(0) = 0$,
(s2) $s(x \oplus y) + s(x \cdot y) = s(x) + s(y)$,
(s3) If $x \vee \neg x = 1$, then $s(x) \in [0, 1]$.

Proposition 5.5 *The following properties hold for hyperstates of IBP*$_0$*-algebras:*

(i) $s(\neg x) = 1 - s(x)$,
(ii) if $x \leq y$, *then* $s(x) \leq s(y)$,
(iii) if $x \cdot y = 0$, $s(x \oplus y) = s(x) + s(y)$,
(iv) if $x \oplus y = 1$, $s(x \cdot y) = s(x) \cdot s(y)$,
(v) $s(x \wedge y) + s(x \vee y) = s(x) + s(y)$,
(vi) the restriction p *of* s *to* $\mathscr{B}(\mathbf{A})$ *is a* $[0, 1]$*-valued and finitely additive probability measure,*
(vii) if $x \in coRad(\mathbf{A})$, *then* $s(x) \in coRad(\mathscr{L}(\mathbb{R}))$. *If* $x \in Rad(\mathbf{A})$, $s(x) \in Rad(\mathscr{L}(\mathbb{R}))$,

(viii) the map $w : \mathscr{H}(\mathbf{A}) \to \mathbb{R}^-$ *defined as*

$$w(x) = \frac{s(x) - 1}{\varepsilon}$$

is a state in the sense of Definition 4.1.

Proof (i). In any MTL-algebra, $x \cdot \neg x = 0$ and $x \oplus \neg x = 1$, thus (s2) and (s1) imply $s(x) + s(\neg x) = s(x \oplus \neg x) + s(x \cdot \neg x) = s(1) + s(0) = 1 + 0 = 1$, whence $s(\neg x) = 1 - s(x)$.

(ii). If $x \leq y$, then $x \cdot \neg y = 0$. Thus from $(s2)$, $s(x \oplus \neg y) = s(x) + s(\neg y) = s(x) + 1 - s(y) \leq 1$. Thus $s(x) \leq s(y)$.

(iii) and (iv) are direct consequences of (s1) and (s2).

(v). As we already observed in the proof of Proposition 5.5, in every MTL-algebra $x \cdot y = (x \wedge y) \cdot (x \vee y)$. Analogously, in every IBP$_0$-algebra, $x \oplus y = (x \wedge y) \oplus (x \vee y)$. Thus, from (s2), $s(x) + s(y) = s((x \wedge y) \cdot (x \vee y)) + s((x \wedge y) \oplus (x \vee y)) = s((x \wedge y) \cdot (x \vee y)) + s(x \wedge y) + s(x \vee y) - s((x \wedge y) \cdot (x \vee y)) = s(x \wedge y) + s(x \vee y)$.

(vi). That the restriction p of s to $\mathscr{B}(\mathbf{A})$ satisfies $p(1) = 1$ and $p(x \wedge y) + p(x \vee y) = p(x) + p(y)$ is ensured by (s1), (s2) together with the fact that, for all $x, y \in \mathscr{B}(\mathbf{A})$, $x \cdot y = x \wedge y$ and $x \oplus y = x \vee y$. Finally, that for every $x \in \mathscr{B}(\mathbf{A})$, $p(x) \in [0, 1]$ is exactly (s3).

(vii). Let $x \in coRad(\mathbf{A})$. Then, for every $n \in \mathbb{N}$, $n.x \leq \neg x$ and, from (ii), $s(n.x) \leq s(\neg x)$. Now, $x \cdot m.x = 0$ for every $m \in \mathbb{N}$, whence, in particular, $s(n.x) = n.s(x)$. Thus, $n.s(x) \leq 1 - s(x)$ for every $n \in \mathbb{N}$, i.e., $s(x) \in coRad(\mathscr{L}(\mathbb{R}))$. The second part of the claim now easily follows since $x \in Rad(\mathbf{A})$ iff $\neg x \in coRad(\mathbf{A})$ and $\alpha \in Rad(\mathscr{L}(\mathbb{R}))$ iff $\neg\alpha \in coRad(\mathscr{L}(\mathbb{R}))$ because both $\mathbf{A}$ and $\mathscr{L}(\mathbb{R})$ are strongly perfect MTL-algebras.

$(viii)$. As we already recalled in Sect. 3, $\mathscr{H}(\mathbf{A}) = Rad(\mathbf{A})$. Thus, if $x \in \mathscr{H}(\mathbf{A})$, from (vii), $s(x) \in Rad(\mathscr{L}(\mathbb{R}))$, whence there is $r_x \in \mathbb{R}^+$ such that $s(x) = 1 - \varepsilon r_x$. Therefore, $w(x) = s(x)/\varepsilon - 1/\varepsilon = -r_x \in \mathbb{R}^-$. It is left to prove that w is a state in the sense of Definition 4.1. First of all, $w(1) = s(1)/\varepsilon - 1/\varepsilon = 0$. Moreover, if $x, y \in \mathscr{H}(\mathbf{A})$, $x \oplus y = 1$, and hence $w(x \cdot y) = (s(x \cdot y) - 1)/\varepsilon = (s(x) + s(y) - 2)/\varepsilon = (s(x) - 1)/\varepsilon + (s(y) - 1)/\varepsilon = w(x) + w(y)$. The monotonicity of w easily follows from the monotonicity of s, (ii) above. Then the claim is settled. □

The next result is the main theorem of this paper and it shows that each hyperstate of an IBP$_0$ algebra $\mathbf{A}$ decomposes in a probability measure on its Boolean skeleton and a state on the maximal prelinear semihoop contained in A.

Theorem 5.6 *For every IBP$_0$-algebra* $\mathbf{A}$ *and every hyperstate* $s : A \to \mathscr{L}(\mathbb{R})$ *there are a probability measure* $p : \mathscr{B}(\mathbf{A}) \to [0, 1]$, *a state* $w \in \mathcal{W}(\mathscr{H}(\mathbf{A}))$ *and an infinitesimal* $\varepsilon > 0$ *such that, for every* $a \in A$,

$$s(a) = p(b_a) + \varepsilon(w(\neg b_a \vee c_a) - w(b_a \vee c_a)).$$

Proof Let p and w respectively be as in Proposition 5.5 (vi) and (viii). Let $a \in A$. Then, by Proposition 5.3 (i), $a = (b_a \vee \neg c_a) \wedge (\neg b_a \vee c_a)$ which equals $(b_a \wedge c_a) \vee (\neg b_a \wedge \neg c_a)$. Thus, since $(b_a \wedge c_a) \wedge (\neg b_a \wedge \neg c_a) = 0$,

$$\begin{aligned} s(a) &= s((b_a \wedge c_a) \vee (\neg b_a \wedge \neg c_a)) \\ &= s(b_a \wedge c_a) + s(\neg b_a \wedge \neg c_a) \\ &= s(b_a \wedge c_a) + s(\neg(b_a \vee c_a)) \\ &= s(b_a \wedge c_a) + 1 - s(b_a \vee c_a) \\ &= s(b_a \wedge c_a) + s(b_a) + s(\neg b_a) - s(b_a \vee c_a) \end{aligned} \tag{1}$$

Now, since $\neg b_a \wedge (b_a \wedge c_a) = 0$, $s(\neg b_a) + s(b_a \wedge c_a) = s(\neg b_a \vee (b_a \wedge c_a)) = s(\neg b_a \vee c_a)$. Therefore, from (1), we get

$$\begin{aligned} s(a) &= s(b_a) + (s(\neg b_a \vee c_a) - s(b_a \vee c_a)) \\ &= p(b_a) + \varepsilon w(\neg b_a \vee c_a) + 1 - \varepsilon w(b_a \vee c_a) - 1 \\ &= p(b_a) + \varepsilon(w(\neg b_a \vee c_a) - w(b_a \vee c_a)). \end{aligned}$$

Thus, the claim is settled. □

The following result is hence a direct consequence of Theorem 5.6 and Flaminio and Kroupa (2015, Corollary 4.0.5).

Corollary 5.7 *For every IBP$_0$-algebra* $\mathbf{A}$ *and every hyperstate* $s : A \to \mathscr{L}(\mathbb{R})$ *there are a regular Borel measure* μ_s *on the Stone space* $Max(\mathscr{B}(\mathbf{A}))$ *of* $\mathscr{B}(\mathbf{A})$*, a state* $w \in \mathcal{W}(\mathscr{H}(\mathbf{A}))$ *and an infinitesimal* $\varepsilon > 0$ *such that, for every* $a \in A$,

$$s(a) = \int_{Max(\mathscr{B}(\mathbf{A}))} (b_a)^* \, \mathrm{d}\mu_s + \varepsilon(w(\neg b_a \vee c_a) - w(b_a \vee c_a)),$$

where $(b_a)^*$ *denotes the characteristic function of the clopen subset of* $Max(\mathscr{B}(\mathbf{A}))$ *corresponding to* b_a *via Stone duality.*

Now, let $\mathbf{A}$ be a IBP$_0$-algebra such that $\mathscr{H}(\mathbf{A})$ is cancellative (i.e., $\mathbf{A}$ belongs to the variety of MV-algebras generated by perfect MV-algebras). Then, from Corollary 3.4 (see also Birkhoff 1967), $\mathscr{H}(\mathbf{A})$ embeds into $\mathbf{K}(\mathscr{H}(\mathbf{A}))$. Therefore, the following easily holds.

Corollary 5.8 *Let* $\mathbf{A}$ *be a IBP$_0$-algebra such that* $\mathscr{H}(\mathbf{A})$ *is cancellative. Then, for a hyperstate* $s : A \to \mathscr{L}(\mathbb{R})$ *there are a probability measure* $p : \mathscr{B}(\mathbf{A}) \to [0, 1]$ *and an* ℓ*-group state* $\sigma : \mathbf{K}(\mathscr{H}(\mathbf{A})) \to \mathbb{R}$ *such that, for every* $a \in A$,

$$s(a) = p(b_a) + \varepsilon \cdot \sigma([\neg b_a \vee c_a, b_a \vee c_a])$$

6 Conclusions and Future Work

The present paper aims at defining a notion of state of prelinear semihoops and hyperstate of IBP_0-algebras. Our investigation was mainly motivated by two key observations:

1. First, states of MV-algebras can be regarded as those mappings that arise from applying Mundici's functor Γ to states of unital ℓ-groups. Thus, with an analogue reasoning, a notion of state (more precisely, hyperstate) of perfect MV-algebras arises applying Di Nola and Lettieri's functor.
2. Second, the variety of IBP_0-algebras is categorically equivalent to a category of triples made of a Boolean algebra, a prelinear semihoop and a special operation which is meant to represent, in this category, the natural algebraic *join*. Thus, IBP_0-algebras can be decomposed in a Boolean algebra and a prelinear semihoop, and a notion of state of IBP_0-algebras can be inspired by this decomposition.

Therefore, we first introduce a notion of state for prelinear semihoops which, in turn, is suggested by Goodearl's definition of state of an ℓ-group and a version of Grothendieck group construction we proved for lattice-ordered monoids. Hyperstates of IBP_0-algebras are then introduced and we prove that, indeed, each hyperstate can be decomposed into a probability function and a state of a prelinear semihoop.

In our future work we plan to deepen the methodologies applied to the present paper to both extend hyperstates to other classes of (not necessarily involutive) MTL-algebras which satisfy the equation $(2x)^2 = 2(x^2)$ and also to provide deeper insights for these mappings. In particular, a strengthening of Theorem 5.6 and Corollary 5.7 to provide a integral representation for hyperstates of IBP_0-algebras.

References

Aguzzoli, S., T. Flaminio, and S. Ugolini. Equivalences between subcategories of MTL-algebras via Boolean algebras and prelinear semihoops. *Journal of Logic and Computation*. https://doi.org/10.1093/logcom/exx014 (in print).

Birkhoff, G. 1967. *Lattice theory*, 3rd ed., vol. XXV. Colloquium Publications, American Mathematical Society.

Cignoli, R., I.M.L. D'Ottaviano, and D. Mundici. 2000. *Algebraic foundations of many-valued reasoning*. Kluwer.

Cignoli, R., and A. Torrens. 2006. Free algebras in varieties of glivenko MTL-algebras satisfying the equation $2(x^2) = (2x)^2$. *Studia Logica* 83: 157–181.

Diaconescu, D., A. Ferraioli, T. Flaminio, and B. Gerla. 2014a. Exploring infinitesimal events through MV-algebras and non-Archimedean states. In *Proceedings of IPMU 2014*, vol. 443, ed. A. Laurent, et al., 385–394, Part II, CCIS.

Diaconescu, D., T. Flaminio, and I. Leuştean. 2014. Lexicographic MV-algebras and lexicographic states. *Fuzzy Sets and Systems* 244: 63–85.

Di Nola, A., G. Georgescu, and I. Leuştean. 2000. States on perfect MV-algebras. In *Discovering the world with fuzzy logic*, ed. V. Novak, I. Perfilieva, , vol. 57, 105–125, Studies in Fuzziness and Soft Computing, Heidelberg: Physica.

Di Nola, A., and A. Lettieri. 1994. Perfect MV-algebras are categorically equivalent to Abelian ℓ-groups. *Studia Logica* 53 (3): 417–432.
Esteva, F., L. Godo, P. Hájek, and F. Montagna. 2003. Hoops and fuzzy logic. *Journal of Logic and Computation* 13 (4): 532–555.
Flaminio, T., and T. Kroupa. 2015. States of MV-algebras. In *Handbook of mathematical fuzzy logic*, ed. P. Cintula, C. Fermüller and C. Noguera, vol. III. London: Studies in Logic, Mathematical Logic and Foundations, College Publications.
Glass, A., and W. Holland. 1989. *Lattice-ordered groups: Advances and techniques*, vol. 48. Springer.
Goodearl, K.R. 1986. *Partially ordered Abelian group with interpolation*, vol. 20. AMS Mathematical Survey and Monographs.
Hájek, P. 1998. *Metamathematics of fuzzy logics*. Dordrecht: Kluwer Academic Publishers.
He, P., B. Zhao, and X. Xin. 2017. States and internal states on semihoops. *Soft Computing* 21 (11): 2941–2957.
Kroupa, T. 2006. Every state on semisimple MV-algebra is integral. *Fuzzy Sets and Systems* 157 (20): 2771–2787.
Mundici, D. 1986. Interpretation of ACF*-algebras in Łukasiewicz sentential calculus. *Journal of Functional Analysis* 65: 15–63.
Mundici, D. 1995. Averaging the Truth-value in Łukasiewicz Logic. *Studia Logica* 55 (1): 113–127.
Mundici, D. 2011. Advanced Łukasiewicz calculus and MV-algebras. In *Trends in Logic*, vol. 35. Springer.
Noguera, C., F. Esteva, and J. Gispert. 2005. Perfect and bipartite IMTL-algebras and disconnected rotations of prelinear semihoops. *Archive for Mathematical Logic* 44: 869–886.
Panti, G. 2009. Invariant measures on free MV-algebras. *Communications in Algebra* 36 (8): 2849–2861.
Weibel, C.A. 2013. The K-book: an introduction to algebraic K-theory. In *Graduate Studies in Mathematics*, vol. 145. AMS.

The Category of Node-and-Choice Forms, with Subcategories for Choice-Sequence Forms and Choice-Set Forms

Peter A. Streufert

Abstract The literature specifies extensive-form games in many styles, but lacks a systematic way of translating across those styles. This paper is a step toward such a translation system. It defines **NCF**, the category of node-and-choice forms. The category's objects are extensive forms in essentially any style, and the category's isomorphisms are made to accord with the literature's small handful of ad hoc style equivalences. Further, this paper develops two full subcategories: **CsqF** for forms whose nodes are choice-sequences, and **CsetF** for forms whose nodes are choice-sets. It is shown that **NCF** is "isomorphically enclosed" in **CsqF** in the sense that each **NCF** form is isomorphic to a **CsqF** form. Similarly, it is shown that $\mathbf{CsqF}_{\tilde{a}}$ is isomorphically enclosed in **CsetF** in the sense that each **CsqF** form with no-absentmindedness is isomorphic to a **CsetF** form. The converses are found to be almost immediate, and the resulting equivalences unify and simplify two ad hoc style equivalences in Kline and Luckraz (2016) and Streufert (2019). Aside from the larger agenda, this paper already makes three practical contributions. Style equivalences are made easier to derive by [1] a natural concept of isomorphic invariance and [2] the composability of isomorphic enclosures. In addition, [3] some new consequences of equivalence are systematically deduced.

Keywords Extensive form · Game form · Isomorphic enclosure

AMS Classification 91A70

JEL Classification C73

The author is grateful to an anonymous reviewer for many probing questions and insights, and to Deanna Walker for valuable suggestions.

P. A. Streufert (✉)
Economics Department, University of Western Ontario,
London, ON N6A 5C2, Canada
e-mail: pstreuf@uwo.ca

S. Ju et al. (eds.), *Nonclassical Logics and Their Applications*,
Logic in Asia: Studia Logica Library,
https://doi.org/10.1007/978-981-15-1342-8_2

1 Introduction

1.1 Some Foundational Issues

There are many styles in which to specify an extensive-form game, and some of these styles are reviewed below in Sect. 1.2. It is widely believed that any definition or result in one style also holds in any other style. This widespread belief might be formally explored. For example, when a result in one style is translated to another style, how would we define the sense in which the translation is correct or incorrect? Also, if a new style were to be proposed, how would we define the sense in which the new style was fully or partially legitimate? This paper is part of a larger agenda to explore such foundational issues by means of category theory. Some aspects of this agenda are explored further in Sect. 1.3. Then Sects. 1.4–1.7 introduce this paper in particular.

1.2 Specification Styles

To set the stage, this subsection recalls that there are many styles in which to specify an extensive-form game. All styles must specify [a] nodes, which are variously called "histories", "vertices", or "states"; and [b] choices, which are variously called "actions", "alternatives", "labels", or "programs". The following paragraphs arrange the styles into five broad groups according to how the styles specify nodes and choices.

[1] Some styles specify nodes and choices abstractly without restriction. Classic examples from economics include the style of Kuhn (1953) and the style of Selten (1975). Examples from computer science and/or logic include the "labeled transition system" style[1] in Blackburn et al. (2001), p. 3, and elsewhere; the style of Shoham and Leyton-Brown (2009), p. 125; and the "epistemic process graph" style of van Benthem (2014), p. 70. A final example is the "node-and-choice" style of this paper (see Fig. 1). Because each of these styles specifies nodes and choices abstractly without restriction, each can be roughly understood to encompass all other styles as special cases.[2]

---

[1]Note 7 more precisely links "labeled transition systems" with "node-and-choice forms".

[2]Accordingly, this paper's "node-and-choice" style essentially encompasses all other extensive-form styles. Several aspects of this claim should be clarified. [1] An extensive-form game specifies a tree. This feature excludes recursively specified stochastic games such as those of Mertens (2002). [2] A node-and-choice form is assumed to be discrete in the sense that every node has a finite number of predecessors. This assumption excludes non-discrete extensive-form games such as those of Dockner et al. (2000), and Alós-Ferrer and Ritzberger (2016). [3] A node-and-choice form assumes that information sets do not share alternatives. This assumption is insubstantial in the sense of note 19 below. [4] A node-and-choice form assumes that exactly one player moves at each information

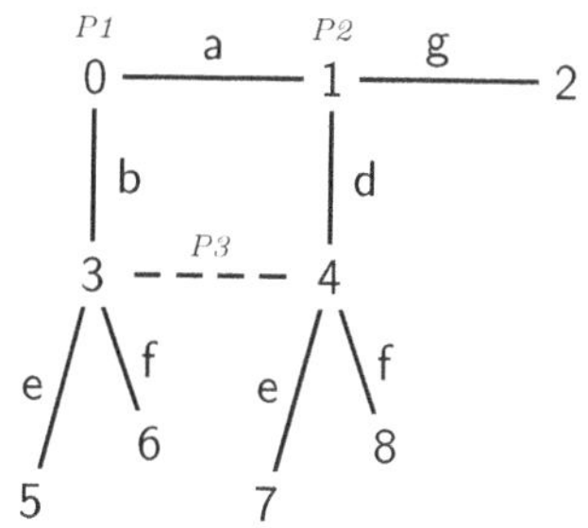

Fig. 1 A node-and-choice form (later called an "**NCF** form"). Player $P3$ selects choice e or choice f without knowing whether she is at node 3 or node 4

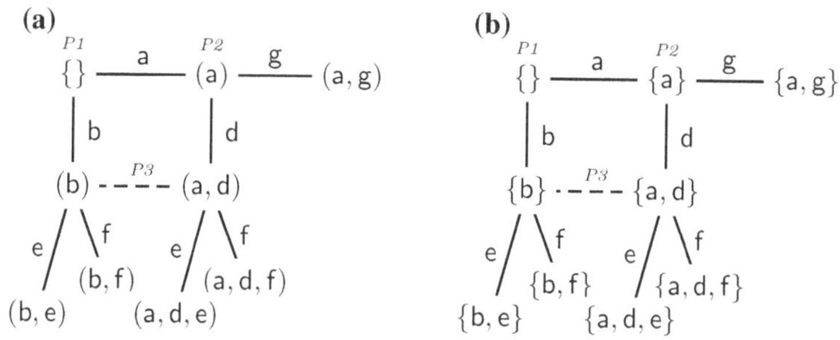

Fig. 2 **a** A choice-sequence form (later called a "**CsqF** form"). **b** A choice-set form (later called a "**CsetF** form"). These special kinds of node-and-choice forms are developed further in this paper

[2] Other styles specify nodes as sequences of choices. A popular example in economics is the style of Osborne and Rubinstein 1994, page 200. Examples from logic include the "logical game" style of Hodges 2013, Sect. 2, and the "epistemic forest model" style of van Benthem 2014, page 130. Examples from computer science include the "protocol" style of Parikh and Ramanujam 1985, the "history-based multi-agent structure" style of Pacuit 2007, and the "sequence-form representation" style of Shoham and Leyton-Brown 2009, page 129. A final example is the "choice-sequence" style of this paper (see Fig. 2a).

[3] Other styles specify nodes as sets of choices. Examples include the "choice-set" style of Streufert (2019) (henceforth "SE"), and also the "choice-set" style of this paper (see Fig. 2b).

There are still other possibilities. [4] Some styles specify choices as sets of nodes, as in the "simple" style of Alós-Ferrer and Ritzberger (2016), Sect. 6.3 (see Fig. 3a). [5] Other styles express both nodes and choices as sets of outcomes, as in the style of von Neumann and Morgenstern (1944), Sect. 10, and the style of Alós-Ferrer and Ritzberger (2016), Sect. 6.2 (see Fig. 3b). Possibilities [1]–[5] are arranged in a spectrum by SE (Streufert 2019), Fig. 2. Further, SE Sect. 7 explains how each possibility has its own advantages and disadvantages.

---

set. Accordingly, simultaneous moves by several players are specified by several information sets, as in Osborne and Rubinstein (1994), p. 202.

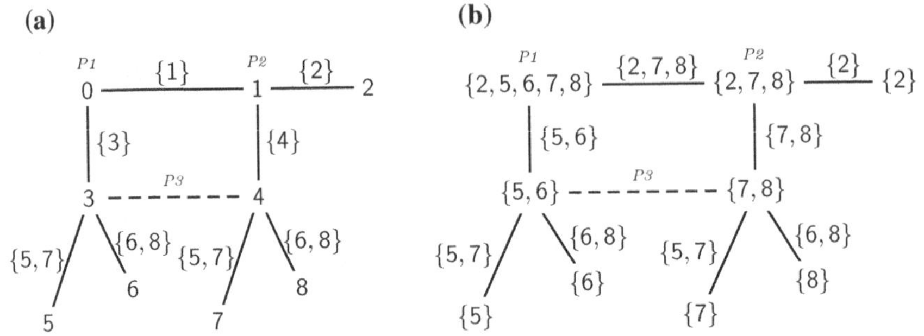

Fig. 3 In **a**, choices are node sets. In **b**, both nodes and choices are outcome sets. These special kinds of node-and-choice forms are not developed further in this paper

1.3 General Motivation

The first rigorous comparisons of specification styles have only recently appeared in Alós-Ferrer and Ritzberger (2016), Sect. 6.3, in Kline and Luckraz (2016), and in SE (whose Fig. 2 provides an overview of all these results). These contributions show, by ad hoc constructions, that the five styles in the above figures are of roughly equal generality. To be somewhat more precise, these papers argue that one style is at least as general as another style, by showing that each game[3] in the first style can be reasonably mapped to a game in the second style. Then two styles are regarded as equivalent if such an argument can be made in both directions. Notice that each such argument hinges upon an ad hoc mapping linking games in one style to games in another style. Lacking is a way to compare styles that is based on a systematic way of comparing games. This paper starts to provide that systematization in a fashion that is compatible with the prior style equivalences.

Further, a larger agenda emerges, for one can hope to do more than systematically translate games from one style to games in another style. One can hope to systematically translate properties, defined for games, from one style to another. One can hope to systematically translate equilibrium concepts from one style to another. So ultimately, one can hope to systematically translate theorems from one style to another.

Such an overarching translation system promises conceptual benefits. Foremost in the author's mind is the formal synthesis of results and questions from the many disciplines and subdisciplines which are currently studying some version of game theory. There might be much to gain because there is so much diversity. In addition, the author has been made aware of another benefit, namely, that categorical translations between games may allow for syntactic translations between the logical

[3]To be meticulous, these papers concern forms rather than games. In other words, they stop before specifying player preferences.

languages that are interpreted in those games. This would accord with the correspondence theory of van Benthem (2001), and Conradie et al. (2014).

Formal translation is a daunting task. Fortunately, category theory promises to be a powerful and natural tool. In order to gain access to this tool, the author's current objective is to construct a category [a] whose objects are extensive-form games in any style, and [b] whose isomorphisms accord with the handful of style equivalences already in the literature. This involves three steps. The first step was Streufert (2018) (henceforth "SP"). That paper defined **NCP**, which is the category of node-and-choice "preforms", where a preform is a rooted tree with choices and information sets. The second step is the present paper. This paper defines **NCF**, which is the category of node-and-choice "forms", where a form augments a preform with players. The third step is a future paper which will augment **NCF** forms with preferences in order to define extensive-form games.

Elsewhere there is relatively little categorical work on game theory. Lapitsky (1999), and Jiménez (2014) define categories for simultaneous-move games. Machover and Terrington (2014) define a category for some cooperative games. Abramsky et al. (2000), Hyland and Ong (2000), and McCusker (2000) develop categories for some specialized games in computer science. Vannucci (2007) defines a category for extensive-form games, but every morphism merely maps a game to itself. Lastly, (Honsell et al. 2012; Abramsky and Winschel 2017; Hedges 2017; Ghani et al. 2018) define categories for various games that assume trivial information sets. Streufert (2018) and the present paper depart from the literature by considering nontrivial information sets.

1.4 This Paper's Categorical Investments

As explained two paragraphs ago, this paper constructs a category of forms [a] whose objects are forms in any style, and [b] whose isomorphisms accord with the style equivalences already in the literature. Goals [a] and [b] are discussed in the next two paragraphs.

Section 2 introduces **NCF**, which is the category of node-and-choice forms, in which both nodes and choices are specified abstractly without restriction. Thereby goal [a] is achieved. Further, one special kind of node-and-choice form is a choice-sequence form, in which nodes are choice-sequences. Correspondingly, Sect. 3 introduces **CsqF**, which is the full **NCF** subcategory for choice-sequence forms. Similarly, another special kind of node-and-choice form is a choice-set form, in which nodes are choice-sets. Correspondingly, Sect. 4 introduces **CsetF**, which is the full **NCF** subcategory for choice-set forms. Finally, consider again the five styles in Sect. 1.2. **NCF** itself corresponds to style [1], **CsqF** corresponds to style [2], and **CsetF** corresponds to style [3]. Left for future research are style [4] with its node-set choices, and style [5] with its outcome-set nodes and outcome-set choices. These two additional styles will correspond to two additional subcategories of **NCF**, as suggested in Sect. 5.2's discussion of future research.

To achieve goal [b], Sect. 2 defines **NCF**'s morphisms in such a way that the category's isomorphisms accord with the style equivalences in the literature. Since this paper does not build subcategories for the node-set and outcome-set styles, only two of the literature's style equivalences remain: [i] Kline and Luckraz (2016) Theorems 1 and 2, which are essentially an equivalence between node-and-choice forms and choice-sequence forms, and [ii] SE Theorems 3.1 and 3.2, which are essentially an equivalence between (no-absentminded) choice-sequence forms and choice-set forms. As discussed earlier, each of these two equivalences is a matching pair of results, in which each result states that each form in one style can be reasonably mapped to a form in the other style. Section 3.2 proposes to strengthen each such result by requiring that each form in one style is **NCF** isomorphic to a form in the other style. This new kind of result is called an "isomorphic enclosure", and a matching pair of isomorphic enclosures is called an "isomorphic equivalence". Equivalence [i] accords with Corollary 3.3(b), which states that **NCF** and **CsqF** are isomorphically equivalent. Similarly, equivalence [ii] accords with Corollary 3.3(b), which states that $\mathbf{CsqF}_{\bar{a}}$ and **CsetF** are isomorphically equivalent. The paragraphs after these two corollaries provide historical context, more details, and more senses in which the two corollaries accord with literature's equivalences [i] and [ii].

Other results show that **NCF** is pleasant in other ways. Theorem 2.3 shows that **NCF** is a well-defined category. Theorem 2.4 shows that an **NCF** isomorphism can be characterized by bijections for nodes, choices, and players. Theorem 2.7 shows that there is a forgetful functor from **NCF** to **NCP**, which is SP's category of node-and-choice preforms. In addition, various results in Sects. 2.1–2.3 show that the category interacts naturally with game-theoretic concepts like the assignment of information sets to players. Also, Sect. 2.4 shows that the properties of no-absentmindedness and perfect-information are invariant to **NCF** isomorphisms. Finally, the paragraph after Corollary 3.5 shows how the negation of isomorphic enclosure formalizes the notion that a property is truly "restrictive" and "substantial" as opposed to merely "notational".

1.5 This Paper's Categorical Dividends

Section 1.4 argues that **NCF** systematizes prior style equivalences and that it is a pleasant category in a variety of other ways. Also, Sects. 1.3 and 5.2 argue that **NCF** promises to be of practical importance in the larger agenda of translating game theory across styles. Further, the following three paragraphs identify three practical ways that **NCF** directly contributes to game theory.

First, isomorphic invariance is a natural and powerful concept. For example, two elementary propositions in Sect. 3.3 use isomorphic invariance to find [1] general circumstances in which one subcategory is strictly isomorphically enclosed by another and [2] general circumstances in which an isomorphic enclosure can be restricted to smaller subcategories. The latter proposition is used by Corollary 3.7(b) to easily construct an isomorphic enclosure for the proof highlighted in the next paragraph.

Further, both propositions are used by Sect. 4.3 to easily derive new results about perfect-information.

Second, isomorphic enclosures can be composed (note 17). Such compositions can make it much easier to derive other isomorphic enclosures. For example, the proof of Corollary 4.3(b)'s reverse direction is just six lines long, and the third paragraph following the corollary's proof explains how this simple argument replaces six difficult pages in SE's proof of its Theorem 3.2. Thus the isomorphic equivalence of Corollary 4.3(b) is much easier to prove than the corresponding ad hoc equivalence of SE Theorems 3.1 and 3.2 (this was called equivalence [ii] in Sect. 1.4).

Third, isomorphic enclosures have consequences for form derivatives, and Sect. 5.1 deduces them simultaneously for all isomorphic enclosures. More specifically, each isomorphic enclosure is defined via isomorphisms, and Proposition 2.6 implies that each such isomorphism has consequences not only for form components (such as nodes, choices, and players) but also for form derivatives (such as the precedence relation among nodes, and each player's collection of information sets). In contrast, the literature's ad hoc style equivalences concern only form components.

1.6 Explicitness for Novice Category Theorists and Others

It is hoped that this paper can be understood by readers from economics, computer science, logic, and mathematics. Many such readers know little category theory, and these readers can benefit from the fact this paper presumes little prior knowledge of the subject. In fact, the paper's results and proofs use just the definitions of category, isomorphism, functor, and full subcategory. These fundamental definitions can be found in the early pages of Simmons (2011), Awodey (2010), and Mac Lane (1998) (arranged in descending order of accessibility).

Because the intended audience includes novice category theorists, this paper tries to avoid the notational and conceptual shortcuts that expert category theorists use to suppress the routine details that are of no interest to them. Thus expert category theorists will find the writing unusually detailed (and perhaps also awkward, pedantic, and tedious). For example, consider Sect. 2.1's concept of a functioned tree (T, p), which consists of a set T with an associated structure p. A category theorist would be apt to make the structure p implicit, to routinely denote a tree by T, and to routinely identify a morphism from one tree T to another tree T' by a function $\tau{:}T \to T'$. This paper keeps the structure p explicit, keeps denoting a functioned tree by (T, p), and keeps distinguishing between the function $\tau{:}T \to T'$ and the morphism $[(T, p), (T', p'), \tau]$. Such explicitness helps novice category theorists and slows expert category theorists.

More generally, because the intended audience spans several disciplines, this paper tries to avoid the specialized shortcuts that any of the disciplines use to suppress routine details. Thus, readers from almost any of the disciplines will probably find the writing uncomfortably detailed at one point or another. For example, consider the fundamental concept of a function. In particular, consider the two functions $f{:}\mathbb{R} \to \mathbb{R}$

defined by $(\forall x{\in}\mathbb{R})\ f(x) = \sin(x)$, and $g{:}\mathbb{R} \to [-1, 1]$ defined by $(\forall x{\in}\mathbb{R})\ g(x) = \sin(x)$. A category theorist, who regards the codomain as an essential part of a function's definition, would think $f \neq g$ (Mac Lane 1998, p. 1). In contrast, a set theorist, who identifies a function with its graph, would think $f = g$ (Halmos 1974, pp. 30–33). Meanwhile, an economist, with applications in mind, would wonder whether understanding this issue is really worth the effort. In order to speak plainly to several disciplines at the same time, this paper will be explicit about the distinction between a function and its graph. Details are in Sect. 2.1.

1.7 Organization

Section 2 develops **NCF**, the category of node-and-choice forms. Less generally, Sect. 3 develops the subcategory **CsqF** for choice-sequence forms, and Sect. 4 develops the subcategory **CsetF** for choice-set forms. Sections 3.2 and 3.3 use the context of **CsqF** to introduce the general concept of isomorphic enclosure, and to introduce general propositions about isomorphic invariance. Further, Sect. 5.1 uses parts of Sects. 3 and 4 to illustrate some general consequences of isomorphic enclosure. Finally, Sect. 5.2 discusses future research.

Although many proofs appear within the text, twelve lengthy proofs and their associated lemmas are relegated to the appendices. Appendix A concerns **NCF**, Appendix B concerns **CsqF**, and Appendix C concerns **CsetF**.

2 The Category of Node-and-Choice Forms

2.1 Objects

Node-and-choice forms are defined and developed in terms of functions, relations, and correspondences. In accord with the last paragraph of Sect. 1.6, the next two paragraphs are explicit about these almost-standard concepts.

Let a *function* $f{:}X \to Y$ be a triple (X, Y, f^{gr}), such that X and Y are sets and such that f^{gr} is a subset of $X{\times}Y$ satisfying $(\forall x{\in}X)(\exists! y{\in}Y)\ (x, y){\in}f^{\mathsf{gr}}$. The sets X, Y, and f^{gr} are, respectively, the domain, codomain, and graph of f. Note that many functions can share the same graph, but that only one of these functions is surjective. Thus a graph, together with the property of surjectivity, can be used to define a function.

Let a *relation* R between X and Y be a triple (X, Y, R^{gr}) such that X and Y are sets and R^{gr} is a subset of $X{\times}Y$. In accord with economics (Mas-Colell et al. 1995, p. 949), let a *correspondence* $F{:}X{\rightrightarrows}Y$ be a triple (X, Y, F^{gr}), such that X and Y are sets and F^{gr} is a subset of $X{\times}Y$. Although relations and correspondences are formally identical, they are used in different ways. Finally, let a relation R *on* X be a relation of the form (X, X, R^{gr}).

Let T be a set of elements t called *nodes*. As in SP Sect. 2.1 (where "SP" abbreviates Streufert 2018), a pair (T, p) is a *functioned tree* iff there are $t^o \in T$ and $X \subseteq T$ such that [T1] p is a nonempty function from $T \setminus \{t^o\}$ onto X and [T2] $(\forall t \in T \setminus \{t^o\})(\exists m \in \mathbb{N}_1)\ p^m(t) = t^o$.[4] Call p the *(immediate) predecessor* function.

A functioned tree (uniquely) determines many entities. First, it determines its *root* node t^o and its set X of *decision* nodes. Second, it determines its *stage* function $k{:}T \rightarrow \mathbb{N}_0$ by [a] $k(t^o) = 0$ and [b] $(\forall t \in T \setminus \{t^o\})\ p^{k(t)}(t) = t^o$. Third, it determines its *(strict) precedence relation* $\prec$ on T by $(\forall t^1 \in T, t^2 \in T)\ t^1 \prec t^2$ iff $(\exists m \in \mathbb{N}_1)\ t^1 = p^m(t^2)$. Relatedly, it determines its *weak precedence relation* $\preccurlyeq$ on T by $(\forall t^1 \in T, t^2 \in T)$ $t^1 \preccurlyeq t^2$ iff $(\exists m \in \mathbb{N}_0)\ t^1 = p^m(t^2)$. Finally, it determines the set $\mathcal{Z}$ of maximal chains in $(T, \preccurlyeq)$. This can be split into the set $\mathcal{Z}_{\mathsf{ft}}$ of finite maximal chains and the (possibly empty) set $\mathcal{Z}_{\mathsf{inft}}$ of infinite maximal chains. These derived entities and their basic properties are developed in SP Sects. 2.1 and 2.2.

Let C be a set of elements c called *choices*. A triple $\Pi = (T, C, \otimes)$ is a *(node-and-choice) preform* (SP Sect. 3.1) iff

[P1] there is a correspondence $F{:}T \rightrightarrows C$ and a $t^o \in T$ such that $\otimes$ is a bijection from F^{gr} onto $T \setminus \{t^o\}$,

[P2] (T, p) is a functioned tree where $p{:}T \setminus \{t^o\} \rightarrow F^{-1}(C)$ [5] is defined by $p^{\mathsf{gr}} = \{(t^\sharp, t) \in T^2 | (\exists c \in C)(t, c, t^\sharp) \in \otimes^{\mathsf{gr}}\}$, [6] and

[P3] $\mathcal{H}$ partitions $F^{-1}(C)$ where $\mathcal{H} \subseteq \mathcal{P}(T)$ is defined by $\mathcal{H} = \{F^{-1}(c) | c \in C\}$.[5,7]

Call[5,6,7] $\otimes$ the *node-and-choice* operator, and let $t \otimes c$ denote its value at $(t, c) \in F^{\mathsf{gr}}$. Call F the *feasibility* correspondence, call t^o the *root* node, call p the *immediate-predecessor* function, and call $\mathcal{H}$ the collection of *information sets*. In addition, let X equal $F^{-1}(C)$ (inconsequentially, SP uses $F^{-1}(C)$ rather than X). Call X the *decision-node set*.[8]

---

[4]Let $\mathbb{N}_0$ be $\{0, 1, 2, \dots\}$, let $\mathbb{N}_1$ be $\{1, 2, \dots\}$, and for any function f, let f^0 be the identity function.

[5][a] For $c \in C$, let $F^{-1}(c) = \{t \in T | c \in F(t)\}$. [b] Let $F^{-1}(C) = \cup_{c \in C} F^{-1}(c)$.

[6]SP Lemma C.1(a) shows that [P1] implies the well-definition and surjectivity of p.

[7]A preform can be regarded as a special kind of labeled transition system (e.g. Blackburn et al. 2001, p. 3; van Benthem 2001, p. 36). More precisely, a *labeled transition system* is a pair $(S, (R_a)_{a \in A})$ consisting of [a] a set S of states s and [b] a collection of binary relations R_a, each defined over S, which is indexed by a set A of labels a. A preform $(T, C, \otimes)$ determines a labeled transition system $(S, (R_a)_{a \in A})$ by setting $S = T$, setting $A = C$, and setting each $R_c^{\mathsf{gr}} = \{(t, t^\sharp) | (t, c, t^\sharp) \in \otimes^{\mathsf{gr}}\}$. Conversely, a labeled transition system $(S, (R_a)_{a \in A})$, which meets certain restrictions, determines a preform $(T, C, \otimes)$ by setting $T = S$, setting $C = A$, and setting $\otimes^{\mathsf{gr}} = \{(s, a, s') | (s, s') \in R_a^{\mathsf{gr}}\}$. Among the restrictions are the following. First, if the derived $(T, C, \otimes)$ is to satisfy [P1], then the original $(S, (R_a)_{a \in A})$ must be deterministic in the sense of Blackburn et al. (2001), pp. 3–4. Second, if the derived $(T, C, \otimes)$ is to satisfy [P2], then $\{(s', s) | (\exists a \in A)(s, s') \in R_a^{\mathsf{gr}}\}$ must be the graph of the immediate-predecessor function of a functioned tree. Both restrictions are substantial, and the second corresponds to [1] in note 2).

[8]SP Lemma C.1(b, c) implies that a preform's t^o and X coincide with the underlying tree's t^o and X. Hence the symbols t^o and X are unambiguous.

A node-and-choice preform Π (uniquely) determines many entities. First, it determines its F, t^o, p, $\mathcal{H}$, and X, as discussed in the previous paragraph. Second, [P2] determines the functioned tree (T, p), which in turn determines k, $\prec$, $\preccurlyeq$, $\mathcal{Z}$, $\mathcal{Z}_{\text{ft}}$, and $\mathcal{Z}_{\text{inft}}$, as discussed in the second-previous paragraph. Third, define the preform's *previous-choice* function $q{:}T\smallsetminus\{t^o\} \rightarrow C$ by $q^{\text{gr}} = \{(t^\sharp, c){\in}T{\times}C|(\exists t{\in}T)$ $(t, c, t^\sharp){\in}{\otimes}^{\text{gr}}\}$. All these entities and their basic properties are developed in SP Sects. 3.1 and 3.2. Among the basic properties is the convenient fact that $(p, q) = \otimes^{-1}$.

Let I be a set of elements i called *players*. A quadruple $\Phi = (I, T, (C_i)_{i\in I}, \otimes)$ is a *(node-and-choice) form* iff

[F1] $$(T, C, \otimes) \text{ is a preform where } C = \cup_{i\in I} C_i,$$

[F2] $$(\forall i{\in}I, j{\in}I\smallsetminus\{i\})\ C_i{\cap}C_j = \varnothing, \text{ and}$$

[F3] $$(\forall t{\in}T)(\exists i{\in}I)\ F(t) \subseteq C_i.$$

Each C_i is the set of choices that are assigned to player i. The definitions in this paragraph are new to this paper (and an earlier version, Streufert 2016).

A node-and-choice form Φ (uniquely) determines many entities. First, [F1] determines C and the preform $(T, C, \otimes)$, which in turn determines F, t^o, p, q, $\mathcal{H}$, X, k, $\prec$, $\preccurlyeq$, $\mathcal{Z}$, $\mathcal{Z}_{\text{ft}}$, and $\mathcal{Z}_{\text{inft}}$, as discussed in the second-previous paragraph. Second, define $(X_i)_{i\in I}$ at each i by $X_i = \cup_{c\in C_i} F^{-1}(c)$. X_i is the set of decision nodes that are assigned to player i. Third, define $(\mathcal{H}_i)_{i\in I}$ at each i by $\mathcal{H}_i = \{F^{-1}(c)|c{\in}C_i\}$. $\mathcal{H}_i$ is the collection of information sets that are assigned to player i.

Proposition 2.1 *Suppose $(I, T, (C_i)_i, \otimes)$ is a node-and-choice form with its X, $\mathcal{H}$, $(X_i)_{i\in I}$, and $(\mathcal{H}_i)_{i\in I}$. Then the following hold.*

(a) $\cup_{i\in I} X_i = X$ *and* $(\forall i{\in}I, j{\in}I\smallsetminus\{i\})\ X_i{\cap}X_j = \varnothing$.
(b) $(\forall i{\in}I)\ \mathcal{H}_i$ *partitions* X_i.
(c) $\cup_{i\in I}\mathcal{H}_i = \mathcal{H}$ *and* $(\forall i{\in}I, j{\in}I\smallsetminus\{i\})\ \mathcal{H}_i{\cap}\mathcal{H}_j = \varnothing$. *(Proof A.3.)*

Here are two minor remarks. [1] A preform can be understood as a one-player form. Specifically, $(T, C, \otimes)$ is a preform iff $(\{1\}, T, (C), \otimes)$ is a form, where $(C_i)_i = (C)$ is taken to mean $C_1 = C$. [2] A player i in a form is said to be *vacuous* iff $C_i = \varnothing$. A vacuous player i necessarily has $X_i = \varnothing$ and $\mathcal{H}_i = \varnothing$. Vacuous players can be convenient. For example, one can posit the existence of a chance player, and yet create a game without chance nodes by letting the chance player be vacuous.

2.2 Morphisms

A (node-and-choice) *preform morphism* (SP Sect. 3.3) is a quadruple $\alpha = [\Pi, \Pi', \tau, \delta]$ such that $\Pi = (T, C, \otimes)$ and $\Pi' = (T', C', \otimes')$ are preforms,

[PM1] $\tau{:}T \rightarrow T'$,

[PM2] $\delta{:}C \rightarrow C'$, and

[PM3] $\{\,(\tau(t), \delta(c), \tau(t^\sharp)) \mid (t, c, t^\sharp){\in}\otimes^{\mathsf{gr}}\,\} \subseteq \otimes'^{\mathsf{gr}}$.

SP Propositions 3.3 and 3.4 give two characterizations of preform morphisms which feel more category-theoretic. A (node-and-choice) *form morphism* is a quintuple $\beta = [\Phi, \Phi', \iota, \tau, \delta]$ s.t. $\Phi = (I, T, (C_i)_{i\in I}, \otimes)$ and $\Phi' = (I', T', (C'_{i'})_{i'\in I'}, \otimes')$ are forms,

[FM1] $[\Pi, \Pi', \tau, \delta]$ is a preform morphism where $\Pi = (T, C, \otimes)$, $C = \cup_{i\in I} C_i$, $\Pi' = (T', C', \otimes')$, and $C' = \cup_{i'\in I'} C'_{i'}$,

[FM2] $\iota{:}I \rightarrow I'$, and

[FM3] $(\forall i{\in}I)\ \delta(C_i) \subseteq C'_{\iota(i)}$.

The first paragraph of Proposition 2.2 rearranges the definition of a morphism. Meanwhile, the second and third paragraphs concern the many derivatives which can be constructed, via Sect. 2.1, from the source and target forms. Parts (k) and (m) are new, while the remainder are obtained by combining [FM1] with various SP results for preforms and trees.

Proposition 2.2 *Suppose* $\Phi = (I, T, (C_i)_{i\in I}, \otimes)$ *and* $\Phi' = (I', T', (C'_{i'})_{i'\in I'}, \otimes')$ *are forms. Let* $C = \cup_{i\in I} C_i$ *and* $C' = \cup_{i'\in I'} C'_{i'}$. *Then* $[\Phi, \Phi', \iota, \tau, \delta]$ *is a morphism iff the following hold.*

(a) $\iota : I \rightarrow I'$.
(b) $\tau : T \rightarrow T'$.
(c) $\delta : C \rightarrow C'$.
(d) $(\forall i{\in}I)\ \delta(C_i) \subseteq C'_{\iota(i)}$.
(e) $\{\,(\tau(t), \delta(c), \tau(t^\sharp)) \mid (t, c, t^\sharp){\in}\otimes^{\mathsf{gr}}\,\} \subseteq \otimes'^{\mathsf{gr}}$.

Further, suppose $[\Phi, \Phi', \iota, \tau, \delta]$ *is a morphism. Let* $\Pi = (T, C, \otimes)$ *and* $\Pi' = (T', C', \otimes')$. *Also, derive* F, t^o, p, q, X, $(X_i)_{i\in I}$, $\mathcal{H}$, *and* $(\mathcal{H}_i)_{i\in I}$ *from* Π *and* Φ. *Also, derive* F', t'^o, p', q', X', $(X'_{i'})_{i'\in I'}$, $\mathcal{H}'$, *and* $(\mathcal{H}'_{i'})_{i'\in I'}$ *from* Π' *and* Φ'. *Then the following hold.*

(f) $\{\,(\tau(t), \delta(c)) \mid (t, c){\in}F^{\mathsf{gr}}\,\} \subseteq F'^{\mathsf{gr}}$.
(g) $t'^o \preccurlyeq' \tau(t^o)$.
(h) $\{\,(\tau(t^\sharp), \tau(t)) \mid (t^\sharp, t){\in}p^{\mathsf{gr}}\,\} \subseteq p'^{\mathsf{gr}}$.
(i) $\{\,(\tau(t^\sharp), \delta(c)) \mid (t^\sharp, c){\in}q^{\mathsf{gr}}\,\} \subseteq q'^{\mathsf{gr}}$.
(j) $\tau(X) \subseteq X'$.
(k) $(\forall i{\in}I)\ \tau(X_i) \subseteq X'_{\iota(i)}$.
(l) $(\forall H{\in}\mathcal{H})(\exists H'{\in}\mathcal{H}')\ \tau(H) \subseteq H'$.
(m) $(\forall i{\in}I, H{\in}\mathcal{H}_i)(\exists H'{\in}\mathcal{H}'_{\iota(i)})\ \tau(H) \subseteq H'$.

Finally, derive k, $\prec$, $\preccurlyeq$, $\mathcal{Z}_{\text{ft}}$, *and* $\mathcal{Z}_{\text{inft}}$ *from* (T, p). *Also, derive* k', $\prec'$, $\preccurlyeq'$, $\mathcal{Z}_{\text{ft}}'$, *and* $\mathcal{Z}_{\text{inft}}'$ *from* (T', p'). *Then the following hold.*

(n) $(\forall t{\in}T)\ k'(\tau(t)) = k(t) + k'(\tau(t^o))$.
(o) $\{\,(\tau(t^1), \tau(t^2)) \mid (t^1, t^2){\in}{\prec^{\text{gr}}}\,\} \subseteq {\prec'^{\text{gr}}}$.
(p) $\{\,(\tau(t^1), \tau(t^2)) \mid (t^1, t^2){\in}{\preccurlyeq^{\text{gr}}}\,\} \subseteq {\preccurlyeq'^{\text{gr}}}$.
(q) $(\forall Z{\in}\mathcal{Z}_{\text{ft}})(\exists Z'{\in}\mathcal{Z}_{\text{ft}}'{\cup}\mathcal{Z}_{\text{inft}}')\ \tau(Z) \subseteq Z'$.
(r) $(\forall Z{\in}\mathcal{Z}_{\text{inft}})(\exists Z'{\in}\mathcal{Z}_{\text{inft}}')\ \tau(Z) \subseteq Z'$. *(Proof A.4.)*

2.3 *The Category* NCF

This paragraph and Theorem 2.3 define the category **NCF**, which is called the *category of node-and-choice forms*. Let an object be a (node-and-choice) form $\Phi = (I, T, (C_i)_{i\in I}, \otimes)$. Let an arrow be a (node-and-choice) form morphism $\beta = [\Phi, \Phi', \iota, \tau, \delta]$. Let source, target, identity, and composition be

$$\beta^{\text{src}} = [\Phi, \Phi', \iota, \tau, \delta]^{\text{src}} = \Phi,$$
$$\beta^{\text{trg}} = [\Phi, \Phi', \iota, \tau, \delta]^{\text{trg}} = \Phi',$$
$$\text{id}_\Phi = \text{id}_{(I,T,(C_i)_{i\in I},\otimes)} = [\Phi, \Phi, \text{id}_I, \text{id}_T, \text{id}_{\cup_{i\in I} C_i}], \text{ and}$$
$$\beta'{\circ}\beta = [\Phi', \Phi'', \iota', \tau', \delta']{\circ}[\Phi, \Phi', \iota, \tau, \delta] = [\Phi, \Phi'', \iota'{\circ}\iota, \tau'{\circ}\tau, \delta'{\circ}\delta],$$

where id_I, id_T, and $\text{id}_{\cup_{i\in I} C_i}$ are identities in **Set**.

Theorem 2.3 **NCF** *is a category. (Proof A.5.)*

Theorem 2.4 *Suppose* $\beta = [\Phi, \Phi', \iota, \tau, \delta]$ *is a morphism. Then (a)* β *is an isomorphism iff* ι, τ, *and* δ *are bijections. Further (b) if* β *is an isomorphism, then* $\beta^{-1} = [\Phi', \Phi, \iota^{-1}, \tau^{-1}, \delta^{-1}]$. *(Proof A.7.)*

Corollary 2.5 *Suppose* $[\Phi, \Phi', \iota, \tau, \delta]$ *is a morphism. Let* Π *be the preform in* Φ, *and let* Π' *be the preform in* Φ'. *Then* $[\Phi, \Phi', \iota, \tau, \delta]$ *is an isomorphism iff [1]* $[\Pi, \Pi', \tau, \delta]$ *is a preform isomorphism and [2]* ι *is a bijection.*

Proof Note $[\Pi, \Pi', \tau, \delta]$ is a preform morphism by [FM1] for $[\Phi, \Phi', \iota, \tau, \delta]$. Thus SP Theorem 3.7(a) shows that [1] is equivalent to the bijectivity of τ and δ. Therefore [1] and [2] together are equivalent to the bijectivity of ι, τ, and δ. By Theorem 2.4(a), this is equivalent to $[\Phi, \Phi', \iota, \tau, \delta]$ being an isomorphism. □

Proposition 2.6 organizes some[9] of the consequences of a form isomorphism. The proposition's first paragraph concerns form components, while the second and third paragraphs concern form derivatives. Consequences (a)–(c) repeat the forward direction of Theorem 2.4(a). Consequences (d), (k), and (m) are new, while the remainder

[9]The proposition's list of consequences is not exhaustive. For example, in the notation of the proposition's second paragraph, Lemma A.2(b) deduces that $(\forall c{\in}C)\ \tau(F^{-1}(c)) = (F')^{-1}(\delta(c))$.

are obtained by combining the forward direction of Corollary 2.5 with SP results about preforms and trees. The entire proposition is comparable to Proposition 2.2 for morphisms, and Sect. 5.1 will discuss how the proposition contributes directly to game theory.

To address a minor technical issue, note that many of the proposition's consequences are formulated by restricting functions. In each case, the codomain of the restriction is defined so that the restriction is surjective. Two other minor technical issues are discussed in notes 10 and 11.

Proposition 2.6 *Suppose* $[\Phi, \Phi', \iota, \tau, \delta]$ *is an isomorphism, where* $\Phi = (I, T, (C_i)_{i\in I}, \otimes)$ *and* $\Phi' = (I', T', (C'_{i'})_{i'\in I'}, \otimes')$. *Let* $C = \cup_{i\in I} C_i$ *and* $C' = \cup_{i'\in I'} C'_{i'}$. *Then the following hold.*

(a) ι *is a bijection from* I *onto* I'.
(b) τ *is a bijection from* T *onto* T'.
(c) δ *is a bijection from* C *onto* C'.
(d) $(\forall i\in I)$ $\delta|_{C_i}$ *is a bijection from* C_i *onto* $C'_{\iota(i)}$.[10]
(e) $(\tau, \delta, \tau)|_{\otimes^{\text{gr}}}$ *is a bijection from* $\otimes^{\text{gr}}$ *onto* $\otimes'^{\text{gr}}$.

Further, let $\Pi = (T, C, \otimes)$ *and* $\Pi' = (T', C', \otimes')$. *Also, derive* F, t^o, p, q, X, $(X_i)_{i\in I}$, $\mathcal{H}$, *and* $(\mathcal{H}_i)_{i\in I}$ *from* Π *and* Φ. *Also, derive* F', t'^o, p', q', X', $(X'_{i'})_{i'\in I'}$, $\mathcal{H}'$, *and* $(\mathcal{H}'_{i'})_{i'\in I'}$ *from* Π' *and* Φ'. *Then the following hold.*

(f) $(\tau, \delta)|_{F^{\text{gr}}}$ *is a bijection from* F^{gr} *onto* F'^{gr}.
(g) $\tau(t^o) = t'^o$.
(h) $(\tau, \tau)|_{p^{\text{gr}}}$ *is a bijection from* p^{gr} *onto* p'^{gr}.
(i) $(\tau, \delta)|_{q^{\text{gr}}}$ *is a bijection from* q^{gr} *onto* q'^{gr}.
(j) $\tau|_X$ *is a bijection from* X *onto* X'.
(k) $(\forall i\in I)$ $\tau|_{X_i}$ *is a bijection from* X_i *onto* $X'_{\iota(i)}$.[10]
(l) $\tau|_{\mathcal{H}}$ *is a bijection from* $\mathcal{H}$ *onto* $\mathcal{H}'$.[11]
(m) $(\forall i\in I)$ $\tau|_{\mathcal{H}_i}$ *is a bijection from* $\mathcal{H}_i$ *onto* $\mathcal{H}'_{\iota(i)}$.[10, 11]

Finally, derive k, $\prec$, $\preccurlyeq$, $\mathcal{Z}$, $\mathcal{Z}_{\text{ft}}$, *and* $\mathcal{Z}_{\text{inft}}$ *from* (T, p). *Also, derive* k', $\prec'$, $\preccurlyeq'$, $\mathcal{Z}'$, $\mathcal{Z}_{\text{ft}}'$, $\mathcal{Z}_{\text{inft}}'$ *from* (T', p'). *Then the following hold.*

(n) $(\forall t\in T)$ $k'(\tau(t)) = k(t)$.
(o) $(\tau, \tau)|_{\prec^{\text{gr}}}$ *is a bijection from* $\prec^{\text{gr}}$ *onto* $\prec'^{\text{gr}}$.
(p) $(\tau, \tau)|_{\preccurlyeq^{\text{gr}}}$ *is a bijection from* $\preccurlyeq^{\text{gr}}$ *onto* $\preccurlyeq'^{\text{gr}}$.
(q) $\tau|_{\mathcal{Z}_{\text{ft}}}$ *is a bijection from* $\mathcal{Z}_{\text{ft}}$ *onto* $\mathcal{Z}_{\text{ft}}'$.[11]
(r) $\tau|_{\mathcal{Z}_{\text{inft}}}$ *is a bijection from* $\mathcal{Z}_{\text{inft}}$ *onto* $\mathcal{Z}_{\text{inft}}$.[11]
(s) $\tau|_{\mathcal{Z}}$ *is a bijection from* $\mathcal{Z}$ *onto* $\mathcal{Z}'$.[11] *(Proof A.9.)*

---

[10]To be clear, parts (d), (k), and (m) do hold when there is a vacuous player i. In this case, C_i is empty, and thus, $\delta|_{C_i}$, $C'_{\iota(i)}$, X_i, $\tau|_{X_i}$, $X'_{\iota(i)}$, $\mathcal{H}_i$, $\tau|_{\mathcal{H}_i}$, and $\mathcal{H}'_{\iota(i)}$ are all empty as well.

[11]In parts (l), (m), (q), (r), and (s), τ is understood to be the function $\mathcal{P}(T) \ni S \mapsto \{\tau(t)|t\in S\} \in \mathcal{P}(T')$. For example, if $H \in \mathcal{H}$, then $\tau(H) = \{\tau(t)|t\in H\}$. Similarly, if $Z \in \mathcal{Z}_{\text{ft}} \cup \mathcal{Z}_{\text{inft}}$, then $\tau(Z) = \{\tau(t)|t\in Z\}$.

As already noted, the definition of a form incorporates a preform, and the definition of a form morphism incorporates a preform morphism. Correspondingly, Theorem 2.7 shows there is a "forgetful" functor P from **NCF** to **NCP**, where **NCP** is SP's category of node-and-choice preforms.[12]

Theorem 2.7 *Define* P *from* **NCF** *to* **NCP** *by*

$$\begin{aligned}\mathsf{P}_0 &: (I, T, (C_i)_{i\in I}, \otimes) \mapsto (T, \cup_{i\in I} C_i, \otimes) \text{ and}\\ \mathsf{P}_1 &: [\Phi, \Phi', \iota, \tau, \delta] \mapsto [\mathsf{P}_0(\Phi), \mathsf{P}_0(\Phi'), \tau, \delta].\end{aligned}$$

Then P *is a well-defined functor. (Proof A.10.)*

2.4 No-Absentmindedness and Perfect-Information

Consider an arbitrary category **Z**, and a property which is defined for the objects of **Z**. The property is said to be *isomorphically invariant* iff, for each object, the object satisfies the property iff all of its isomorphs satisfy the property. This section explores two isomorphically invariant properties: [1] no-absentmindedness and [2] perfect-information. Both properties restrict information sets.

No-absentmindedness is a standard property which is widely regarded as being very weak (see, for example, Alós-Ferrer and Ritzberger 2016, Sect. 4.2.3). To define this property in **NCP**, consider an **NCP** preform with its $\prec$ and $\mathcal{H}$. Then the preform is said to have *no-absentmindedness* iff $(\nexists H\in\mathcal{H}, t^A\in H, t^B\in H)\; t^A \prec t^B$.[13] Further, consider an **NCF** form with its preform. Then the form is said to have *no-absentmindedness* iff its preform has no-absentmindedness.

Proposition 2.8 *(a°) If* $[\Pi, \Pi', \tau, \delta]$ *is an* **NCP** *morphism and* Π' *has no-absentmindedness, then* Π *has no-absentmindedness. (a) No-absentmindedness is isomorphically invariant in* **NCP**. *(b°) If* $[\Phi, \Phi', \iota, \tau, \delta]$ *is an* **NCF** *morphism and* Φ' *has no-absentmindedness, then* Φ *has no-absentmindedness. (b) No-absentmindedness is isomorphically invariant in* **NCF**. *(Proof A.11.)*

Let $\mathbf{NCP}_{\bar{\mathsf{a}}}$ be the full subcategory of **NCP** whose objects are preforms with no-absentmindedness. (Subscripts are being used for isomorphically invariant properties.) Similarly, let $\mathbf{NCF}_{\bar{\mathsf{a}}}$ be the full subcategory of **NCF** whose objects are forms with no-absentmindedness. No-absentmindedness will appear again in Sect. 3.3.

---

[12] Incidentally, SP Theorem 3.9 shows there is a similar functor F from **NCP** to **Tree**, where **Tree** is SP's category of functioned trees. Hence $\mathsf{F}\circ\mathsf{P}$ is a functor from **NCF** to **Tree**. By SP Theorem 2.8, **Tree** is isomorphic to the full subcategory of **Grph** for converging arborescences.

[13] Piccione and Rubinstein (1997), Fig. 1, provides an example of absentmindedness. A corresponding **NCP** preform $\Pi = (T, C, \otimes)$ can be defined by $T = \{(), (\mathsf{a}), (\mathsf{b}), (\mathsf{a}, \mathsf{a}), (\mathsf{a}, \mathsf{b})\}$, $C = \{\mathsf{a}, \mathsf{b}\}$, and $\otimes = \{((), \mathsf{a}, (\mathsf{a})),\ ((), \mathsf{b}, (\mathsf{b})),\ ((\mathsf{a}), \mathsf{a}, (\mathsf{a}, \mathsf{a})),\ ((\mathsf{a}), \mathsf{b}, (\mathsf{a}, \mathsf{b}))\}$. No-absentmindedness fails because $\mathcal{H}$ contains $H = \{(), (\mathsf{a})\}$ and $() \prec (\mathsf{a})$. A corresponding **NCF** form $\Phi = (I, T, (C_i)_i, \otimes)$ can be defined by setting T and $\otimes$ as above, setting $I = \{1\}$, and setting $C_1 = \{\mathsf{a}, \mathsf{b}\}$. The existence of this example is used in the proof of Corollary 3.5.

Perfect-information is another standard property. It is restrictive, and at the same time, there are many interesting games which satisfy it (see, for example, Osborne and Rubinstein 1994 Part II). As in SP Sect. 3.5, an **NCP** preform, with its collection $\mathcal{H}$ of information sets H, is said to have *perfect-information* iff $(\forall H \in \mathcal{H})\ |H| = 1$. Perfect-information is strictly stronger than no-absentmindedness.[14,15] Further, an **NCF** form is said to have *perfect-information* iff the form's preform has perfect-information. (In spite of Proposition 2.9, the existence of a *morphism* does not lead to any logical relationship between the source's perfect-information and the target's perfect-information. Thus there is no "Proposition 2.9(a^o)" to mimic Proposition 2.8(a^o).)

Proposition 2.9 *(a) Perfect-information is isomorphically invariant in* **NCP**. *(b) Perfect-information is isomorphically invariant in* **NCF**. *(Proof A.12.)*

Let $\mathbf{NCP_p}$ be the full subcategory of **NCP** whose objects are preforms with perfect-information. (The subscript $_{\tilde{\mathrm{a}}\mathrm{p}}$ would be equivalent to the subscript $_\mathrm{p}$, because no-absentmindedness is implied by perfect-information, as shown in note 14.) Further, let $\mathbf{NCF_p}$ be the full subcategory of **NCF** whose objects are forms with perfect-information. Perfect-information will appear again in Sect. 4.3.

3 The Subcategory of Choice-Sequence Forms

3.1 Objects

Let a *(finite) sequence* be the graph of a function from $\{1, 2, \ldots m\}$ for some non-negative integer m (to be clear, the empty sequence[16] with empty domain is admitted by $m = 0$). Thus this paper regards a sequence as a set of ordered pairs. For example, $t^* = \{(1, \mathsf{g}), (2, \mathsf{f}), (3, \mathsf{f})\}$ is a sequence with domain $\{1, 2, 3\}$. An alternative notation for the same entity is $t^* = (\mathsf{g}, \mathsf{f}, \mathsf{f})$. Yet another is $t^* = (t_n^*)_{n=1}^3$ where $t_1^* = \mathsf{g}$ and $t_2^* = t_3^* = \mathsf{f}$.

Let the *length* of a sequence t be $|t|$. For instance, the length of the example sequence is $|t^*| = |\{(1, \mathsf{g}), (2, \mathsf{f}), (3, \mathsf{f})\}| = 3$, which is consistent with the observation

---

[14] To see that perfect-information implies no-absentmindedness, assume no-absentmindedness is violated. Then there is $H \in \mathcal{H}$, $t^A \in H$, and $t^B \in H$ such that $t^A \prec t^B$. Thus $t^A \neq t^B$. So $|H| > 1$ and perfect-information is violated.

[15] A simple example of a form which satisfies no-absentmindedness but not perfect-information is a form corresponding to a two-person simultaneous-move game. Specifically, define the **NCP** preform $\Pi = (T, C, \otimes)$ by $T = \{\{\}, (\mathsf{a}), (\mathsf{b}), (\mathsf{a}, \mathsf{c}), (\mathsf{a}, \mathsf{d}), (\mathsf{b}, \mathsf{c}), (\mathsf{b}, \mathsf{d})\}$, $C = \{\mathsf{a}, \mathsf{b}, \mathsf{c}, \mathsf{d}\}$, and $\otimes = \{(\{\}, \mathsf{a}, (\mathsf{a})), (\{\}, \mathsf{b}, (\mathsf{b})), ((\mathsf{a}), \mathsf{c}, (\mathsf{a}, \mathsf{c})), ((\mathsf{a}), \mathsf{d}, (\mathsf{a}, \mathsf{d})), ((\mathsf{b}), \mathsf{c}, (\mathsf{b}, \mathsf{c})), ((\mathsf{b}), \mathsf{d}, (\mathsf{b}, \mathsf{d}))\}$. Note that $\mathcal{H}$ consists of $H = \{\{\}\}$ and $H' = \{(\mathsf{a}), (\mathsf{b})\}$. No-absentmindedness holds because [i] H is a singleton and [ii] neither $(\mathsf{a}) \prec (\mathsf{b})$ nor $(\mathsf{a}) \succ (\mathsf{b})$. Perfect-information fails because $|H'| \neq 1$. A corresponding **NCF** form $\Phi = (I, T, (C_i)_i, \otimes)$ can be defined by setting T and $\otimes$ as above, setting $I = \{1, 2\}$, and setting $C_1 = \{\mathsf{a}, \mathsf{b}\}$ and $C_2 = \{\mathsf{c}, \mathsf{d}\}$. The existence of this example is used in the proof of Corollary 4.4. [A slightly more complicated example with the same combination of properties can be obtained from any of the five figures in Sect. 1.2].

[16] The empty sequence is the empty set. Further, $\{\}$ and $\varnothing$ are alternative notations for the empty set. This paper uses $\{\}$ for a root node, and uses $\varnothing$ for all other purposes.

that $(2, \mathsf{f}) \neq (3, \mathsf{f})$. Note that the length of the empty sequence $\{\}$ is $|\{\}| = 0$. Next, let the *range* of a sequence t be $R(t) = \{\, t_n \mid n{\in}\{1, 2, \ldots |t|\}\,\}$. For instance, the range of the example sequence is $R(t^*) = \{\, t^*_n \mid n{\in}\{1, 2, 3\}\,\} = \{\mathsf{g}, \mathsf{f}, \mathsf{f}\} = \{\mathsf{g}, \mathsf{f}\}$. Note that the range of the empty sequence $\{\}$ is $R(\{\}) = \varnothing$.

Let the *concatenation* $t{\oplus}s$ of two sequences t and s be $\{(1, t_1), \ldots (|t|, t_{|t|}), (|t|{+}1, s_1), \ldots (|t|{+}|s|, s_{|s|})\}$. Thus the concatenation of a sequence $t = (t_1, t_2, \ldots t_{|t|})$ with a one-element sequence (c) is $t{\oplus}(c) = (t_1, t_2, \ldots t_{|t|}, c)$. Next, for any sequence t and any $\ell \in \{0, 1, 2, \ldots |t|\}$, let ${}_1t_\ell$ denote the *initial segment* $(t_1, t_2, \ldots t_\ell)$. Thus for any sequence t, ${}_1t_0 = \{\}$.

A *choice-sequence* **NCP** preform is an **NCP** preform $(T, C, \otimes)$ such that

[Csq1] T is a collection of (finite) sequences which contains $\{\}$,

[Csq2] $(\forall\, (t, c, t^\sharp) \in \otimes^{\mathsf{gr}})\ t{\oplus}(c) = t^\sharp$.

Let **CsqP** be the full subcategory of **NCP** whose objects are choice-sequence preforms. Proposition 3.1 lists some of the special properties of **CsqP** preforms. Incidentally, property (i) and assumption [Csq1] together imply that each node in a **CsqP** preform is actually a choice sequence, as the terminology suggests.

Proposition 3.1 *Suppose* $(T, C, \otimes)$ *is a* **CsqP** *preform. Derive its* F, t^o, p, q, k, $\prec$, *and* $\preccurlyeq$. *Then the following hold.*

(a) $t^o = \{\}$.
(b) $(\forall t^\sharp{\in}T{\setminus}\{\{\}\})\ p(t^\sharp) = {}_1t^\sharp_{|t^\sharp|-1}$ *and* $q(t^\sharp) = t^\sharp_{|t^\sharp|}$.
(c) $\otimes^{\mathsf{gr}} = \{\, (t, c, t^\sharp){\in}T{\times}C{\times}T \mid t{\oplus}(c){=}t^\sharp \,\}$.
(d) $F^{\mathsf{gr}} = \{\, (t, c){\in}T{\times}C \mid t{\oplus}(c){\in}T \,\}$.
(e) $(\forall t{\in}T, m{\in}\{0, 1, \ldots |t|\})\ p^m(t) = {}_1t_{|t|-m}$.
(f) $(\forall t{\in}T)\ k(t) = |t|$.
(g) $(\forall t{\in}T)\ t = (q{\circ}p^{|t|-\ell}(t))_{\ell=1}^{|t|}$.
(h) $C = \cup_{t\in T} R(t)$.
(i) $(\forall t^A{\in}T, t^B{\in}T)\ t^A \prec t^B$ *iff* $(|t^A| < |t^B|$ *and* $t^A = {}_1t^B_{|t^A|})$.
(j) $(\forall t^A{\in}T, t^B{\in}T)\ t^A \preccurlyeq t^B$ *iff* $(|t^A| \le |t^B|$ *and* $t^A = {}_1t^B_{|t^A|})$. *(Proof B.1.)*

Finally, let a *choice-sequence* **NCF** form be an **NCF** form whose preform is a **CsqP** preform. Then let **CsqF** be the full subcategory of **NCF** whose objects are choice-sequence **NCF** forms.

3.2 Isomorphic Enclosure

Consider two full subcategories **A** and **B** of some overarching category **Z**. Say that **A** is *isomorphically enclosed* in **B** (in symbols, $\mathbf{A} \dashrightarrow \mathbf{B}$) iff every object of **A** is isomorphic to an object of **B**. Note that $\mathbf{A} \dashrightarrow \mathbf{B}$ concerns not only the subcategories

A and **B** but also, implicitly, the overarching category **Z** within which isomorphisms are defined. Further note that isomorphic enclosures can be composed in the sense that **A** $\underset{\cdot}{\rightarrow}$ **B** and **B** $\underset{\cdot}{\rightarrow}$ **C** imply **A** $\underset{\cdot}{\rightarrow}$ **C**.[17] Finally, let **A** $\underset{\cdot}{\leftrightarrow}$ **B** mean that both **A** $\underset{\cdot}{\rightarrow}$ **B** and **A** $\underset{\cdot}{\leftarrow}$ **B** hold. Call $\underset{\cdot}{\leftrightarrow}$ *isomorphic equivalence*. Isomorphic equivalence implies the standard categorical concept of equivalence in Mac Lane (1998) p. 18.

Theorem 3.2 *(a)* **NCP** $\underset{\cdot}{\rightarrow}$ **CsqP**. *In particular, suppose* $\Pi = (T, C, \otimes)$ *is an* **NCP** *preform with its* p, q, *and* k. *Define* $\bar{T} = \{\, (q \circ p^{k(t)-\ell}(t))_{\ell=1}^{k(t)} \,|\, t \in T \,\}$, *define* $\bar{\tau}{:}T \rightarrow \bar{T}$ *by* $\bar{\tau}(t) = (q \circ p^{k(t)-\ell}(t))_{\ell=1}^{k(t)}$, *and define* $\bar{\otimes}$ *by surjectivity and* $\bar{\otimes}^{\mathsf{gr}} = \{\, (\bar{\tau}(t), c, \bar{\tau}(t^\sharp)) \,|\, (t, c, t^\sharp) \in \otimes^{\mathsf{gr}} \,\}$. *Then* $\bar{\Pi} = (\bar{T}, C, \bar{\otimes})$ *is an* **CsqP** *preform,* $\bar{\tau}$ *is a bijection, and* $[\Pi, \bar{\Pi}, \bar{\tau}, \mathsf{id}_C]$ *is an* **NCP** *isomorphism. (b)* **NCF** $\underset{\cdot}{\rightarrow}$ **CsqF**. *In particular, suppose* $\Phi = (I, T, (C_i)_{i \in I}, \otimes)$ *is an* **NCF** *form. Define* $\bar{T}$, $\bar{\tau}$, *and* $\bar{\otimes}$ *as in part (a). Then* $\bar{\Phi} = (I, \bar{T}, (C_i)_{i \in I}, \bar{\otimes})$ *is a* **CsqF** *form and* $[\Phi, \bar{\Phi}, \mathsf{id}_I, \bar{\tau}, \mathsf{id}_{\cup_{i \in I} C_i}]$ *is an* **NCF** *isomorphism. (Proof B.3.)*[18]

Corollary 3.3 *(a)* **NCP** $\underset{\cdot}{\leftrightarrow}$ **CsqP**. *(b)* **NCF** $\underset{\cdot}{\leftrightarrow}$ **CsqF**.

Proof *(a)*. **NCP** $\underset{\cdot}{\rightarrow}$ **CsqP** by Theorem 3.2(a). Conversely, each **CsqP** preform is an **NCP** preform by definition. *(b)*. This is very similar to (a). Change "preform" to "form", **P** to **F**, and (a) to (b). □

This equivalence has a long history. In the more distant past, it was informally understood that game trees could be specified in terms of either [i] a collection of nodes and a collection of edges or [ii] a collection of sequences. Harris (1985) p. 617 provides an example of this informal understanding. Specification style [i] uses the nomenclature of graph theory (e.g., Tutte 1984), and style-[i] trees were the basis on which Kuhn (1953) and Selten (1975) built game forms. Later, style-[ii] trees became the basis on which Osborne and Rubinstein (1994) built game forms.

Kline and Luckraz (2016)[19] (henceforth "KL16") develop this equivalence by a pair of theorems. In recognition of the above authors, they call style-[i] forms "KS forms" and call style-[ii] forms "OR forms". Then, one of their theorems (their Theorem 2) shows that a KS form can be derived from each OR form, while the other theorem (their Theorem 1) shows that each KS form can be mapped to an OR

[17]To prove composability, recall **A** $\underset{\cdot}{\rightarrow}$ **B** means that [a] each **A** form is isomorphic to a **B** form. Similarly, **B** $\underset{\cdot}{\rightarrow}$ **C** means that [b] each **B** form is isomorphic to a **C** form. [a] and [b] imply that each **A** form is isomorphic to a **C** form, and this is what is meant by **A** $\underset{\cdot}{\rightarrow}$ **C**.

[18]Theorems 3.2 and 4.2 draw upon Lemmas A.14 and A.15. These nontrivial lemmas show how to construct isomorphisms in **NCP** and **NCF** from bijections for nodes, choices, and players. These lemmas appear to have application beyond this paper.

[19]The terms "choice", "action", and "alternative" are fundamentally synonymous. However, the literature tends to use "choice" when it is assumed that information sets do not share alternatives, and conversely, to use "action" when the assumption is relaxed. The assumption itself is insubstantial in the sense that one can always introduce more alternatives until each information set has its own alternatives (see SE Sect. 5.2, first paragraph, for more discussion). This paper makes the assumption for notational convenience, and correspondingly, uses "choice" (see SP Proposition 3.2(16b) and the paragraphs beforehand). In contrast, KL16 relaxes the assumption and uses "action".

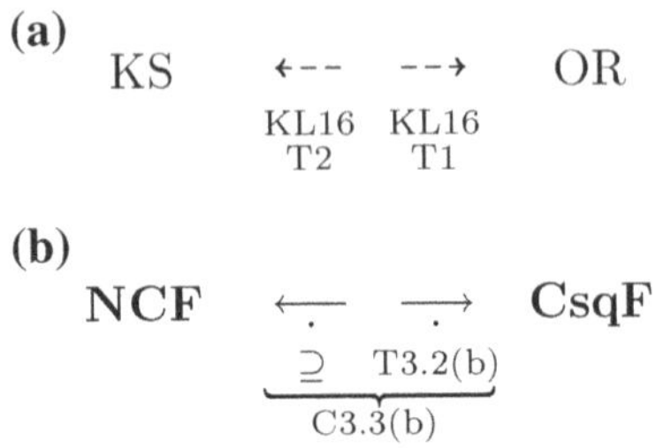

Fig. 4 **a** The ad hoc equivalence of Kline and Luckraz (2016). **b** The isomorphic equivalence of Corollary 3.3(b). T = Theorem. C = Corollary

form.[20] These two theorems are depicted by the two arrows in Fig. 4a. The arrows are dashed to convey that the equivalence is ad hoc.

Corollary 3.3(b) develops the equivalence further. Style-[i] forms are written as **NCF** forms, and style-[ii] forms are written as **CsqF** forms. Corollary 3.3(b) is then a pair of results: one half (the very easy half) shows that an **NCF** form is isomorphic to each **CsqF** form, while the other half (Theorem 3.2) shows that each **NCF** form is isomorphic to a **CsqF** form. Thus the corollary's isomorphic equivalence strengthens the KL16 equivalence by introducing isomorphisms.

There are further senses in which the corollary's isomorphic equivalence accords with the KL16 equivalence. In the backward direction, KL16 Theorem 2 is appealing because the nodes in the constructed KS form are identical to the sequences in the given OR form. This is possible because KS nodes admit OR sequences as special cases. Nonetheless KL16 Theorem 2 is nontrivial because KS forms do not admit OR forms as special cases. Here the analogous result is cleaner: **NCF** forms have been defined so that **NCF** forms admit **CsqF** forms as special cases. In the forward direction, KL16 Theorem 1 is made appealing by KL16 Lemma 2, which shows that there is a bijection α from the "vertex histories" in the given KS form to the nodes in the constructed OR form. That bijection is closely related to Theorem 3.2's bijection $\bar{\tau}$, which maps from the nodes of the given **NCF** form to the nodes in the constructed **CsqF** form.

3.3 More About No-Absentmindedness

3.3.1. Proposition 3.4 describes a general situation in which one subcategory strictly isomorphically encloses another. In the proposition, w and s are two properties defined for the objects of **Z**. Further, $w \Leftarrow\!\not\Rightarrow s$ means that w is strictly weaker than s. In other words, $w \Leftarrow\!\not\Rightarrow s$ means that [a] each object of **Z** satisfies w if it satisfies s, and [b] there is an object of **Z** that satisfies w but not s. Corollary 3.5 applies Proposition 3.4 to the nonvacuous property of no-absentmindedness.

[20]SE Theorems 3.2 and 3.1 adapt and slightly extend KL16 Theorems 2 and 1.

Proposition 3.4 *Suppose w and s are properties defined for the objects of* $\mathbf{Z}$, *and that s is isomorphically invariant. Let* $\mathbf{Z}_{\mathrm{w}}$ *be the full subcategory of* $\mathbf{Z}$ *whose objects satisfy w, and let* $\mathbf{Z}_{\mathrm{s}}$ *be the full subcategory of* $\mathbf{Z}$ *whose objects satisfy s. Then $w \Leftarrow\!\!\not\Rightarrow s$ implies* $\mathbf{Z}_{\mathrm{w}} \underset{\cdot}{\leftarrow\!\!\not\rightarrow} \mathbf{Z}_{\mathrm{s}}$.

Proof Suppose $w \Leftarrow\!\!\not\Rightarrow s$. To see $\mathbf{Z}_{\mathrm{w}} \underset{\cdot}{\leftarrow} \mathbf{Z}_{\mathrm{s}}$, take an object of $\mathbf{Z}_{\mathrm{s}}$. Since $w \Leftarrow s$, the object is also an object of $\mathbf{Z}_{\mathrm{w}}$. Thus (trivially) the object is isomorphic to an object of $\mathbf{Z}_{\mathrm{w}}$. To see $\mathbf{Z}_{\mathrm{w}} \underset{\cdot}{\not\rightarrow} \mathbf{Z}_{\mathrm{s}}$, note the assumption $w \Leftarrow\!\!\not\Rightarrow s$ implies that there is an object of $\mathbf{Z}$ that satisfies w and violates s. Thus there is an object of $\mathbf{Z}_{\mathrm{w}}$ that violates s. Thus since s is isomorphically invariant, this object does not have an isomorph that satisfies s. Thus the object does not have an isomorph in $\mathbf{Z}_{\mathrm{s}}$. □

Corollary 3.5 *(a)* $\mathbf{NCP} \underset{\cdot}{\leftarrow\!\!\not\rightarrow} \mathbf{NCP}_{\tilde{\mathrm{a}}}$. *(b)* $\mathbf{NCF} \underset{\cdot}{\leftarrow\!\!\not\rightarrow} \mathbf{NCF}_{\tilde{\mathrm{a}}}$.

Proof (a). Consider Proposition 3.4 at $\mathbf{Z}$ equal to $\mathbf{NCP}$, when w is the vacuous property satisfied by all objects of $\mathbf{NCP}$, and s is the property of no-absentmindedness. No-absentmindedness is invariant by Proposition 2.8(a). Further the vacuous property is strictly weaker than no-absentmindedness because there exists an absentminded preform (recall note 13). Thus Proposition 3.4 implies that $\mathbf{NCP}_{\mathrm{w}} = \mathbf{NCP}$ strictly isomorphically encloses $\mathbf{NCP}_{\mathrm{s}} = \mathbf{NCP}_{\tilde{\mathrm{a}}}$. *(b).* This is very similar to (a). Change "preform" to "form", **P** to **F**, and (a) to (b). □

To better interpret Corollary 3.5, recall Theorem 3.2(b) which states $\mathbf{NCF} \underset{\cdot}{\rightarrow} \mathbf{CsqF}$. Formally, this means each **NCF** form is isomorphic to a **CsqF** form. This can be interpreted to mean that the property of having choice-sequence nodes is not "restrictive". In contrast, Corollary 3.5(b) implies $\mathbf{NCF} \underset{\cdot}{\not\rightarrow} \mathbf{NCF}_{\tilde{\mathrm{a}}}$. Formally, this means there is at least one **NCF** form (such as the one in note 13) that is not isomorphic to an $\mathbf{NCF}_{\tilde{\mathrm{a}}}$ form. This can be interpreted to mean that the property of no-absentmindedness is "restrictive". Informally, the first result states that choice-sequence-ness is "purely notational". In contrast, the second result states that no-absentmindedness is "substantial", "significant", and "real", and that it "limits the range of decision processes and social interactions that can be modelled". The categorical concept of isomorphic enclosure ($\underset{\cdot}{\rightarrow}$) serves to formalize and to standardize these important terms. Note that both an isomorphic enclosure, and the negation of an isomorphic enclosure, are meaningful.

3.3.2. Next, Proposition 3.6 shows that an isomorphic enclosure can be restricted by any isomorphically invariant property. Corollary 3.7 uses this result to restrict Corollary 3.3 by no-absentmindedness. Corollary 3.7 will in turn be used in the remarkably quick proof of Corollary 4.3.

Proposition 3.6 *Suppose that* $\mathbf{A}$ *and* $\mathbf{B}$ *are full subcategories of* $\mathbf{Z}$, *and that w is an isomorphically invariant property defined for the objects of* $\mathbf{Z}$. *Let* $\mathbf{A}_{\mathrm{w}}$ *be the full subcategory of* $\mathbf{A}$ *whose objects satisfy w, and let* $\mathbf{B}_{\mathrm{w}}$ *be the full subcategory of* $\mathbf{B}$ *whose objects satisfy w. Then* $\mathbf{A} \underset{\cdot}{\rightarrow} \mathbf{B}$ *implies* $\mathbf{A}_{\mathrm{w}} \underset{\cdot}{\rightarrow} \mathbf{B}_{\mathrm{w}}$.

Proof Suppose $\mathbf{A} \underset{\cdot}{\rightarrow} \mathbf{B}$. To show $\mathbf{A}_{\mathrm{w}} \underset{\cdot}{\rightarrow} \mathbf{B}_{\mathrm{w}}$, take an object of $\mathbf{A}_{\mathrm{w}}$. Then [1] the object is an object of $\mathbf{A}$ and [2] the object satisfies w. By [1] and $\mathbf{A} \underset{\cdot}{\rightarrow} \mathbf{B}$, the object has an

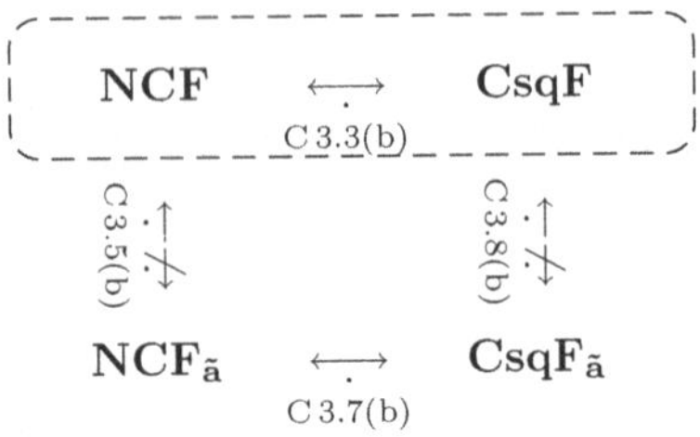

Fig. 5 Half of the previous figure, augmented with some results about no-absentmindedness. C = Corollary

isomorph in **B**. By [2] and the isomorphic invariance of w, the isomorph satisfies w. The conclusions of the previous two sentences imply that the isomorph is in $\mathbf{B_w}$. □

Corollary 3.7 *(a)* $\mathbf{NCP}_{\tilde{a}} \underset{\cdot}{\leftrightarrow} \mathbf{CsqP}_{\tilde{a}}$. *(b)* $\mathbf{NCF}_{\tilde{a}} \underset{\cdot}{\leftrightarrow} \mathbf{CsqF}_{\tilde{a}}$.

Proof (a) follows from Corollary 3.3(a), Proposition 3.6, and Proposition 2.8(a). (b) is very similar to (a). Just change (a) to (b). □

3.3.3. Finally, Corollary 3.8 could be proved by mimicking the proof of Corollary 3.5, in which case Proposition 3.4 would be employed once for part (a) at $\mathbf{Z} = \mathbf{CsqP}$, and again for part (b) at $\mathbf{Z} = \mathbf{CsqF}$. Instead, Corollary 3.8 is proved by composing isomorphic enclosures (note 17), and the proof of the corollary's part (b) is illustrated by Fig. 5. Both proof techniques are straightforward, and a more interesting example of composition will soon appear in the proof of Corollary 4.3.

Corollary 3.8 *(a)* $\mathbf{CsqP} \underset{\cdot}{\leftarrow\!\not\rightarrow} \mathbf{CsqP}_{\tilde{a}}$. *(b)* $\mathbf{CsqF} \underset{\cdot}{\leftarrow\!\not\rightarrow} \mathbf{CsqF}_{\tilde{a}}$.

Proof *(a)*. This is very similar to (b). Change **F** to **P**, and (b) to (a). *(b)*. To see $\mathbf{CsqF} \underset{\cdot}{\leftarrow} \mathbf{CsqF}_{\tilde{a}}$, note that $\mathbf{CsqF} \underset{\cdot}{\leftarrow} \mathbf{NCF} \underset{\cdot}{\leftarrow} \mathbf{NCF}_{\tilde{a}} \underset{\cdot}{\leftarrow} \mathbf{CsqF}_{\tilde{a}}$ by, respectively, Corollary 3.3(b), Corollary 3.5(b), and Corollary 3.7(b). To see $\mathbf{CsqF} \underset{\cdot}{\not\rightarrow} \mathbf{CsqF}_{\tilde{a}}$, suppose it were. Then $\mathbf{NCF} \underset{\cdot}{\rightarrow} \mathbf{CsqF} \underset{\cdot}{\rightarrow} \mathbf{CsqF}_{\tilde{a}} \underset{\cdot}{\rightarrow} \mathbf{NCF}_{\tilde{a}}$ by, respectively, Corollary 3.3(b), the supposition of the previous sentence, and Corollary 3.7(b). This contradicts Corollary 3.5(b), which states that $\mathbf{NCF} \underset{\cdot}{\not\rightarrow} \mathbf{NCF}_{\tilde{a}}$. □

4 The Subcategory of Choice-Set Forms

4.1 Objects

Let a *choice-set* **NCP** preform be an **NCP** preform $(T, C, \otimes)$ such that

[Cset1] T is a collection of finite sets which contains {} and

[Cset2] $(\forall (t, c, t^\sharp) \in \otimes^{\mathrm{gr}})\ t \cup \{c\} = t^\sharp$.

Then let **CsetP** be the full subcategory of **NCP** whose objects are choice-set **NCP** preforms. Proposition 4.1 lists some of the special properties of **CsetP** preforms.[21] Incidentally, property (f) and assumption [Cset1] together imply that each node in a **CsetP** preform is actually a choice set, in accord with the terminology. More significantly, property (g) shows that every **CsetP** preform has no-absentmindedness. In this sense the combination of [Cset1] and [Cset2] is restrictive.

Proposition 4.1 *Suppose* $(T, C, \otimes)$ *is a* **CsetP** *preform with its* F, t^o, p, q, k, $\prec$, $\preccurlyeq$, *and* $\mathcal{H}$. *Then the following hold.*

(a) $t^o = \{\}$.
(b) $(\forall t^\sharp \in T \setminus \{\{\}\})\ q(t^\sharp) \notin p(t^\sharp)$ *and* $p(t^\sharp) \cup \{q(t^\sharp)\} = t^\sharp$.
(c) $(\forall t \in T)\ k(t) = |t|$.
(d) $(\forall t \in T, m \in \{0, 1, \ldots |t|\})\ p^m(t) \subseteq t$ and $t \setminus p^m(t) = \{\, q \circ p^n(t) \mid m > n \geq 0 \,\}$.
(e) $(\forall t \in T)\ t = \{\, q \circ p^n(t) \mid |t| > n \geq 0 \,\}$.
(f) $C = \cup T$.
(g) $(T, C, \otimes)$ *has no-absentmindedness.*
(h) $(\forall t \in T, H \in \mathcal{H})\ |t \cap F(H)| \leq 1$.
(i) $(\forall t^A \in T, t^B \in T)\ t^A \subseteq t^B$ *implies* $t^A = p^{|t^B| - |t^A|}(t^B)$.
(j) $(\forall t^A \in T, t^B \in T)\ t^A \prec t^B$ *iff* $t^A \subset t^B$.
(k) $(\forall t^A \in T, t^B \in T)\ t^A \preccurlyeq t^B$ *iff* $t^A \subseteq t^B$.
(l) $\otimes^{\mathsf{gr}} = \{\ (t, c, t^\sharp) \in T \times C \times T \mid c \notin t,\ t \cup \{c\} = t^\sharp\ \}$.[22]
(m) $F^{\mathsf{gr}} = \{\ (t, c) \in T \times C \mid c \notin t,\ t \cup \{c\} \in T\ \}$. *(Proof C.2.)*

Finally, let a *choice-set* **NCF** form be an **NCF** form whose preform is a **CsetP** preform. Then let **CsetF** be the full subcategory of **NCF** whose objects are choice-set **NCF** forms.

4.2 *Isomorphic Enclosure*

Theorem 4.2 *(a)* $\mathbf{CsqP}_{\tilde{\mathrm{a}}} \underset{\cdot}{\rightarrow}$ **CsetP**. *In particular, suppose* $\bar{\Pi} = (\bar{T}, \bar{C}, \bar{\otimes})$ *is a* $\mathbf{CsqP}_{\tilde{\mathrm{a}}}$ *preform. Define* $T = R(\bar{T})$, *and define* $\otimes$ *by surjectivity and* $\otimes^{\mathsf{gr}} = \{\, (R(\bar{t}), \bar{c}, R(\bar{t}^\sharp)) \mid (\bar{t}, \bar{c}, \bar{t}^\sharp) \in \bar{\otimes}^{\mathsf{gr}}\, \}$. *Then* $\Pi = (T, \bar{C}, \otimes)$ *is a* **CsetP** *preform,* $R|_{\bar{T}}$ *is a bijection, and* $[\bar{\Pi}, \Pi, R|_{\bar{T}}, \mathsf{id}_{\bar{C}}]$ *is an* **NCP** *isomorphism. (b)* $\mathbf{CsqF}_{\tilde{\mathrm{a}}} \underset{\cdot}{\rightarrow}$ **CsetF**. *In particular, suppose* $\bar{\Phi} = (\bar{I}, \bar{T}, (\bar{C}_{\bar{i}})_{\bar{i} \in \bar{I}}, \bar{\otimes})$ *is a* $\mathbf{CsqF}_{\tilde{\mathrm{a}}}$ *form. Define* T *and* $\otimes$ *as in part (a). Then* $\Phi = (\bar{I}, T, (\bar{C}_{\bar{i}})_{\bar{i} \in \bar{I}}, \otimes)$ *is a* **CsetF** *form and* $[\bar{\Phi}, \Phi, \mathsf{id}_{\bar{I}}, R|_{\bar{T}}, \mathsf{id}_{\cup_{\bar{i} \in \bar{I}} \bar{C}_{\bar{i}}}]$ *is an* **NCF** *isomorphism. (Proof C.3.)*

---

[21] Almost every **CsetP** property in Proposition 4.1 has a **CsqP** analog in Proposition 3.1. The properties are merely presented in different orders because they are proved in different orders. The exceptions are that properties (g)–(i) have no **CsqP** analogs in Proposition 3.1.

[22] Lemma C.1 shows the following are equivalent: [a] $c \notin t$ and $t \cup \{c\} = t^\sharp$. [b] $t \neq t^\sharp$ and $t \cup \{c\} = t^\sharp$. [c] $t \neq t^\sharp$ and $t = t^\sharp \setminus \{c\}$. [d] $t \subseteq t^\sharp$ and $\{c\} = t^\sharp \setminus t$.

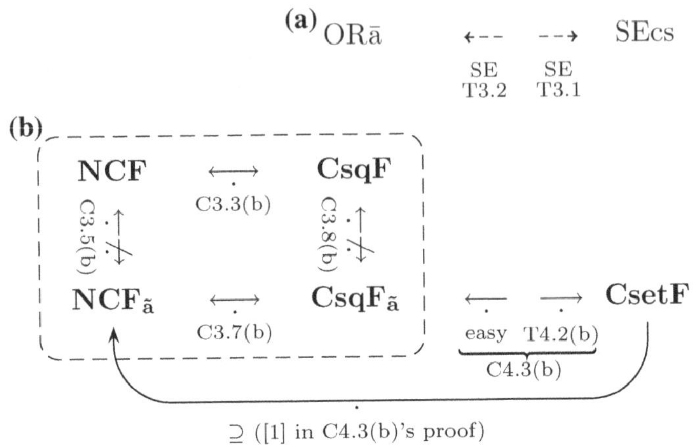

Fig. 6 **a** An ad hoc equivalence from SE. **b** The previous figure, augmented with Corollary 4.3(b) and its proof. T = Theorem. C = Corollary

Corollary 4.3 *(a)* $\mathbf{CsqP}_{\bar{a}} \underset{\cdot}{\leftrightarrow} \mathbf{CsetP}$. *(b)* $\mathbf{CsqF}_{\bar{a}} \underset{\cdot}{\leftrightarrow} \mathbf{CsetF}$.

Proof *(a)*. This is very similar to (b). Change "form" to "preform", **F** to **P**, (b) to (a), and the last phrase to "because it has no-absentmindedness by Proposition 4.1(g)".

(b). Theorem 4.2(b) shows $\mathbf{CsqF}_{\bar{a}} \underset{\cdot}{\rightarrow} \mathbf{CsetF}$. Thus it remains to show $\mathbf{CsqF}_{\bar{a}} \underset{\cdot}{\leftarrow} \mathbf{CsetF}$. Since isomorphic enclosures can be composed, it suffices to show [1] $\mathbf{CsetF} \underset{\cdot}{\rightarrow} \mathbf{NCF}_{\bar{a}}$ and [2] $\mathbf{NCF}_{\bar{a}} \underset{\cdot}{\rightarrow} \mathbf{CsqF}_{\bar{a}}$. [2] is the forward direction of Corollary 3.7(b). [1] holds simply because any **CsetF** form is a $\mathbf{NCF}_{\bar{a}}$ form. To see this, take a **CsetF** form. It is an **NCF** form by construction. It has no-absentmindedness because its preform has no-absentmindedness by Proposition 4.1(g). □

Corollary 4.3(b) is analogous to an ad hoc style equivalence in SE. There, a pair of results argues that no-absentminded OR forms ("ORā forms" in this subsection) are equivalent to SE-choice-set forms ("SEcs forms" in this subsection). One of the results (SE Theorem 3.2) shows that an ORā form can be reasonably derived from each SEcs form, and the other result (SE Theorem 3.1) shows that each ORā form can be reasonably mapped to an SEcs form. These two theorems are depicted by the two dashed arrows in Fig. 6a.

Corollary 4.3(b) strengthens this equivalence. $\mathbf{CsqF}_{\bar{a}}$ forms are like ORā forms in that both specify nodes as choice-sequences, and **CsetF** forms are like SEcs forms in that both specify nodes as choice-sets. Then, Corollary 4.3(b)'s isomorphic equivalence is a matching pair of results: one half (labelled "easy" in Fig. 6b) shows that a $\mathbf{CsqF}_{\bar{a}}$ form is isomorphic to each **CsetF** form, while the other half (Theorem 4.2) shows that each $\mathbf{CsqF}_{\tilde{a}}$ form is isomorphic to a **CsetF** form. Thus Corollary 4.3(b) strengthens the SE equivalence by introducing isomorphisms.[23]

Corollary 4.3(b)'s proof highlights how useful it is to compose isomorphic enclosures. In particular, consider the reverse direction of Corollary 4.3(b), which is $\mathbf{CsqF}_{\tilde{a}} \underset{\cdot}{\leftarrow} \mathbf{CsetF}$ in Fig. 6b, and compare it with SE Theorem 3.2, which is $\text{OR}\tilde{a} \dashleftarrow \text{SEcs}$ in Fig. 6a. The lemmas and proof for SE Theorem 3.2 span six difficult pages. In contrast, the reverse direction of Corollary 4.3(b) is proved in six lines by composing an easily-proved enclosure ($\mathbf{CsetF} \underset{\cdot}{\rightarrow} \mathbf{NCF}_{\tilde{a}}$ in part [1] of proof) with a previously-proved enclosure ($\mathbf{NCF}_{\tilde{a}} \underset{\cdot}{\rightarrow} \mathbf{CsqF}_{\tilde{a}}$ from the forward half of Corollary 3.7(b)). Figure 6b shows this composition as the curved arrow followed by the forward direction of Corollary 3.7(b).

4.3 More About Perfect-Information

Corollaries 4.4 and 4.5 are additional applications of Sect. 3.3's general propositions using isomorphic invariance.

Corollary 4.4 *(a)* $\mathbf{NCP}_{\tilde{a}} \underset{\cdot\ \cdot}{\leftarrow\!\!\not\rightarrow} \mathbf{NCP}_{\mathrm{p}}$. *(b)* $\mathbf{NCF}_{\tilde{a}} \underset{\cdot\ \cdot}{\leftarrow\!\!\not\rightarrow} \mathbf{NCF}_{\mathrm{p}}$.

Proof *(a)*. Consider Proposition 3.4 at **Z** equal to **NCP**, when w is the property of no-absentmindedness **ã**, and s is the property of perfect-information p. Perfect-information is isomorphically invariant by Proposition 2.9(a). Further no-absentmindedness is strictly weaker than perfect-information by notes 14 and 15. Thus Proposition 3.4 implies that $\mathbf{NCP}_{\tilde{a}}$ strictly isomorphically encloses $\mathbf{NCP}_{\mathrm{p}}$. *(b)*. This is very similar to (a). Change **P** to **F**, and (a) to (b). □

Corollary 4.5 *(a)* $\mathbf{NCP}_{\mathrm{p}} \underset{\cdot}{\leftrightarrow} \mathbf{CsqP}_{\mathrm{p}} \underset{\cdot}{\leftrightarrow} \mathbf{CsetP}_{\mathrm{p}}$. *(b)* $\mathbf{NCF}_{\mathrm{p}} \underset{\cdot}{\leftrightarrow} \mathbf{CsqF}_{\mathrm{p}} \underset{\cdot}{\leftrightarrow} \mathbf{CsetF}_{\mathrm{p}}$.

Proof *(a)*. Corollary 3.7(a) and Corollary 4.3(a) imply $\mathbf{NCP}_{\tilde{a}} \underset{\cdot}{\leftrightarrow} \mathbf{CsqP}_{\tilde{a}} \underset{\cdot}{\leftrightarrow} \mathbf{CsetP}$. Thus, Propositions 3.6 and 2.9(a) imply that $\mathbf{NCP}_{\tilde{a}\mathrm{p}} \underset{\cdot}{\leftrightarrow} \mathbf{CsqP}_{\tilde{a}\mathrm{p}} \underset{\cdot}{\leftrightarrow} \mathbf{CsetP}_{\mathrm{p}}$, where $\mathbf{NCP}_{\tilde{a}\mathrm{p}}$ is the full subcategory of **NCP** consisting of those objects that satisfy both no-absentmindedness and perfect-information, and where similarly $\mathbf{CsqP}_{\tilde{a}\mathrm{p}}$ is the full subcategory of **CsqP** consisting of those objects that satisfy both no-absentmindedness and perfect-information. Since no-absentmindedness is weaker than perfect-information (note 14), $\mathbf{NCP}_{\tilde{a}\mathrm{p}} = \mathbf{NCP}_{\mathrm{p}}$ and $\mathbf{CsqP}_{\tilde{a}\mathrm{p}} = \mathbf{CsqP}_{\mathrm{p}}$. *(b)*. This is very similar to (a). Change **P** to **F**, and (a) to (b). □

Incidentally, since isomorphic equivalence implies categorical equivalence, Corollary 4.5(a) implies $\mathbf{NCP}_{\mathrm{p}}$, $\mathbf{CsqP}_{\mathrm{p}}$, and $\mathbf{CsetP}_{\mathrm{p}}$ are categorically equivalent. Further, SP Theorem 3.13 and Corollary 3.14 show that $\mathbf{NCP}_{\mathrm{p}}$, **Tree**, and $\mathbf{Grph}_{\mathrm{ca}}$ are categorically equivalent, where **Tree** is the category of functioned trees which SP uses

---

[23]There is also another sense in which Corollary 4.3(b) accords with the SE equivalence. The forward half of the corollary is Theorem 4.2, and that theorem transforms choice-sequence nodes to choice-set nodes via the bijection $R|_{\bar{T}}$. That same bijection is used in SE Theorem 3.1.

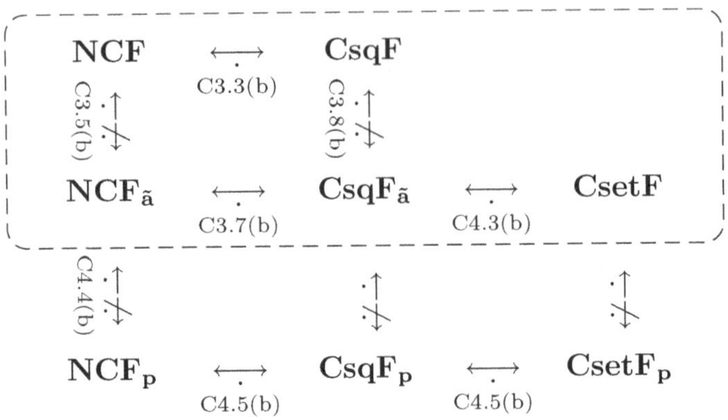

Fig. 7 Most of the previous figure, augmented with some results about perfect-information. C = Corollary

in its development of **NCP**, and where $\mathbf{Grph_{ca}}$ is the full subcategory of **Grph** whose objects are converging arborescences. Thus, $\mathbf{NCP_p}$, $\mathbf{CsqP_p}$, $\mathbf{CsetP_p}$, **Tree**, and $\mathbf{Grph_{ca}}$ are categorically equivalent.

Figure 7's arrow diagram illustrates most of the isomorphic-enclosure results from Sects. 3.2 and following. In addition, the diagram has some unlabelled arrows. They are derived by composing arrows as in the proof of Corollary 3.8. Many diagonal arrows could be similarly derived.

5 Further Remarks

5.1 Deducing Consequences from an Isomorphic Enclosure

Consider this paper's first isomorphic enclosure. Theorem 3.2 shows that each **NCF** form Φ is isomorphic to a **CsqF** form $\bar{\Phi}$ by means of an isomorphism which transforms nodes via the bijection $\bar{\tau}$. Proposition 2.6 deduces many consequences from such an isomorphism. For example, its part (o) implies that $(\forall t^1{\in}T, t^2{\in}T)\ t^1 \prec t^2$ iff $\bar{\tau}(t^1) \bar{\prec} \bar{\tau}(t^2)$, where T is the node set of Φ, $\prec$ is derived from Φ, and $\bar{\prec}$ is derived from $\bar{\Phi}$. Although such consequences about form derivatives like $\prec$ and $\bar{\prec}$ are tantalizingly natural, the consequences about form derivatives in Proposition 2.6(f)–(s) take about 10 pages to prove. That work is important because such consequences are fundamental to drawing more conclusions from the isomorphic enclosure of **NCF** in **CsqF**.

As Sect. 3.2 explained, the isomorphic enclosure of **NCF** in **CsqF** is analogous to KL16 Theorem 1. No consequences about form derivatives have been deduced from that ad hoc theorem, and an analog of Proposition 2.6(f)–(s) would likely require about 10 pages to prove. Moreover, like KL16 Theorem 1, no consequences about form derivatives have been deduced from KL16 Theorem 2 or from SE Theorems 3.1 and 3.2. Each of these ad hoc theorems has its own formulation, so deriving analogs of Proposition 2.6(f)–(s) for the three of them would likely require another $3 \times 10 = 30$ pages.

In contrast, Proposition 2.6(f)–(s) applies not only to the isomorphic enclosure of **NCF** in **CsqF**. It applies to any isomorphic enclosure. Thus it applies to all the arrows in Fig. 7, as well as to all isomorphic enclosures in the future.

5.2 *Future Research*

As discussed in Sect. 1.3, this paper is part of a larger agenda to translate game theory across specification styles. In this larger context, isomorphic enclosures can be seen as a way to translate form components from one style to another, and on the basis of these isomorphic enclosures, Proposition 2.6(f)–(s) (discussed just above) can be seen as a way of translating form derivatives from one style to another.

The results of this paper wait to be expanded in three orthogonal directions.

[1] There is more to translate beyond forms and their derivatives. This would include properties that forms might satisfy, and theorems that might relate these properties to one another. (This paper makes some limited progress in this direction by exploring the isomorphically invariant properties of no-absentmindedness and perfect-information, and by identifying some special properties of **CsqF** forms and **CsetF** forms via Propositions 3.1 and 4.1.) Expanding in this direction would correspond to expanding the three substantive sections of this paper.

[2] This paper concerns only three styles: **NCF**, **CsqF**, and **CsetF**. There are other styles to explore, including the two neglected styles mentioned at the start of this paper, namely, the "node-set" style of Alós-Ferrer and Ritzberger (2016), Sect. 6.3, and the "outcome-set" style of von Neumann and Morgenstern (1944), Alós-Ferrer and Ritzberger (2016) Sect. 6.2. Expanding in this direction will require defining new **NCF** subcategories for "node-set" forms and "outcome-set" forms, and will correspond to adding, to the present paper, two new sections for the two new subcategories.

[3] This paper concerns only forms, which need to be augmented with preferences in order to define games. At the higher level of games, many more issues emerge. To return to [1], there is more to translate, including equilibrium concepts and the theorems which might relate one equilibrium concept to another. To return to [2], there will be more than five styles because there are alternative ways to specify preferences over the same form. Expanding in this third direction will require building a new category for games which incorporates this paper's category for forms.

Appendix A: NCF

Lemma A.1 *Suppose* $(T, C, \otimes)$ *is an* **NCP** *preform with its* F, t^o, p, q, *and* $\mathcal{H}$. *Then the following hold.*

(a) $|T| \geq 2$, $|C| \geq 1$, $|\otimes^{\mathrm{gr}}| \geq 1$.
(b) $(\forall H{\in}\mathcal{H}, c{\in}C)$ $c \in F(H)$ *iff* $F^{-1}(c) = H$.
(c) $(\forall H{\in}\mathcal{H}, t^\sharp{\in}T{\smallsetminus}\{t^o\})$ $q(t^\sharp) \in F(H)$ *iff* $p(t^\sharp) \in H$.

Proof *(a)*. In the paragraph after SP Eq. (1), remark [ii] shows that $(\nexists t{\in}T)$ $p(t) = t$. Thus, since p is nonempty by [T1], there are distinct $t^1 \in T$ and $t^2 \in T$ such that $t^1 = p(t^2)$. Thus, by the definition of p in [P2], there is $c \in C$ such that $(t^1, c, t^2) \in \otimes^{\mathrm{gr}}$.

(b). (Forward direction). Suppose [a] $c \in C$, [b] $H \in \mathcal{H}$, and [c] $c \in F(H)$. [c] implies there is [d] $t \in H$ such that [e] $c \in F(t)$. [e] implies [f] $t \in F^{-1}(c)$. Meanwhile, [a] and [P3] imply [g] $F^{-1}(c) \in \mathcal{H}$. Since $\mathcal{H}$ is a partition by [P3], [b] and [g] imply H and $F^{-1}(c)$ are elements of the same partition. Hence [d] and [f] imply $H = F^{-1}(c)$.

(Reverse direction). Suppose $c \in C$, [a] $H \in \mathcal{H}$, and [b] $F^{-1}(c) = H$. Since H belongs to a partition by [a] and [P3], there is [c] $t \in H$. [c] and [b] implies $t \in F^{-1}(c)$, which implies $c \in F(t)$. This and [c] imply $c \in F(H)$.

(c). (Forward direction). Suppose $H \in \mathcal{H}$, [a] $t^\sharp \in T{\smallsetminus}\{t^o\}$ and [b] $q(t^\sharp) \in F(H)$. [b] and the forward direction of part (b) imply [c] $F^{-1}(q(t^\sharp)) = H$. Meanwhile, [a] and SP Proposition 3.1(b) imply $p(t^\sharp){\otimes}q(t^\sharp) = t^\sharp$. This and [P1] imply $(p(t^\sharp), q(t^\sharp)) \in F^{\mathrm{gr}}$. This implies $p(t^\sharp) \in F^{-1}(q(t^\sharp))$, which equals H by [c].

(Reverse direction). Suppose $H \in \mathcal{H}$, [a] $t^\sharp \in T{\smallsetminus}\{t^o\}$ and [b] $p(t^\sharp) \in H$. [a] and SP Proposition 3.1(b) imply $p(t^\sharp){\otimes}q(t^\sharp) = t^\sharp$. This and [P1] imply $(p(t^\sharp), q(t^\sharp)) \in F^{\mathrm{gr}}$. This implies $q(t^\sharp) \in F(p(t^\sharp))$. This and [b] imply $q(t^\sharp) \in F(H)$. □

Lemma A.2 [24]*Suppose* $\alpha = [\Pi, \Pi', \tau, \delta]$ *is a preform morphism, where* $\Pi = (T, C, \otimes)$ *determines* F *and where* $\Pi' = (T', C', \otimes')$ *determines* F'. *Then the following hold. (a)* $(\forall c{\in}C)$ $\tau(F^{-1}(c)) \subseteq (F')^{-1}(\delta(c))$. *(b) Suppose* α *is an isomorphism. Then* $(\forall c{\in}C)$ $\tau(F^{-1}(c)) = (F')^{-1}(\delta(c))$.

Proof *(a)*. Take c. It will be argued that

$$\begin{aligned}
\tau(F^{-1}(c)) &= \{\, t'{\in}T' \mid (\exists t{\in}T)\ t'{=}\tau(t) \text{ and } t{\in}F^{-1}(c) \,\} \\
&= \{\, t'{\in}T' \mid (\exists t{\in}T)\ t'{=}\tau(t) \text{ and } (t, c){\in}F^{\mathrm{gr}} \,\} \\
&\subseteq \{\, t'{\in}T' \mid (\exists t{\in}T)\ t'{=}\tau(t) \text{ and } (\tau(t), \delta(c)){\in}F'^{\mathrm{gr}} \,\} \\
&= \{\, t'{\in}T' \mid (\exists t{\in}T)\ t'{=}\tau(t) \text{ and } (t', \delta(c)){\in}F'^{\mathrm{gr}} \,\} \\
&\subseteq \{\, t'{\in}T' \mid (t', \delta(c)){\in}F'^{\mathrm{gr}} \,\} \\
&= (F')^{-1}(\delta(c)).
\end{aligned}$$

[24]This lemma excerpts parts of proofs from SP. In particular, the proof of part (a) rearranges part of SP Proof C.12's argument for SP Proposition 3.5, and the proof of the part (b) rearranges part of the argument for SP Lemma C.17(e).

The first inclusion holds by (18a) of SP Lemma C.6. The second inclusion holds because $\tau(T) \subseteq T'$ by [PM1]. The equalities are rearrangements.

(b). Take c. It will be argued that

$$\begin{aligned}\tau(F^{-1}(c)) &= \{\, t'{\in}T' \mid (\exists t{\in}T)\ t'{=}\tau(t) \text{ and } t{\in}F^{-1}(c)\,\}\\ &= \{\, t'{\in}T' \mid (\exists t{\in}T)\ t'{=}\tau(t) \text{ and } (t,c){\in}F^{\mathrm{gr}}\,\}\\ &= \{\, t'{\in}T' \mid (\exists t{\in}T)\ t'{=}\tau(t) \text{ and } (\tau(t),\delta(c)){\in}F'^{\mathrm{gr}}\,\}\\ &= \{\, t'{\in}T' \mid (\exists t{\in}T)\ t'{=}\tau(t) \text{ and } (t',\delta(c)){\in}F'^{\mathrm{gr}}\,\}\\ &= \{\, t'{\in}T' \mid (t',\delta(c)){\in}F'^{\mathrm{gr}}\,\}\\ &= (F')^{-1}(\delta(c)).\end{aligned}$$

The third equality holds by SP Proposition 3.8(c). The fifth holds because τ is a bijection by SP Theorem 3.7 (second sentence). The remaining equalities are rearrangements. □

Proof A.3 (for Proposition 2.1). *(a)*. First it is shown that [1] $\cup_{i\in I}X_i = X$ by arguing in steps that $\cup_{i\in I}X_i$ by the definition of $(X_i)_{i\in I}$ equals $\cup_{i\in I}(\cup_{c\in C_i}F^{-1}(c))$; which by rearrangement equals $\cup_{c\in\cup_{i\in I}C_i}F^{-1}(c)$; which by the definition of C equals $\cup_{c\in C}F^{-1}(c)$; which by definition equals $F^{-1}(C)$; which by definition (in Sect. 2.1) equals X. Thus it remains to show that $(\forall i{\in}I,\ j{\in}I\smallsetminus\{i\})\ X_i{\cap}X_j = \varnothing$. Toward that end, suppose there are $i^1 \in I$ and $i^2 \in I$ such that [2] $i^1 \neq i^2$ and $X_{i^1}{\cap}X_{i^2} \neq \varnothing$. This nonemptiness and [1] imply there is [3] $t \in X$ such that [4] $t \in X_{i^1}{\cap}X_{i^2}$. [3] and [F3] imply there is $i^* \in I$ such that [5] $F(t) \subseteq C_{i^*}$. [4] implies $t \in X_{i^1}$, which by the definition of X_{i^1} implies there is [6^1] $c^1 \in C_{i^1}$ such that $t \in F^{-1}(c^1)$. The previous set membership is equivalent to [7^1] $c^1 \in F(t)$. [7^1] and [5] imply $c^1 \in C_{i^*}$, and thus [6^1] and [F2] imply [8^1] $i^1 = i^*$. Similarly, [4] implies $t \in X_{i^2}$, which by the definition of X_{i^2} implies there is [6^2] $c^2 \in C_{i^2}$ such that $t \in F^{-1}(c^2)$. The previous set membership is equivalent to [7^2] $c^2 \in F(t)$. [7^2] and [5] imply $c^2 \in C_{i^*}$, and thus [6^2] and [F2] imply [8^2] $i^2 = i^*$. [8^1] and [8^2] imply $i^1 = i^2$, which contradicts [2].

(b). Take i. First it is shown that [1] $\mathcal{H}_i \subseteq \mathcal{H}$. This is done by arguing, in steps, that $\mathcal{H}_i$ by definition equals $\{F^{-1}(c)|c{\in}C_i\}$; which by the definition of C is a subset of $\{F^{-1}(c)|c{\in}C\}$; which by definition (in [P3]) equals $\mathcal{H}$. Since $\mathcal{H}$ is a partition by [P3], [1] implies that the elements of $\mathcal{H}_i$ are nonempty and disjoint. Thus it remains to show that $\cup\mathcal{H}_i = X_i$. This is done by arguing, in steps, that $\cup\mathcal{H}_i$ by the definition of $\mathcal{H}_i$ equals $\cup\{F^{-1}(c)|c{\in}C_i\}$; which by the definition of X_i equals X_i.

(c). First it is shown that $\cup_{i\in I}\mathcal{H}_i = \mathcal{H}$. This is done by arguing, in steps, that $\cup_{i\in I}\mathcal{H}_i$ by the definition of $(\mathcal{H}_i)_{i\in I}$ equals $\cup_{i\in I}\{F^{-1}(c)|c{\in}C_i\}$; which by rearrangement equals $\{F^{-1}(c)|c{\in}\cup_{i\in I}C_i\}$; which by the definition of C equals $\{F^{-1}(c)|c{\in}C\}$; which by definition (in [P3]) equals $\mathcal{H}$. Thus it remains to show $(\forall i{\in}I,\ j{\in}I\smallsetminus\{i\})\ \mathcal{H}_i{\cap}\mathcal{H}_j = \varnothing$. Toward that end, suppose $i^1 \in I$ and $i^2 \in I$ satisfy [1] $i^1 \neq i^2$ and $\mathcal{H}_{i^1}{\cap}\mathcal{H}_{i^2} \neq \varnothing$. This nonemptiness implies there is $H \in \mathcal{H}_{i^1}{\cap}\mathcal{H}_{i^2}$. $H \in \mathcal{H}_{i^1}$ and part (b) implies H is a nonempty subset of X_{i^1}. Similarly, $H \in \mathcal{H}_{i^2}$ and part (b) implies H is a nonempty subset of X_{i^2}. The previous two sentences imply $X_{i^1}{\cap}X_{i^2} \neq \varnothing$. Hence part (a) implies $i^1 = i^2$, which contradicts [1]. □

Proof A.4 (for Proposition 2.2). The next two paragraphs prove the first paragraph of the proposition. In particular, the next two paragraphs show that $[\Phi, \Phi', \iota, \tau, \delta]$ is a morphism iff (a)–(e) hold.

Forward Direction. Assume $[\Phi, \Phi', \iota, \tau, \delta]$ is a morphism. Then [FM1] implies $[(T, C, \otimes), (T', C', \otimes'), \tau, \delta]$ is a preform morphism, so [PM1] implies (b), [PM2] implies (c), and [PM3] implies (e). Further, [FM2] implies (a), and [FM3] implies (d).

Reverse Direction. Assume (a)–(e). Since Φ and Φ' are forms by assumption, it suffices to show [FM1]–[FM3]. [FM3] holds by (d). [FM2] holds by (a). For [FM1], note that Π and Π' are preforms by [F1] and the assumption that Φ and Φ' are forms. Thus it suffices to show [PM1]–[PM3]. [PM1] holds by (b), [PM2] holds by (c), and [PM3] holds by (e).

Henceforth assume that $[\Phi, \Phi', \iota, \tau, \delta]$ is a morphism. The remaining two paragraphs of the proposition follow from Claims 1, 2, 4, and 5 below.

Claim 1: *(k) holds.* Take i. It is argued, in steps, that $\tau(X_i)$ by definition equals $\tau(\cup_{c \in C_i} F^{-1}(c))$, which by rearrangement equals $\cup\{\, \tau(F^{-1}(c)) \mid c \in C_i \,\}$, which by Lemma A.2(a) is included in $\cup\{\, (F')^{-1}(\delta(c)) \mid c \in C_i \,\}$, which by rearrangement is $\cup\{\, (F')^{-1}(c') \mid c' \in \delta(C_i) \,\}$, which by [FM3] is included in $\cup\{\, (F')^{-1}(c') \mid c' \in C'_{\iota(i)} \,\}$, which by definition is $X'_{\iota(i)}$.

Claim 2: *(m) holds.* Take i and $H \in \mathcal{H}_i$. By the definition of $\mathcal{H}_i$, there exists [1] $c \in C_i$ such that [2] $H = F^{-1}(c)$. Let [3] $H' = (F')^{-1}(\delta(c))$. [1] and [FM3] imply $\delta(c) \in C'_{\iota(i)}$. Thus the definition of $\mathcal{H}'_{\iota(i)}$ implies $(F')^{-1}(\delta(c)) \in \mathcal{H}'_{\iota(i)}$. This and [3] imply $H' \in \mathcal{H}'_{\iota(i)}$. Thus it remains to show that $\tau(H) \subseteq H'$. It is argued, in steps, that $\tau(H)$ by [2] equals $\tau(F^{-1}(c))$, which by Lemma A.2(a) is included in $(F')^{-1}(\delta(c))$, which by [3] equals H'.

Claim 3: *(a) $[\Pi, \Pi, \tau, \delta]$ is an* **NCP** *morphism. (b) $[(T, p), (T', p'), \tau]$ is a* **Tree** *morphism.* (a) follows from [FM1]. For (b), note that (a) and SP Theorem 3.9 imply that $\mathsf{F}_1([\Pi, \Pi', \tau, \delta])$ is a **Tree** morphism. By that theorem's definition of F, $\mathsf{F}_1([\Pi, \Pi', \tau, \delta]) = [\mathsf{F}_0(\Pi), \mathsf{F}_0(\Pi'), \tau] = [(T, p), (T', p'), \tau]$.

Claim 4: *(f), (h), (i), (j), and (l) hold.* Because of Claim 3(a), these parts follow from various results in SP. In particular, (f) follows from SP Lemma C.6(18a). (h) follows from SP Lemma C.9(20a). (i) follows from SP Lemma C.9(20b). (j) follows from SP Proposition 3.4(22a) since Sect. 2.1 defines X equal to $F^{-1}(C)$ and thus X' equal to $(F')^{-1}(C')$. (l) follows from SP Proposition 3.5.

Claim 5: *(g) and (n)–(r) hold.* Because of Claim 3(b), these parts of the proposition follow from various parts of SP Proposition 2.4. In particular, (g) follows from SP Proposition 2.4(a). (n)–(p) follow from SP Proposition 2.4(c)–(e). (q) follows from SP Proposition 2.4(h). (r) follows from SP Proposition 2.4(g).

□

Proof A.5 (for Theorem 2.3). The next two paragraphs draw upon SP Theorem 3.6, which showed that **NCP** is a well-defined category.

This paragraph shows that, for each form Φ, id_Φ is a form morphism. Toward this end, take a form $\Phi = [I, T, (C_i)_{i\in I}, \otimes]$. By [F1], let $\Pi = (T, \cup_{i\in I} C_i, \otimes)$ be its **NCP** preform. It must be shown that $\mathsf{id}_\Phi = [\Phi, \Phi, \mathsf{id}_I, \mathsf{id}_T, \mathsf{id}_{\cup_{i\in I} C_i}]$ satisfies [FM1]–[FM3]. [FM1] holds because $[\Pi, \Pi, \mathsf{id}_T, \mathsf{id}_{\cup_{i\in I} C_i}]$ is an **NCP** identity, and hence, an **NCP** morphism. [FM2] holds because $\mathsf{id}_I{:}I \to I$. [FM3] holds because $(\forall j{\in}I)$ $\mathsf{id}_{\cup_{i\in I} C_i}(C_j) = C_j = C_{\mathsf{id}_I(j)}$.

This paragraph shows that, for any two form morphisms β and β', $\beta'{\circ}\beta$ is a form morphism. Toward this end, take form morphisms $\beta = [\Phi, \Phi', \iota, \tau, \delta]$ and $\beta' = [\Phi', \Phi'', \iota', \tau', \delta']$, where $\Phi = (I, T, (C_i)_{i\in I}, \otimes)$, $\Phi' = (I', T', (C'_{i'})_{i'\in I'}, \otimes')$, and $\Phi'' = (I'', T'', (C''_{i''})_{i''\in I''}, \otimes'')$. By [F1], let Π, Π', and Π'' be the **NCP** preforms underlying Φ, Φ', and Φ''. It must be shown that $\beta'{\circ}\beta = [\Phi, \Phi'', \tau'{\circ}\iota, \tau'{\circ}\iota, \delta'{\circ}\delta]$ satisfies [FM1]–[FM3]. For [FM1], it must be shown that the quadruple $[\Pi, \Pi'', \tau'{\circ}\tau, \delta'{\circ}\delta]$ is an **NCP** morphism. This holds because [a] the quadruple equals $[\Pi', \Pi'', \tau', \delta']{\circ}$ $[\Pi, \Pi', \tau, \delta]$ in **NCP**, and because [b] $[\Pi, \Pi', \tau, \delta]$ and $[\Pi', \Pi'', \tau', \delta']$ are **NCP** morphisms by [FM1] for β and β'. For [FM2], it must be shown that $\iota'{\circ}\iota{:}I \to I''$. This holds because $\iota{:}I \to I'$ by [FM2] for β, and because $\iota'{:}I' \to I''$ by [FM2] for β'. For [FM3], it must be shown that $(\forall i{\in}I)$ $(\delta'{\circ}\delta)(C_i) \subseteq C''_{\iota'\circ\iota(i)}$. To prove this, take i. It is argued that $\delta'(\delta(C_i)) \subseteq \delta'(C'_{\iota(i)}) \subseteq C''_{\iota'\circ\iota(i)}$, where the first inclusion holds because $\delta(C_i) \subseteq C'_{\iota(i)}$ by [FM3] for β, applied at i, and where the second inclusion holds by [FM3] for β', applied at $i' = \iota(i)$.

The previous two paragraphs have established the well-definition of identity and composition. The unit and associative laws are immediate. Thus **NCF** is a category (e.g. Mac Lane 1998, p. 10). □

Lemma A.6 *Suppose* $[\Phi, \Phi', \iota, \tau, \delta]$ *is a morphism, where* $\Phi = (I, T, (C_i)_{i\in I}, \otimes)$ *and* $\Phi' = (I', T', (C'_{i'})_{i'\in I'}, \otimes')$. *Further suppose that* ι *and* δ *are bijections. Then the following hold.*
(a) $(\forall i{\in}I)$ $\delta|_{C_i}$ *is a bijection from* C_i *onto* $C'_{\iota(i)}$.
(b) $(\forall i'{\in}I')$ $\delta^{-1}|_{C'_{i'}}$ *is a bijection from* $C'_{i'}$ *onto* $C_{\iota^{-1}(i')}$.

Proof Define $C = \cup_{i\in I} C_i$ and $C' = \cup_{i'\in I'} C'_{i'}$. The lemma follows from Claims 3 and 4.

Claim 1: *δ is a bijection from $\cup_{i\in I} C_i$ onto $\cup_{i'\in I'} C'_{i'}$.* [FM1] implies [PM2], which implies δ is a function from C to C'. Thus the definitions of C and C' imply δ is a function from $\cup_{i\in I} C_i$ to $\cup_{i'\in I'} C'_{i'}$. δ is a bijection by assumption.

Claim 2: $(\forall i{\in}I)$ $\delta(C_i) = C'_{\iota(i)}$. Take i. [FM3] implies $\delta(C_i) \subseteq C'_{\iota(i)}$. Thus it remains to show that $C'_{\iota(i)} \setminus \delta(C_i) = \varnothing$. Toward that end, suppose contrariwise there is c' such that [a] $c' \in C'_{\iota(i)}$ and [b] $c' \notin \delta(C_i)$. [a] and Claim 1 implies that $\delta^{-1}(c')$ is a well-defined element of $\cup_{k\in I} C_k$. Thus there is $j \in I$ such that $\delta^{-1}(c') \in C_j$. This implies [c] $c' \in \delta(C_j)$. [c] and [b] imply [d] $i \neq j$. Also, [c] and [FM3] imply $c' \in C'_{\iota(j)}$. This and

[a] imply [e] $c' \in C'_{\iota(i)} \cap C'_{\iota(j)}$. Meanwhile, [d] and the bijectivity of ι imply [f] $\iota(i) \neq \iota(j)$. [e] and [f] contradict [F2] for Φ'.

Claim 3: *(a) holds.* This follows from the bijectivity of δ and Claim 2.

Claim 4: *(b) holds.* Since ι is bijective, it suffices to prove that $(\forall i \in I)\ \delta^{-1}|_{C'_{\iota(i)}}$ is a bijection from $C'_{\iota(i)}$ onto C_i. By Claim 2, this is equivalent to proving that $(\forall i \in I)\ \delta^{-1}|_{\delta(C_i)}$ is a bijection from $\delta(C_i)$ onto C_i. This follows from part (a). □

Proof A.7 (for Theorem 2.4). Let the components of Φ be $(I, T, (C_i)_{i \in I}, \otimes)$, define $C = \cup_i C_i$, let the components of Φ' be $(I', T', (C'_{i'})_{i' \in I'}, \otimes')$, and define $C' = \cup_{i'} C'_{i'}$.

The forward half of (a) and all of (b). Suppose that β is an isomorphism (Awodey 2010, p. 12, Definition 1.3). Recall that $\beta = [\Phi, \Phi', \iota, \tau, \delta]$ and let $\beta^{-1} = [\Phi^*, \Phi^{**}, \iota^*, \tau^*, \delta^*]$. Then

$$[1]\ \ [\Phi^*, \Phi^{**}, \iota^*, \tau^*, \delta^*] \circ [\Phi, \Phi', \iota, \tau, \delta] = \mathsf{id}_\Phi = [\Phi, \Phi, \mathsf{id}_I, \mathsf{id}_T, \mathsf{id}_C] \text{ and}$$
$$[2]\ \ [\Phi, \Phi', \iota, \tau, \delta] \circ [\Phi^*, \Phi^{**}, \iota^*, \tau^*, \delta^*] = \mathsf{id}_{\Phi'} = [\Phi', \Phi', \mathsf{id}_{I'}, \mathsf{id}_{T'}, \mathsf{id}_{C'}],$$

where the first equality in both lines holds by the definition of β^{-1}, and the second equality in both lines holds by the definition of id. The well definition of $\circ$ in [1] implies [a] $\Phi^* = \Phi'$. Analogously, the well definition of $\circ$ in [2] implies [b] $\Phi^{**} = \Phi$. The third component of [1] implies $\iota^* \circ \iota = \mathsf{id}_I$, and the third component of [2] implies $\iota \circ \iota^* = \mathsf{id}_{I'}$. Thus ι is a bijection from I onto I' and [c] $\iota^* = \iota^{-1}$. Similarly, the fourth components of [1] and [2] imply τ is a bijection from T onto T' and [d] $\tau^* = \tau^{-1}$. Similarly again, the fifth components of [1] and [2] imply δ is a bijection from C onto C' and [e] $\delta^* = \delta^{-1}$. To conclude, the previous three sentences have shown that ι, τ, and δ are bijections. Further,

$$\beta^{-1} = [\Phi^*, \Phi^{**}, \iota^*, \tau^*, \delta^*] = [\Phi', \Phi, \iota^{-1}, \tau^{-1}, \delta^{-1}],$$

where the first equality follows from the definition of β^{-1}, and where the second equality follows from [a]–[e].

The reverse half of (a). Suppose that ι, τ, and δ are bijections. Define $\beta^* = [\Phi', \Phi, \iota^{-1}, \tau^{-1}, \delta^{-1}]$. Derive Π from Φ and Π' from Φ'. The remainder of this paragraph will show that β^* is a form morphism by showing that it satisfies

[FM1′] $\quad [\Pi', \Pi, \tau^{-1}, \delta^{-1}]$ is a preform morphism,

[FM2′] $\quad \iota^{-1}{:}I' \to I$, and

[FM3′] $\quad (\forall i' \in I')\ \delta^{-1}(C'_{i'}) \subseteq C_{\iota^{-1}(i')}$.

To see [FM1′], first note that $[\Pi, \Pi', \tau, \delta]$ is a preform morphism by [FM1] for β. Thus the bijectivity of τ and δ, together with SP Theorem 3.7(a), imply that $[\Pi', \Pi, \tau^{-1}, \delta^{-1}]$ is an **NCP** isomorphism. Hence a fortiori, it is a preform morphism. To see [FM2′], first note that $\iota{:}I \to I'$ by [FM2] for β. Thus the bijectivity of

ι implies that $\iota^{-1}{:}I' \to I$. Finally, to see [FM3′], consider Lemma A.6. The lemma's assumptions are met because the theorem assumes that $\beta = [\Phi, \Phi', \iota, \tau, \delta]$ is a morphism and because the start of this paragraph assumes that ι and δ are bijections. Thus the lemma's part (b) implies that $(\forall i'{\in}I')\ \delta^{-1}(C'_{i'}) = C_{\iota^{-1}(i')}$.

To conclude, β^* is a form morphism by the previous paragraph. Further,

$$\beta^*{\circ}\beta = [\Phi', \Phi, \iota^{-1}, \tau^{-1}, \delta^{-1}]{\circ}[\Phi, \Phi', \iota, \tau, \delta] = \mathsf{id}_\Phi \text{ and}$$
$$\beta{\circ}\beta^* = [\Phi, \Phi', \iota, \tau, \delta]{\circ}[\Phi', \Phi, \iota^{-1}, \tau^{-1}, \delta^{-1}] = \mathsf{id}_{\Phi'}.$$

Hence β is an isomorphism (and incidentally, $\beta^{-1} = \beta^*$). □

Lemma A.8 *Suppose* $[\Phi, \Phi', \iota, \tau, \delta]$ *is a morphism, where* $\Phi = (I, T, (C_i)_{i\in I}, \otimes)$ *determines* $(\mathcal{H}_i)_{i\in I}$, *and where* $\Phi' = (I', T', (C'_{i'})_{i'\in I'}, \otimes')$ *determines* $(\mathcal{H}'_{i'})_{i'\in I'}$. *Further suppose that* $[\Pi, \Pi', \tau, \delta]$ *is an isomorphism, where* $\Pi = (T, C, \otimes)$, $C = \cup_{i\in I} C_i$, $\Pi' = (T', C', \otimes')$, *and* $C' = \cup_{i'\in I'} C'_{i'}$. *Then* $(\forall i{\in}I, H{\in}\mathcal{H}_i)\ \tau(H) \in \mathcal{H}'_{\iota(i)}$.

Proof Derive F from Π, and F' from Π'. Since $[\Pi, \Pi', \tau, \delta]$ is an isomorphism, Lemma A.2(b) implies [1] $(\forall c{\in}C)\ \tau(F^{-1}(c)) = (F')^{-1}(\delta(c))$.

Now take i and $H \in \mathcal{H}_i$. Then there is [2] $c^* \in C_i$ such that [3] $H = F^{-1}(c^*)$. First it is shown that [4] $\delta(c^*) \in C'_{\iota(i)}$ by arguing, in steps, that $\delta(c^*)$ by [2] belongs to $\delta(C_i)$, which by [FM3] is included in $C'_{\iota(i)}$. Finally, it is argued, in steps, that $\tau(H)$ by [3] equals $\tau(F^{-1}(c^*))$, which by [1] equals $(F')^{-1}(\delta(c^*))$, which by [4] belongs to $\mathcal{H}'_{\iota(i)}$. □

Proof A.9 (for Proposition 2.6). The proposition follows from Claims 1–4 and 6–7.

Claim 1: *(a)–(c) hold.* The forward direction of Theorem 2.4(a) implies that ι, τ, and δ are bijections.

Claim 2: *(d) holds.* This follows from Lemma A.6(a).

Claim 3: *(k) holds.* Take i. Since τ is a bijection by Claim 1 (part (b)), it suffices to argue that

$$\begin{aligned}\tau(X_i) &= \cup\{\, \tau(F^{-1}(c)) \mid c{\in}C_i \,\}\\ &= \cup\{\, (F')^{-1}(\delta(c)) \mid c{\in}C_i \,\}\\ &= \cup\{\, (F')^{-1}(c') \mid c'{\in}C'_{\iota(i)} \,\} = X'_{\iota(i)}.\end{aligned}$$

The first equality holds by the definition of X_i and a rearrangement. The second equality follows from Lemma A.2(b) because $[\Pi, \Pi', \tau, \delta]$ is an isomorphism by Corollary 2.5. The third equality holds by Claim 2 (part (d)). The fourth equality holds by the definition of $X'_{\iota(i)}$.

Claim 4: *(m) holds.* Take i. Since $[\Pi, \Pi', \tau, \delta]$ is an isomorphism by Corollary 2.5, Lemma A.8 implies that $\tau|_{\mathcal{H}_i}$ is a well-defined function from $\mathcal{H}_i$ into $\mathcal{H}'_{\iota(i)}$. It is injective because τ is injective by Claim 1 (part (b)). To show that it is surjective, take $H' \in \mathcal{H}_{\iota(i)}$. Since $[\Phi', \Phi, \iota^{-1}, \tau^{-1}, \delta^{-1}]$ is an isomorphism by Theorem 2.4(b), $[\Pi', \Pi, \tau^{-1}, \delta^{-1}]$ is an isomorphism by Corollary 2.5. Thus Lemma A.8 can be applied to $[\Phi', \Phi, \tau^{-1}, \iota^{-1}, \delta^{-1}]$. Therefore $H' \in \mathcal{H}_{\iota(i)}$ implies $\tau^{-1}(H') \in \mathcal{H}_{\iota^{-1}\circ\iota(i)}$. Hence $\tau^{-1}(H') \in \mathcal{H}_i$. This implies that $\tau(\tau^{-1}(H')) = H'$ is in the range of $\tau|_{\mathcal{H}_i}$.

Claim 5: *(a).* $[\Pi, \Pi', \tau, \delta]$ *is an* **NCP** *isomorphism, where* $\Pi = (T, C, \otimes)$ *and* $\Pi = (T', C', \otimes')$. *(b)* $[(T, p), (T', p'), \tau]$ *is a* **Tree** *isomorphism.* (a) holds by Corollary 2.5. For (b), note that (a) and SP Theorem 3.9 imply $\mathsf{F}_1([\Pi, \Pi', \tau, \delta])$ is a **Tree** isomorphism. By that theorem's definition of F, $\mathsf{F}_1([\Pi, \Pi', \tau, \delta]) = [\mathsf{F}_0(\Pi), \mathsf{F}_0(\Pi'), \tau] = [(T, p), (T', p'), \tau]$.

Claim 6: *(e), (f), (i), (j), and (l) hold.* These hold by Claim 5(a) and parts of SP Proposition 3.8. In particular, (e) holds by SP Proposition 3.8(b). (f) holds by SP Proposition 3.8(c). (i) holds by SP Proposition 3.8(d). (j) holds by SP Proposition 3.8(a) since Sect. 2.1 defines X as $F^{-1}(C)$ and thus X' as $(F')^{-1}(C')$. (l) holds by SP Proposition 3.8(e).

Claim 7: *(g), (h), and (n)–(s) hold.* These hold by Claim 5(b) and various parts of SP Proposition 2.7. In particular, (g) holds by SP Proposition 2.7(c). (h) holds by SP Proposition 2.7(e). (n) holds by SP Proposition 2.7(d). (o)–(r) hold by SP Proposition 2.7(f)–(i). Finally, (s) follows from (q) and (r). □

Proof A.10 (for Theorem 2.7). By [F1], P_0 maps any form into a preform. By [FM1], P_1 maps any form morphism into a preform morphism. Thus it suffices to show that P preserves source, target, identity, and composition (Mac Lane 1998 p. 13). This is done in the following four claims.

Claim 1: $\mathsf{P}_1(\beta)^{\mathsf{src}} = \mathsf{P}_0(\beta^{\mathsf{src}})$. Take $\beta = [\Phi, \Phi', \iota, \tau, \delta]$. Then it is argued, in steps, that $\mathsf{P}_1(\beta)^{\mathsf{src}}$ by the definition of β is equal to $\mathsf{P}_1([\Phi, \Phi', \iota, \tau, \delta])^{\mathsf{src}}$, which by the definition of P_1 is equal to $[\mathsf{P}_0(\Phi), \mathsf{P}_0(\Phi'), \tau, \delta]^{\mathsf{src}}$, which by the definition of src in **NCP** is equal to $\mathsf{P}_0(\Phi)$, which by the definition of src in **NCF** is equal to $\mathsf{P}_0([\Phi, \Phi', \iota, \tau, \delta]^{\mathsf{src}})$, which by the definition of β is equal to $\mathsf{P}_0(\beta^{\mathsf{src}})$.

Claim 2: $\mathsf{P}_1(\beta)^{\mathsf{trg}} = \mathsf{P}_0(\beta^{\mathsf{trg}})$. This is very similar to Claim 1. Simply change src to trg.

Claim 3: $\mathsf{P}_1(\mathsf{id}_\Phi) = \mathsf{id}_{\mathsf{P}_0(\Phi)}$. Take $\Phi = (I, T, (C_i)_{i\in I}, \otimes)$ and let $C = \cup_i C_i$. First it is shown that [a] $\mathsf{P}_0(\Phi) = (T, C, \otimes)$ by arguing, in steps, that $\mathsf{P}_0(\Phi)$ by the definition of Φ is $\mathsf{P}_0(I, T, (C_i)_{i\in I}, \otimes)$, which by the definition of P_0 is $(T, \cup_{i\in I} C_i, \otimes)$, which by the definition of C is $(T, C, \otimes)$. Then it is argued, in steps, that $\mathsf{P}_1(\mathsf{id}_\Phi)$ by the definition of id in **NCF** is equal to $\mathsf{P}_1([\Phi, \Phi, \mathsf{id}_I, \mathsf{id}_T, \mathsf{id}_C])$, which by the definition of P_1 is equal to $[\mathsf{P}_0(\Phi), \mathsf{P}_0(\Phi), \mathsf{id}_T, \mathsf{id}_C]$, which by [a] is equal to $[(T, C, \otimes), (T, C, \otimes), \mathsf{id}_T, \mathsf{id}_C]$, which by the definition of id in **NCP** is equal to $\mathsf{id}_{(T,C,\otimes)}$, which by [a] is equal to $\mathsf{id}_{\mathsf{P}_0(\Phi)}$.

Claim 4: $\mathsf{P}_1(\beta'\circ\beta) = \mathsf{P}_1(\beta')\circ\mathsf{P}_1(\beta)$. Take $\beta = [\Phi, \Phi', \iota, \tau, \delta]$ and $\beta' = [\Phi', \Phi'', \iota', \tau', \delta']$. First note that, since P_1 is well-defined by the first paragraph, $\mathsf{P}_1([\Phi, \Phi', \iota, \tau, \delta]) = [\mathsf{P}_0(\Phi), \mathsf{P}_0(\Phi'), \tau, \delta]$ and $\mathsf{P}_1([\Phi', \Phi'', \iota', \tau', \delta']) = [\mathsf{P}_0(\Phi'), \mathsf{P}_0(\Phi''), \tau', \delta']$ are preform morphisms. Then it is argued that

$$\begin{aligned}
\mathsf{P}_1(\beta'\circ\beta) &= \mathsf{P}_1([\Phi',\Phi'',\iota',\tau',\delta']\circ[\Phi,\Phi',\iota,\tau,\delta]) \\
&= \mathsf{P}_1([\Phi,\Phi'',\iota'\circ\iota,\tau'\circ\tau,\delta'\circ\delta]) \\
&= [\mathsf{P}_0(\Phi),\mathsf{P}_0(\Phi''),\tau'\circ\tau,\delta'\circ\delta] \\
&= [\mathsf{P}_0(\Phi'),\mathsf{P}_0(\Phi''),\tau',\delta']\circ[\mathsf{P}_0(\Phi),\mathsf{P}_0(\Phi'),\tau,\delta] \\
&= \mathsf{P}_1([\Phi',\Phi'',\iota',\tau',\delta'])\circ\mathsf{P}_1([\Phi,\Phi',\iota,\tau,\delta]) \\
&= \mathsf{P}_1(\beta')\circ\mathsf{P}_1(\beta),
\end{aligned}$$

where the first equality holds by the definitions of β and β', the second by the definition of $\circ$ in **NCF**, the third by the definition of P_1, the fourth by the previous sentence and by the definition of $\circ$ in **NCP**, the fifth by the definition of P_1, and the sixth by the definitions of β and β'. □

Proof A.11 (for Proposition 2.8). *(a°)*. Suppose $[\Pi, \Pi', \tau, \delta]$ is a preform morphism, with $\Pi = (T, C, \otimes)$ determining $\prec$ and $\mathcal{H}$, and with $\Pi' = (T', C', \otimes')$ determining $\prec'$ and $\mathcal{H}'$. It suffices to show that the absentmindedness of Π implies the absentmindedness of Π'. Toward that end, suppose Π has absentmindedness. Then there are [1] $H \in \mathcal{H}$, [2] $t^A \in H$, and [3] $t^B \in H$ such that [4] $t^A \prec t^B$. [1] and SP Proposition 3.5 imply there exists [5] $H' \in \mathcal{H}'$ such that [6] $\tau(H) \subseteq H'$. [2] implies $\tau(t^A) \in \tau(H)$ and thus [6] implies [7] $\tau(t^A) \in H'$. Similarly, [3] implies $\tau(t^B) \in \tau(H)$ and thus [6] implies [8] $\tau(t^B) \in H'$. In addition, [4] and SP Proposition 2.4(d) (via SP Corollary 3.10) imply [9] $\tau(t^A) \prec' \tau(t^B)$. [5], [7], [8], and [9] imply Π' has absentmindedness.

(a). Suppose Π and Π' are isomorphic. Then (a fortiori) there is a morphism to Π from Π' and also a morphism from Π to Π'. By part (a°) and the first morphism, the no-absentmindedness of Π implies the no-absentmindedness of Π'. Similarly, by part (a°) and the second morphism, the no-absentmindedness of Π is implied by the no-absentmindedness of Π'.

(b°). This follows from part (a°) and the definition of no-absentmindedness for forms.

(b). This follows from part (b°) just as part (a) follows from part (a°). □

Proof A.12 (for Proposition 2.9). *Claim 1. If $[\Pi, \Pi', \tau, \delta]$ is an isomorphism and Π' has perfect-information, then Π has perfect-information.* Suppose $[\Pi, \Pi', \tau, \delta]$ is an isomorphism, with $\Pi = (T, C, \otimes)$ determining $\mathcal{H}$ and $\Pi' = (T', C', \otimes)$ determining $\mathcal{H}'$. Further suppose Π does not have perfect-information. It suffices to show that Π' does not have perfect-information. Because Π does not have perfect-information, there are $t^1 \in T$, $t^2 \in T$, and [a] $H \in \mathcal{H}$ such that [b] $t^1 \neq t^2$ and [c] $\{t^1, t^2\} \subseteq H$. SP Proposition 3.8(e) implies $\tau|_{\mathcal{H}}$ is a bijection from $\mathcal{H}$ onto $\mathcal{H}'$. Hence [a] implies [d] $\tau(H) \in \mathcal{H}'$. Further, SP Theorem 3.7 implies that τ is a bijection from T onto T'. Hence [b] implies [e] $\tau(t^1) \neq \tau(t^2)$. Yet further, [c] implies [f] $\{\tau(t^1), \tau(t^2)\} \subseteq \tau(H)$. [d], [e], and [f] imply that Π' does not have perfect-information.

(a). This follows from Claim 1.

(b). This follows from part (a) and the definition of perfect-information for forms. □

Lemma A.13 [25]*Suppose that* (T, p) *is a functioned tree and that* $\tau{:}T \to T'$ *is a bijection. Define the function* p' *by surjectivity and* $p'^{\mathsf{gr}} = \{\, (\tau(t^\sharp), \tau(t)) \mid (t^\sharp, t){\in}p^{\mathsf{gr}} \,\}$. *Then* (T', p') *is a functioned tree.*

Proof Since (T, p) is a functioned tree, there exist $t^o \in T$ and $X \subseteq T$ to satisfy [T1]–[T2]. Define $t'^o = \tau(t^o)$ and $X' = \tau(X)$. It suffices to show

[T1′] $\quad p'$ is a nonempty function from $T' \smallsetminus \{t'^o\}$ onto X', and

[T2′] $\quad (\forall t'{\in}T'{\smallsetminus}\{t'^o\})(\exists m{\ge}1)\ (p')^m(t') = t'^o.$

These two statements are shown by Claims 6 and 8.

Claim 1 : $\tau|_{T\smallsetminus\{t^o\}}{:}T{\smallsetminus}\{t^o\} \to T'{\smallsetminus}\{t'^o\}$ *is a bijection*. This follows from the bijectivity of τ and the definition of t'^o.

Claim 2 : $\tau|_X{:}X \to X'$ *is a bijection*. This follows from the bijectivity of τ and the definition of X'.

Claim 3 : $\tau|_X{\circ}p{\circ}(\tau|_{T\smallsetminus\{t^o\}})^{-1}$ *is a nonempty function from* $T'{\smallsetminus}\{t'^o\}$ *onto* X'. The claim follows from composition. In particular, $(\tau|_{T\smallsetminus\{t^o\}})^{-1}{:}T'{\smallsetminus}\{t'^o\} \to T{\smallsetminus}\{t^o\}$ is a bijection by Claim 1, $p{:}T{\smallsetminus}\{t^o\} \to X$ is nonempty and surjective by [T1], and $\tau|_X{:}X \to X'$ is a bijection by Claim 2. These functions appear on the bottom, left, and top of Fig. 8.

Claim 4 : $p'^{\mathsf{gr}} = (\tau|_X{\circ}p{\circ}(\tau|_{T\smallsetminus\{t^o\}})^{-1})^{\mathsf{gr}}$. It will be argued that

$$\begin{aligned}
p'^{\mathsf{gr}} &= \{\, (\tau(t^\sharp), \tau(t)) \mid (t^\sharp, t){\in}p^{\mathsf{gr}} \,\} \\
&= \{\, (\tau(t^\sharp), \tau(t)) \mid t^\sharp{\in}T{\smallsetminus}\{t^o\},\ t{=}p(t^\sharp) \,\} \\
&= \{\, (\tau(t^\sharp), \tau{\circ}p(t^\sharp)) \mid t^\sharp{\in}T{\smallsetminus}\{t^o\} \,\} \\
&= \{\, (\tau(t^\sharp), \tau{\circ}p(t^\sharp)) \mid t'^\sharp{\in}T'{\smallsetminus}\{t'^o\},\ t^\sharp{=}(\tau|_{T\smallsetminus\{t'^o\}})^{-1}(t'^\sharp) \,\} \\
&= \{\, (\tau{\circ}(\tau|_{T\smallsetminus\{t'^o\}})^{-1}(t'^\sharp), \tau{\circ}p{\circ}(\tau|_{T\smallsetminus\{t'^o\}})^{-1}(t'^\sharp)) \mid t'^\sharp{\in}T'{\smallsetminus}\{t'^o\} \,\} \\
&= \{\, (t'^\sharp, \tau{\circ}p{\circ}(\tau|_{T\smallsetminus\{t'^o\}})^{-1}(t'^\sharp)) \mid t'^\sharp{\in}T'{\smallsetminus}\{t'^o\} \,\} \\
&= (\tau{\circ}p{\circ}(\tau|_{T\smallsetminus\{t'^o\}})^{-1})^{\mathsf{gr}}.
\end{aligned}$$

The first equality holds by the lemma's definition of p'^{gr}. The second holds since the domain of p is $T{\smallsetminus}\{t^o\}$ by [T1]. The third is a rearrangement. The fourth holds by Claim 1. The fifth and sixth are

[25]Lemmas A.13–A.15 use bijections to construct isomorphisms. Experienced category theorists may find these results straightforward. In fact, portions of Lemma A.13 could be proved by appealing to SP Theorem 2.8's isomorphism between **Tree** and **Grph$_{\mathbf{ca}}$**.

Incidentally, the constructions of Lemmas A.13–A.15 are used in Appendices B and C to prove results about **CsqF** and **CsetF**.

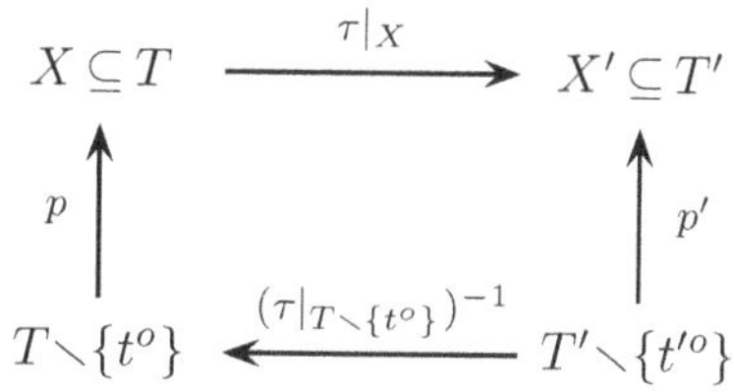

Fig. 8 **Set** diagram for Claims 3 and 5

rearrangements. The last holds because the domain of $(\tau|_{T\setminus\{t'^o\}})^{-1}$ is $T'\setminus\{t'^o\}$ by Claim 1.

Claim 5 : $p' = \tau|_X \circ p \circ (\tau|_{t\setminus\{t'^o\}})^{-1}$, *that is, Fig.* 8 *commutes.* This follows from Claim 4 because [a] p' is surjective by assumption and [b] $\tau|_X \circ p \circ (\tau|_{t\setminus\{t'^o\}})^{-1}$ is surjective by Claim 3.

Claim 6 : *[T1′] holds.* This follows from Claims 3 and 5.

Claim 7 : $(\forall t \in T\setminus\{t^o\})(\exists m \geq 1)\ t^o = p \circ [(\tau|_{T\setminus\{t'^o\}})^{-1} \circ \tau|_X \circ p]^{m-1}(t)$. Take $t \neq t^o$. By [T2] there exists $m \geq 1$ such that $t^o = p^m(t)$. On the one hand, suppose $m = 1$. Then the claim holds by the definition of m. On the other hand, suppose $m \geq 2$. Then proving the claim requires several steps. First, it will be shown that

(a) $(\forall n \mid m-1 \geq n \geq 1)\ \ p^n(t) = (\tau|_{T\setminus\{t'^o\}})^{-1} \circ \tau|_X \circ p^n(t).$

Take any such n. Since τ is bijective, it suffices to show that the composition $(\tau|_{T\setminus\{t'^o\}})^{-1} \circ \tau|_X \circ p^n(t)$ is well-defined. In other words, it suffices to show [i] that $p^n(t) \in X$ and [ii] that $\tau|_X \circ p^n(t)$ is in the domain of $(\tau|_{T\setminus\{t'^o\}})^{-1}$. [i] holds because the codomain of p is X by [T1]. To see [ii], note that $t^o = p^m(t)$ and $m-1 \geq n \geq 1$ imply that $p^n(t)$ is in the domain of p. Thus, since the domain of p is $T\setminus\{t^o\}$ by [T1], we have $p^n(t) \in T\setminus\{t^o\}$. Hence the definition of t'^o and the bijectivity of τ imply $\tau|_X \circ p^n(t) \in T'\setminus\{t'^o\}$. This and Claim 1 imply [ii]. Second, it will be argued that

(b) $(\forall n \mid m-1 \geq n \geq 1)\ \ p^n(t) = [(\tau|_{T\setminus\{t'^o\}})^{-1} \circ \tau|_X \circ p] \circ p^{n-1}(t).$

This holds because the right-hand side of (b) is a rearrangement of the right-hand side of (a). Third, it will be argued that

(c)
$$\begin{aligned} p^{m-1}(t) &= [(\tau_{T\setminus\{t'^o\}})^{-1} \circ \tau|_X \circ p] \circ p^{m-2}(t) \\ &= [(\tau_{T\setminus\{t'^o\}})^{-1} \circ \tau|_X \circ p]^2 \circ p^{m-3}(t) \\ \ldots &= [(\tau_{T\setminus\{t'^o\}})^{-1} \circ \tau|_X \circ p]^{m-2} \circ p(t) \end{aligned}$$

$$= [(\tau_{T \setminus \{t^o\}})^{-1} \circ \tau|_X \circ p]^{m-1}(t),$$

where the first equality holds by (b) at $n{=}m{-}1$, the second by (b) at $n{=}m{-}2$, ..., and the last by (b) at $n{=}1$. Finally, the claim holds because

$$t^o = p^m(t) = p \circ p^{m-1}(t) = p \circ [(\tau|_{T \setminus \{t^o\}})^{-1} \circ \tau|_X \circ p]^{m-1}(t),$$

where the first equality holds by the definition of m, the second is a rearrangement, and the third holds by (c).

Claim 8 : *[T2′] holds*. Take $t' \in T' \setminus \{t'^o\}$. Then Claim 1 implies $(\tau|_{T \setminus \{t^o\}})^{-1}(t') \in T \setminus \{t^o\}$. Thus by Claim 7, there exists $m \geq 1$ such that

$$t^o = p \circ [(\tau|_{T \setminus \{t^o\}})^{-1} \circ \tau|_X \circ p]^{m-1} \circ (\tau|_{T \setminus \{t^o\}})^{-1}(t').$$

It can now be argued that

$$\begin{aligned} t'^o &= \tau|_X(t^o) \\ &= \tau|_X \circ p \circ [(\tau|_{T \setminus \{t^o\}})^{-1} \circ \tau|_X \circ p]^{m-1} \circ (\tau|_{T \setminus \{t^o\}})^{-1}(t') \\ &= [\tau|_X \circ p \circ (\tau^{-1}|_{T \setminus \{t^o\}})^{-1}]^m(t') \\ &= (p')^m(t'). \end{aligned}$$

The first equation holds by the definition of t'^o and the fact that $t^o \in X$ in any functioned tree (by remark [iv] in the paragraph following SP Eq. (1)). The second equation holds by the definition of m, the third is a rearrangement, and the fourth holds by Claim 5.

□

Lemma A.14 *Suppose* $\Pi = (T, C, \otimes)$ *is an* **NCP** *preform. Also suppose* $\tau{:}T \to T'$ *and* $\delta{:}C \to C'$ *are bijections. Define* $\otimes'$ *by surjectivity and* $\otimes'^{\text{gr}} = \{\, (\tau(t), \delta(c), \tau(t^\sharp)) \mid (t, c, t^\sharp) \in \otimes^{\text{gr}} \,\}$. *Also define* $\Pi' = (T', C', \otimes')$. *Then (a)* Π' *is an* **NCP** *preform and (b)* $[\Pi, \Pi', \tau, \delta]$ *is an* **NCP** *isomorphism.*

Proof *(a)*. By [P1] there exist $F{:}T \rightrightarrows C$ and $t^o \in T$ such that $\otimes$ is a bijection from F^{gr} onto $T \setminus \{t^o\}$. Define $F'{:}T' \rightrightarrows C'$ by $F'^{\text{gr}} = \{(\tau(t), \delta(c)) | (t, c) \in F^{\text{gr}}\}$. Also define $t'^o = \tau(t^o)$. It suffices to show that

[P1′] $\otimes'$ is a bijection from F'^{gr} onto $T' \setminus \{t'^o\}$,

[P2′] (T', p') is a functioned tree where $p'{:}T' \setminus \{t'^o\} \to (F')^{-1}(C')$ is defined by $p'^{\text{gr}} = \{(t'^\sharp, t') \in (T')^2 | (\exists c' \in C')(t', c', t'^\sharp) \in \otimes'^{\text{gr}}\}$, and

[P3′] $\{(F')^{-1}(c') | c' \in C'\}$ partitions $(F')^{-1}(C')$.

This is done by Claims 6, 7, and 9.

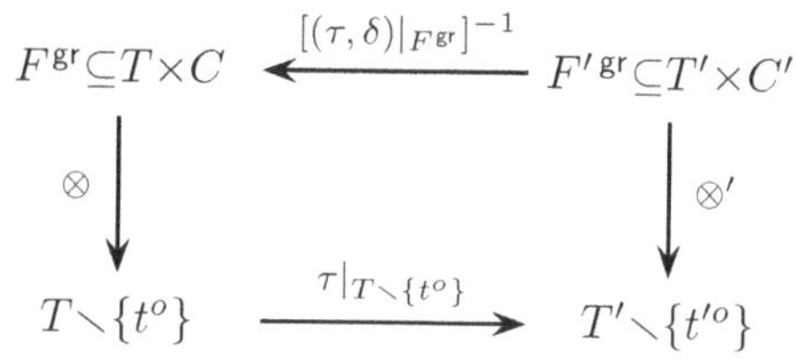

Fig. 9 **Set** diagram for Claims 3 and 5

Claim 1: $(\tau, \delta)|_{F^{\mathrm{gr}}}{:}F^{\mathrm{gr}} \to F'^{\mathrm{gr}}$ *is a bijection*. This follows from the bijectivity of τ, the bijectivity of δ, and the definition of F'.

Claim 2: $\tau|_{T \setminus \{t^o\}}{:}T \setminus \{t^o\} \to T' \setminus \{t'^o\}$ *is a bijection*. This follows from the bijectivity of τ and the definition of t'^o.

Claim 3: $\tau|_{T \setminus \{t^o\}} \circ \otimes \circ [(\tau, \delta)|_{F^{\mathrm{gr}}}]^{-1}$ *is a bijection from* F'^{gr} *onto* $T' \setminus \{t'^o\}$. The claim follows from composition. In particular, $((\tau, \delta)|_{F^{\mathrm{gr}}})^{-1}{:}F'^{\mathrm{gr}} \to F^{\mathrm{gr}}$ is a bijection by Claim 1, $\otimes{:}F^{\mathrm{gr}} \to T \setminus \{t^o\}$ is a bijection by the definitions of F and t^o, and $\tau_{T \setminus \{t^o\}}{:}T \setminus \{t^o\} \to T' \setminus \{t'^o\}$ is a bijection by Claim 2. These three functions appear on the top, left, and bottom of Fig. 9.

Claim 4: $\otimes'^{\mathrm{gr}} = (\tau|_{T \setminus \{t^o\}} \circ \otimes \circ [(\tau, \delta)|_{F^{\mathrm{gr}}}]^{-1})^{\mathrm{gr}}$. It will be argued that

$$\begin{aligned}
\otimes'^{\mathrm{gr}} &= \{\, (\tau(t), \delta(c), \tau(t^\sharp)) \mid (t, c, t^\sharp) \in \otimes^{\mathrm{gr}} \,\} \\
&= \{\, (\tau(t), \delta(c), \tau(t^\sharp)) \mid (t, c) \in F^{\mathrm{gr}},\ t^\sharp = \otimes(t, c) \,\} \\
&= \{\, ((\tau, \delta)(t, c), \tau|_{T \setminus \{t^o\}} \circ \otimes(t, c)) \mid (t, c) \in F^{\mathrm{gr}} \,\} \\
&= \{\, ((\tau, \delta)(t, c), \tau|_{T \setminus \{t^o\}} \circ \otimes(t, c)) \mid (t', c') \in F'^{\mathrm{gr}},\ (t, c) = [(\tau, \delta)|_{F^{\mathrm{gr}}}]^{-1}(t', c') \,\} \\
&= \{\, ((\tau, \delta) \circ [(\tau, \delta)|_{F^{\mathrm{gr}}}]^{-1}(t', c'), \tau|_{T \setminus \{t^o\}} \circ \otimes \circ [(\tau, \delta)|_{F^{\mathrm{gr}}}]^{-1}(t', c') \mid (t', c') \in F'^{\mathrm{gr}} \,\} \\
&= \{\, ((t', c'), \tau|_{T \setminus \{t^o\}} \circ \otimes \circ [(\tau, \delta)|_{F^{\mathrm{gr}}}]^{-1}(t', c') \mid (t', c') \in F'^{\mathrm{gr}} \,\} \\
&= (\tau|_{T \setminus \{t^o\}} \circ \otimes \circ [(\tau, \delta)|_{F^{\mathrm{gr}}}]^{-1})^{\mathrm{gr}}.
\end{aligned}$$

The first equality holds by the lemma's definition of $\otimes'$. The second holds by the definition of F, and the third by the definition of t^o. The fourth holds by Claim 1. The fifth and sixth are rearrangements. The seventh holds by Claim 1.

Claim 5: $\otimes' = \tau|_{T \setminus \{t^o\}} \circ \otimes \circ [(\tau, \delta)|_{F^{\mathrm{gr}}}]^{-1}$, *that is, Fig.* 9 *commutes*. This follows from Claim 4 because [a] $\otimes'$ is surjective by definition and [b] $\tau|_{T \setminus \{t^o\}}$ is surjective by Claim 2.

Claim 6: *[P1′] holds*. This follows from Claims 3 and 5.

Claim 7: *[P2′] holds*. Define p by [P2]. [P2] implies that [a] (T, p) is a functioned tree. Define p' by [P2′]. Claim 6 and SP Lemma C.1(a) implies [b] p' is well-defined and [c] p' is surjective. Because of [b], it suffices to show that (T', p') is a functioned tree.

Toward that end, consider Lemma A.13. Lemma A.13's assumptions are met by [a] and the bijectivity of τ. Thus Lemma A.13 implies that $(T', p^\star)$ is a functioned tree, where the function $p^\star$ is defined by [d] $p^\star$ being surjective and [e] $p^{\star\,\mathrm{gr}} = \{\, (\tau(t^\sharp), \tau(t)) \mid (t^\sharp, t) \in p^{\mathrm{gr}} \,\}$. Thus it suffices to show that $p' = p^\star$.

Toward that end, note [c] and [d] imply that both p' and $p^\star$ are surjective. Thus it suffices to show $p'^{\mathrm{gr}} = p^{\star\,\mathrm{gr}}$. It will be argued that

$$\begin{aligned}
p'^{\mathrm{gr}} &= \{\, (t'^\sharp, t') \in (T')^2 \mid (\exists c' \in C')(t', c', t'^\sharp) \in \otimes'^{\mathrm{gr}} \,\} \\
&= \{\, (t'^\sharp, t') \in (T')^2 \mid (\exists c' \in C')(\exists (t, c, t^\sharp) \in \otimes^{\mathrm{gr}})\ (t', c', t'^\sharp) = (\tau(t), \delta(c), \tau(t^\sharp)) \,\} \\
&= \{\, (t'^\sharp, t') \in (T')^2 \mid (\exists (t, c, t^\sharp) \in \otimes^{\mathrm{gr}})\ (t', t'^\sharp) = (\tau(t), \tau(t^\sharp)) \,\} \\
&= \{\, (\tau(t^\sharp), \tau(t)) \mid (\exists c \in C)(t, c, t^\sharp) \in \otimes^{\mathrm{gr}} \,\} \\
&= \{\, (\tau(t^\sharp), \tau(t)) \mid (t^\sharp, t) \in p^{\mathrm{gr}} \,\} \\
&= p^{\star\,\mathrm{gr}}.
\end{aligned}$$

The first equality holds by the definition of p' two paragraphs ago, and the second equality holds by the definition of $\otimes'$ in the lemma statement. The $\subseteq$ direction of the third equality holds simply because the variable c' does not appear in the right-hand side. The $\supseteq$ direction follows from $\otimes^{\mathrm{gr}} \subseteq T \times C \times T$ and $\delta{:}C \rightarrow C'$. The fourth equality holds because the codomain of τ is T'. The fifth equality follows from the definition of p two paragraphs ago, and the sixth equality follows from [e].

Claim 8: $(\forall c' \in C')\ (F')^{-1}(c') = \tau(F^{-1}(\delta^{-1}(c')))$. Take $c' \in C'$. It is argued, in seven steps, that $(F')^{-1}(c')$ by definition is $\{t' \in T' | c' \in F'(t')\}$, which by rearrangement is $\{t' \in T' | (t', c') \in F'^{\mathrm{gr}}\}$, which, by the definition of F', the bijectivity of τ, and the bijectivity of δ, is $\{t' \in T' | (\tau^{-1}(t'), \delta^{-1}(c')) \in F^{\mathrm{gr}}\}$, which by the bijectivity of τ is $\{t' | (\exists t \in T)\, t' = \tau(t), (\tau^{-1}(t'), \delta^{-1}(c')) \in F^{\mathrm{gr}}\}$, which by rearrangement is $\{\tau(t) | (\exists t \in T)(\tau^{-1} \circ \tau(t), \delta^{-1}(c')) \in F^{\mathrm{gr}}\}$, which by rearrangement is $\tau(\{t \in T | (t, \delta^{-1}(c')) \in F^{\mathrm{gr}}\}$, which by rearrangement is $\tau(F^{-1}(\delta^{-1}(c')))$.

Claim 9: *[P3′] holds.* It must be shown that

[a] $(\forall c' \in C')\ (F')^{-1}(c') \neq \varnothing$,

[b] $(\forall c'^A \in C', c'^B \in C')\ (F')^{-1}(c'^A) \cap (F')^{-1}(c'^B) \neq \varnothing$
$\Rightarrow (F')^{-1}(c'^A) = (F')^{-1}(c'^B)$, and

[c] $\cup_{c' \in C'} (F')^{-1}(c') = (F')^{-1}(C')$.

To show [a], take c'. By the bijectivity of δ, $\delta^{-1}(c') \in C$. Thus by [P3], $F^{-1}(\delta^{-1}(c')) \neq \varnothing$. Thus $\tau(F^{-1}(\delta^{-1}(c'))) \neq \varnothing$. Thus by Claim 8, $(F')^{-1}(c') \neq \varnothing$. To show [b], suppose that [b] were false. Then there would be c'^A and c'^B such that $(F')^{-1}(c'^A)$ and $(F')^{-1}(c'^B)$ intersect and are unequal. Hence by Claim 8, $\tau(F^{-1}(\delta^{-1}(c'^A)))$ and $\tau(F^{-1}(\delta^{-1}(c'^A)))$ intersect and are unequal. Hence by the bijectivity of τ, $F^{-1}(\delta^{-1}(c'^A))$ and $F^{-1}(\delta^{-1}(c'^B))$ intersect and are unequal. This contradicts [P3] because both $\delta^{-1}(c'^A)$ and $\delta^{-1}(c'^B)$ belong to C by the bijectivity of δ. Finally, [c] holds by definition (recall the last sentence of note 5).

(b). This paragraph shows that $[\Pi, \Pi', \tau, \delta]$ is a morphism. Π is a preform by assumption and Π' is a preform by part (a). [PM1] and [PM2] hold by assumption (a fortiori). [PM3] holds with equality by the definition of $\otimes'$.

Finally, SP Theorem 3.7 implies that $[\Pi, \Pi', \tau, \delta]$ is an isomorphism because [a] it is a morphism by the previous paragraph and [b] τ and δ are bijective by assumption. □

Lemma A.15 *Suppose* $\Phi = (I, T, (C_i)_{i\in I}, \otimes)$ *is an* **NCF** *form. Also suppose* $\iota{:}I \to I'$, $\tau{:}T \to T'$, *and* $\delta{:}\cup_{i\in I} C_i \to C'$ *are bijections. Define* $\otimes'$ *by surjectivity and* $\otimes'^{\text{gr}} = \{(\tau(t), \delta(c), \tau(t^\sharp))|(t, c, t^\sharp)\in\otimes^{\text{gr}}\}$. *Also define* $(C'_{i'})_{i'\in I'}$ *at each* i' *by* $C'_{i'} = \delta(C_{\iota^{-1}(i')})$. *Also define* $\Phi' = (I', T', (C'_{i'})_{i'\in I'}, \otimes')$. *Then (a)* Φ' *is an* **NCF** *form and (b)* $[\Phi, \Phi', \iota, \tau, \delta]$ *is an* **NCF** *isomorphism.*

Proof Define $C = \cup_{i\in I} C_i$. Define $\Pi = (T, C, \otimes)$. Define $\Pi' = (T', C', \otimes')$.

Claim 1 : *(a)* Π' *is an* **NCP** *preform and (b)* $[\Pi, \Pi', \tau, \delta]$ *is an* **NCP** *isomorphism.* Consider Lemma A.14. The assumptions of Lemma A.14 are met because [i] Π is an **NCP** preform by [F1], [ii] $\tau{:}T \to T'$ is a bijection by assumption, and [iii] $\delta{:}C \to C'$ is a bijection because $C = \cup_{i\in I} C_i$ by definition and $\delta{:}\cup_{i\in I} C_i \to C'$ is a bijection by assumption. Further, Lemma A.14's definitions of $\otimes'$ and Π' coincide with the present definitions of $\otimes'$ and Π'. Thus Lemma A.14 implies this claim's two conclusions.

Claim 2 : $C' = \cup_{i'\in I'} C'_{i'}$. It is argued, in four steps, that C' by the bijectivity of δ equals $\delta(\cup_{i\in I} C_i)$, which by rearrangement equals $\cup_{i\in I} \delta(C_i)$, which by the bijectivity of ι equals $\cup_{i'\in I'} \delta(C_{\iota^{-1}(i')})$, which by the definition of $(C'_{i'})_{i'\in I'}$ equals $\cup_{i'\in I'} C'_{i'}$.

Claim 3 : Φ' *satisfies [F1]*. It must be shown that $(T', C^\star, \otimes')$ is a preform where $C^\star$ is defined as $\cup_{i'\in I'} C'_{i'}$. Claim 2 implies that $C^\star = C'$. Hence $\Pi' = (T', C^\star, \otimes')$. Hence Claim 1(a) implies that $(T', C^\star, \otimes')$ is a preform.

Claim 4 : Φ' *satisfies [F2]*. Take $i' \in I'$ and $j' \in I' \setminus \{i'\}$. The bijectivity of ι implies $\iota^{-1}(i') \in I$ and $\iota^{-1}(j') \in I \setminus \{\iota^{-1}(i')\}$. Thus [F2] for Φ implies $C_{\iota^{-1}(i')} \cap C_{\iota^{-1}(j')} = \varnothing$. Hence the bijectivity of δ implies $\delta(C_{\iota^{-1}(i')}) \cap \delta(C_{\iota^{-1}(j')}) = \varnothing$. Hence the definition of $(C'_{i'})_{i'\in I'}$ implies $C'_{i'} \cap C'_{j'} = \varnothing$.

Claim 5 : Φ' *satisfies [F3]*. Take $t' \in T'$. The bijectivity of τ implies $\tau^{-1}(t') \in T$. Hence [F3] for Φ implies there is $i \in I$ such that $F(\tau^{-1}(t')) \subseteq C_i$. Hence the bijectivity of ι implies there is $i' \in I'$ such that [a] $F(\tau^{-1}(t')) \subseteq C_{\iota^{-1}(i')}$. Also, it will be shown [b] $F(\tau^{-1}(t')) = \delta^{-1}(F'(t'))$ by arguing, in steps, that $F(\tau^{-1}(t'))$ by rearrangement equals $\{c{\in}C|(\tau^{-1}(t'), c){\in}F^{\text{gr}}\}$, which by Claim 1(b) and SP Proposition 3.8(c) equals $\{c{\in}C|(t', \delta(c)){\in}F'^{\text{gr}}\}$, which by the bijectivity of δ equals $\{c|(\exists c'{\in}C')\ c{=}\delta^{-1}(c'),\ (t', \delta(c)){\in}F'^{\text{gr}}\}$, which by rearrangement equals $\{\delta^{-1}(c')|(\exists c'{\in}C')(t', c'){\in}F'^{\text{gr}}\}$, which by rearrangement equals $\delta^{-1}(\{c'{\in}C'|(t', c'){\in}F'^{\text{gr}}\})$, which by rearrangement equals $\delta^{-1}(F'(t'))$. [a] and [b] imply $\delta^{-1}(F'(t')) \subseteq C_{\iota^{-1}(i')}$. Hence the bijectivity of δ implies $F'(t') \subseteq \delta(C_{\iota^{-1}(i')})$. Hence the definition of $C'_{i'}$ implies $F'(t') \subseteq C'_{i'}$.

Claim 6 : *Φ' is an* **NCF** *form.* This follows from Claims 3–5.

Claim 7 : *$[\Phi, \Phi', \iota, \tau, \delta]$ is an* **NCF** *morphism.* Φ is an **NCF** form by assumption, and Φ' is an **NCF** form by Claim 6. [FM1] holds because $[\Pi, \Pi', \tau, \delta]$ is an **NCP** morphism a fortiori by Claim 1(b). [FM2] holds by assumption. For [FM3], take $i \in I$. It is argued, in two steps, that $\delta(C_i)$ by the bijectivity of ι equals $\delta(C_{\iota^{-1}\circ\iota(i)})$, which by definition of $C'_{\iota(i)}$ equals $C'_{\iota(i)}$.

Claim 8 : *$[\Phi, \Phi', \iota, \tau, \delta]$ is an* **NCF** *isomorphism.* This follows from the reverse direction of Corollary 2.5 because [a] $[\Phi, \Phi', \iota, \tau, \delta]$ is an **NCF** morphism by Claim 7, [b] $[\Pi, \Pi', \tau, \delta]$ is an **NCP** isomorphism by Claim 1(b), and [c] ι is a bijection by assumption.

Conclusion. The lemma's conclusions follow from Claims 6 and 8. □

Appendix B: CsqF

Proof B.1 (for Proposition 3.1). The proposition follows from Claims 1–8 and 13–14.

Claim 1 : *(a) holds.* Suppose [a] $t^o \neq \{\}$. [Csq1] states [b] $\{\} \in T$. [a] and [b] imply $\{\} \in T\smallsetminus\{t^o\}$. Thus by [P1], there are $t \in T$ and $c \in C$ such that $(t, c, \{\}) \in \otimes^{\text{gr}}$. Thus by [Csq2], $p(\{\})\oplus(c) = \{\}$. This is impossible because the left-hand sequence has positive length and the right-hand sequence has zero length.

Claim 2 : *(b) holds.* Take $t^\sharp \in T\smallsetminus\{\{\}\}$. Claim 1 (a) implies $t^\sharp \in T\smallsetminus\{t^o\}$. Thus the reverse direction of SP Proposition 3.1(a) implies $(p(t^\sharp), q(t^\sharp), t^\sharp) \in \otimes$. Thus [Csq2] implies $p(t^\sharp)\oplus(q(t^\sharp)) = t^\sharp$. Thus $p(t^\sharp) = {}_1t^\sharp_{|t^\sharp|-1}$ and $q(t^\sharp) = t^\sharp_{|t^\sharp|}$.

Claim 3 : *(c) holds.* Assume $(t, c, t^\sharp) \in \otimes^{\text{gr}}$. Then [P1] yields $(t, c, t^\sharp) \in T{\times}C{\times}T$, and [Csq2] yields $t\oplus(c) = t^\sharp$. Conversely, suppose [1] $(t, c, t^\sharp) \in T{\times}C{\times}T$ and [2] $t\oplus(c) = t^\sharp$. [2] implies [3] $t = {}_1t^\sharp_{|t^\sharp|-1}$ and [4] $c = t^\sharp_{|t^\sharp|}$. Further, [4] implies $t^\sharp \neq \{\}$. This and [1] implies [5] $t^\sharp \in T\smallsetminus\{\{\}\}$. [5] and Claim 2 (b) imply [6] $p(t^\sharp) = {}_1t^\sharp_{|t^\sharp|-1}$ and [7] $q(t^\sharp) = t^\sharp_{|t^\sharp|}$. [3] and [6] imply [8] $t = p(t^\sharp)$. [4] and [7] imply [9]

$c = q(t^\sharp)$. Further, [5] and Claim 1 (a) imply $t^\sharp \neq t^o$, and thus SP Proposition 3.1(a) implies [10] $(p(t^\sharp), q(t^\sharp), t^\sharp) \in \otimes^{\mathsf{gr}}$. [8]–[10] imply $(t, c, t^\sharp) \in \otimes^{\mathsf{gr}}$.

Claim 4 : *(d) holds.* By [P1], $F^{\mathsf{gr}} \subseteq T \times C$. Thus it suffices to show $(\forall t \in T, c \in C)$ $(t, c) \in F^{\mathsf{gr}}$ iff $t \oplus c \in T$. Suppose $(t, c) \in F^{\mathsf{gr}}$. Then [P1] implies there is [1] $t^\sharp \in T$ such that [2] $(t, c, t^\sharp) \in \otimes^{\mathsf{gr}}$. [2] and Claim 3 (c) imply $t \oplus (c) = t^\sharp$. This and [1] imply $t \oplus (c) \in T$. Conversely, suppose $t \oplus (c) \in T$. There there is $t^\sharp \in T$ such that $t \oplus (c) = t^\sharp$. Thus Claim 3 (c) implies $(t, c, t^\sharp) \in \otimes^{\mathsf{gr}}$. This and [P1] imply $(t, c) \in F^{\mathsf{gr}}$.

Claim 5 : *(e) holds.* Take $t \in T$. The following induction is on $m \in \{0, 1, \ldots |t|\}$. For the initial step, assume $m = 0$. Then $p^0(t) = t = {}_1t_{|t|} = {}_1t_{|t|-0} = {}_1t_{|t|-m}$ by inspection. For the inductive step, assume $m > 0$. Note $m \leq |t|$ implies $|t| - m \geq 0$, which implies $|t| - (m-1) > 0$, which implies [1] ${}_1t_{|t|-(m-1)} \neq \{\}$. Then it is argued, in steps, that $p^m(t)$ by $m > 0$ equals $p(p^{m-1}(t))$, which by the inductive hypothesis equals $p({}_1t_{|t|-(m-1)})$, which by [1] and Claim 2 (b) at $t^\sharp = {}_1t_{|t|-(m-1)}$ equals ${}_1t_{|t|-(m-1)-1}$, which by rearrangement equals ${}_1t_{|t|-m}$.

Claim 6 : *(f) holds.* Take $t \in T$. It will be shown that $p^{|t|}(t) = t^0$ by arguing, in steps, that $p^{|t|}(t)$ by Claim 5 (e) at $m = |t|$ equals ${}_1t_{|t|-|t|}$, which equals ${}_1t_0$, which equals $\{\}$, which by Claim 1 (a) equals t^o. This and the definition of k imply $k(t) = |t|$.

Claim 7 : *(g) holds.* Take $t \in T$. By inspection, the result is equivalent to $(\forall \ell \in \{1, 2, \ldots |t|\})$ $t_\ell = q \circ p^{|t|-\ell}(t)$. On the one hand, take $t = \{\}$. Then $|t| = 0$ so the result is vacuously true. On the other hand, take $t \neq \{\}$. Then take [1] $\ell \in \{1, 2, \ldots |t|\}$. First it will be shown that [2] $p^{|t|-\ell}(t) = {}_1t_\ell$ by arguing, in steps, that $p^{|t|-\ell}(t)$ by Claim 5 (e) at $m = |t| - \ell$ equals ${}_1t_{|t|-(|t|-\ell)}$, which by rearrangement equals ${}_1t_\ell$. Then it is argued, in steps, that $q \circ p^{|t|-\ell}(t)$ by [2] equals $q({}_1t_\ell)$, which by [1] and Claim 2 (b) equals t_ℓ.

Claim 8 : *(h) holds.* Suppose $c \in C$. This and [P3] imply $F^{-1}(c) \neq \varnothing$. Thus there is $t^\star \in T$ such that $(t^\star, c) \in F^{\mathsf{gr}}$. This and Claim 4 (d) imply $t^* \oplus (c) \in T$. Thus $c \in R(t^* \oplus (c)) \subseteq \cup\{R(t) | t \in T\}$. Conversely, suppose $b \in \cup\{R(t) | t \in T\}$. There there is [1] $t^* \in T$ such that [2] $b \in R(t^*)$. [1] and Claim 7 (g) imply that $t^* = (q \circ p^{|t^*|-\ell}(t^*))_{\ell=1}^{|t^*|}$. This and [2] imply there is $\ell^* \in \{1, 2, \ldots |t^*|\}$ such that $b = q \circ p^{|t^*|-\ell^*}(t^*)$. This implies $b \in C$ since the codomain of q is C by the definition of q.

Claim 9 : $(\forall t^A \in T, t^B \in T)$ $(|t^A| < |t^B|$ and $t^A = {}_1t^B_{|t^A|})$ *iff* $t^A \subset t^B$. Take $t^A \in T$ and $t^B \in T$. First, suppose [1] $|t^A| < |t^B|$ and [2] $t^A = {}_1t^B_{|t^A|}$. [1] and the definition of ${}_1t^B_{|t^A|}$ imply [3] ${}_1t^B_{|t^A|} \subset t^B$. [2] and [3] imply $t^A \subset t^B$. Conversely, suppose [4] $t^A \subset t^B$. [Csq1] implies [5] $t^A = \{(1, t^A_1), (2, t^A_2), \ldots (|t^A|, t^A_{|t^A|})\}$ and [6] $t^B = \{(1, t^B_1), (2, t^B_2), \ldots (|t^B|, t^B_{|t^B|})\}$. By inspection, [4]–[6] imply $|t^A| < |t^B|$ and $t^A = {}_1t^B_{|t^A|}$.

Claim 10 : $(\forall t^A{\in}T, t^B{\in}T)$ $(|t^A| \leq |t^B|$ and $t^A = {}_1t^B_{|t^A|})$ *iff* $t^A \subseteq t^B$.[26] In the proof of Claim 9, change $<$ to $\leq$, and $\subset$ to $\subseteq$.

Claim 11 : $(\forall t^A{\in}T, t^B{\in}T)$ $t^A \prec t^B$ *iff* $t^A \subset t^B$. Take $t^A \in T$ and $t^B \in T$. First, suppose $t^A \prec t^B$. This and the definition of $\prec$ imply there is [1] $m \in \{1, 2, ... k(t^B)\}$ such that [2] $t^A = p^m(t^B)$. [1] and Claim 6 (f) imply [3] $m \in \{1, 2, ... |t^B|\}$. Finally, it is argued, in steps, that t^A by [2] equals $p^m(t^B)$, which by [3] and Claim 5 (e) equals ${}_1t^B_{|t^B|-m}$, which by [3] is a strict subset of ${}_1t^B_{|t^B|}$, which by inspection equals t^B. Conversely, suppose $t^A \subset t^B$. This and Claim 9 imply [1] $|t^A| < |t^B|$ and [2] $t^A = {}_1t^B_{|t^A|}$. For convenience, let [3] $m = |t^B| - |t^A|$. Note [1] and $|t^A| \geq 0$ imply [4] $m \in \{1, 2, ... |t^B|\}$. It will now be shown that [5] $t^A = p^m(t^B)$ by arguing, in steps, that t^A by [2] equals ${}_1t^B_{|t^A|}$, which by [3] equals ${}_1t^B_{|t^B|-m}$, which by [4] and Claim 5 (e) at $t = t^B$ equals $p^m(t^B)$. Finally, [5], [4], and the definition of $\prec$ imply $t^A \prec t^B$.

Claim 12 : $(\forall t^A{\in}T, t^B{\in}T)$ $t^A \preccurlyeq t^B$ *iff* $t^A \subseteq t^B$. Take $t^A \in T$ and $t^B \in T$. First, suppose $t^A \preccurlyeq t^B$. Then by the definition of $\preccurlyeq$, either $t^A \prec t^B$ or $t^A = t^B$. In the first case, Claim 11 implies $t^A \subset t^B$. Thus $t^A \subseteq t^B$ in both cases. Conversely, suppose $t^A \subseteq t^B$. Then either $t^A \subset t^B$ or $t^A = t^B$. In the first case, Claim 11 implies $t^A \prec t^B$. Thus the definition of $\preccurlyeq$ implies $t^A \preccurlyeq t^B$ in both cases.

Claim 13 : *(i) holds.* Combine Claims 9 and 11.

Claim 14 : *(j) holds.* Combine Claims 10 and 12.

□

Lemma B.2 *Suppose* $(T, C, \otimes)$ *is a node-and-choice preform with its* t^o, p, q, *and* k . *Let* $\bar{T} = \{\, (q \circ p^{k(t)-\ell}(t))_{\ell=1}^{k(t)} \mid t{\in}T \,\}$. *Then*

$$T \ni t \;\mapsto\; (q \circ p^{k(t)-\ell}(t))_{\ell=1}^{k(t)} \in \bar{T}$$

is a well-defined bijection. Its inverse is

$$T \ni ((...((t^o \otimes \bar{t}_1) \otimes \bar{t}_2) \;\ldots) \otimes \bar{t}_{|\bar{t}|-1}) \otimes \bar{t}_{|\bar{t}|} \;\leftarrow\!\!\mid\; \bar{t} \in \bar{T}$$

(to be clear, $T \ni t^o \leftarrow\!\!\mid \{\} \in \bar{T}$*).*

Proof Let α be the function from T to $\bar{T}$, and conversely, let β be the function to T from $\bar{T}$.

This paragraph shows that $\beta \circ \alpha$ is the identity function on T. The composition is well-defined because [1] the domain of β is $\bar{T}$ and [2] the range of α is $\bar{T}$ by the

---

[26] Claim 10 says that one sequence is an initial segment of another sequence iff the former is a restriction of the latter. This may appear implausible. For example, $\{(2, \mathsf{f})\}$ is not an initial sequence of $t^* = \{(1, \mathsf{g}), (2, \mathsf{f}), (3, \mathsf{f})\}$ even though $\{(2, \mathsf{f})\}$ is a restriction of t^*. This is consistent with Claim 10, because $\{(2, \mathsf{f})\}$ is not a sequence and thus not an element of T by [Csq1].

definition of $\bar{T}$. Thus it suffices to show $(\forall t{\in}T)\ \beta{\circ}\alpha(t) = t$. Toward that end, take $t \in T$. First, suppose $k(t) = 0$. It is argued, in steps, that $\beta{\circ}\alpha(t)$ by the definition of α equals $\beta(\{\})$, which by the definition of β equals t^o, which by $k(t) = 0$ equals t. Second, suppose $k(t) = 1$. It is argued, in steps, that $\beta{\circ}\alpha(t)$ by the definition of α equals $\beta[(q(t))]$, which by the definition of β equals $t^o{\otimes}q(t)$, which by $k(t) = 1$ equals $p(t){\otimes}q(t)$, which by SP Proposition 3.1(b) equals t. Third and finally, suppose $k(t) \geq 2$. It will be argued that

$$\begin{aligned}
\beta{\circ}\alpha(t) &= \beta(\ (q{\circ}p^{k(t)-\ell}(t))_{\ell=1}^{k(t)}\ )\\
&= [[...[[t^o{\otimes}q{\circ}p^{k(t)-1}(t)]{\otimes}q{\circ}p^{k(t)-2}(t)]\ ...\ ]{\otimes}q{\circ}p(t)]{\otimes}q(t)\\
&= [[...[[p^{k(t)}(t){\otimes}q{\circ}p^{k(t)-1}(t)]{\otimes}q{\circ}p^{k(t)-2}(t)]\ ...\ ]{\otimes}q{\circ}p(t)]{\otimes}q(t)\\
&= [[...[[p{\circ}p^{k(t)-1}(t){\otimes}q{\circ}p^{k(t)-1}(t)]{\otimes}q{\circ}p^{k(t)-2}(t)]\ ...\ ]{\otimes}q{\circ}p(t)]{\otimes}q(t)\\
&= [[...[p^{k(t)-1}(t){\otimes}q{\circ}p^{k(t)-2}]\ ...\ ]{\otimes}q{\circ}p(t)]{\otimes}q(t)\\
&\quad ...\\
&= p(t){\otimes}q(t) = t.
\end{aligned}$$

The first equality holds by the definition of α, the second by the definition of β, and the third by the definition of k. The fourth and fifth equalities hold by a rearrangement and SP Proposition 3.1(b). The sixth equality holds by $k(t){-}2$ similar applications of SP Proposition 3.1(b), and the final equality holds by a final application of SP Proposition 3.1(b).

This paragraph shows that $\alpha{\circ}\beta$ is the identity function on $\bar{T}$. The composition is well-defined because [a] the domain of α is T and [b] each value of β is a value of $\otimes$ and the codomain of $\otimes$ is a subset of T. Thus it suffices to show $(\forall \bar{t}{\in}\bar{T})\ \alpha{\circ}\beta(\bar{t}) = \bar{t}$. Toward that end, take $\bar{t}$. First, suppose $\bar{t} = \{\}$. It will be argued, in steps, that $\alpha{\circ}\beta(\{\})$ by the definition of β equals $\alpha(t^o)$, which by the definition of α equals $\{\}$. Second, suppose $\bar{t} \neq \{\}$. Then it suffices to show that $(\forall \ell{\in}\{1, 2, ...|\bar{t}|\})\ (\alpha{\circ}\beta(\bar{t}))_\ell = \bar{t}_\ell$. Toward this end, take ℓ. [i] First assume $\ell < |\bar{t}|$. It will be argued that

$$\begin{aligned}
(\alpha{\circ}\beta(\bar{t}))_\ell &= q{\circ}p^{k(\beta(\bar{t}))-\ell}(\beta(\bar{t}))\\
&= q{\circ}p^{|\bar{t}|-\ell}(\beta(\bar{t}))\\
&= q{\circ}p^{|\bar{t}|-\ell}[\ ((...((t^o{\otimes}\bar{t}_1){\otimes}\bar{t}_2)\ \ldots){\otimes}\bar{t}_{|\bar{t}|-1}){\otimes}\bar{t}_{|\bar{t}|}\ ]\\
&= q{\circ}p^{|\bar{t}|-\ell-1}[\ ((...((t^o{\otimes}\bar{t}_1){\otimes}\bar{t}_2)\ \ldots){\otimes}\bar{t}_{|\bar{t}|-2}){\otimes}\bar{t}_{|\bar{t}|-1}\ ]\\
&\quad ...\\
&= q{\circ}p^{|\bar{t}|-\ell-(|\bar{t}|-\ell)}[\ ((...((t^o{\otimes}\bar{t}_1){\otimes}\bar{t}_2)\ \ldots){\otimes}\bar{t}_{|\bar{t}|-(|\bar{t}|-\ell)-1}){\otimes}\bar{t}_{|\bar{t}|-(|\bar{t}|-\ell)}\ ]\\
&= q{\circ}p^0[\ ((...((t^o{\otimes}\bar{t}_1){\otimes}\bar{t}_2)\ \ldots){\otimes}\bar{t}_{\ell-1}){\otimes}\bar{t}_\ell\ ] = \bar{t}_\ell.
\end{aligned}$$

The first equality holds by the definition of α. The second equality holds because $k(\beta(\bar{t})) = |\bar{t}|$ by inspecting the definitions of k and β. The third holds by the definition of β. The fourth holds by the definition of p. The fifth holds by $|t|{-}\ell{-}1$ similar applications of the definition of p. The sixth is a rearrangement. The seventh holds by the definition of q. [ii] Second assume $\ell = |\bar{t}|$. Then it will be argued that

$$\begin{aligned}(\alpha\circ\beta(\bar{t}))_{|\bar{t}|} &= q\circ p^{k(\beta(\bar{t}))-|\bar{t}|}(\beta(\bar{t})) \\ &= q\circ p^{|\bar{t}|-|\bar{t}|}(\beta(\bar{t})) = q(\beta(\bar{t})) = \bar{t}_{|\bar{t}|} \ ,\end{aligned}$$

The first equality holds by the definition of α and $\ell = |\bar{t}|$. The second equality holds because $k(\beta(\bar{t})) = |\bar{t}|$ by inspecting the definitions of k and β. The third is trivial. The fourth holds by the definitions of q and β. □

Proof B.3 (for Theorem 3.2).

(a). Lemma B.2 implies $\bar{\tau}{:}T \to \bar{T}$ is a bijection. Thus the assumptions of Lemma A.14 are met at $T' = \bar{T}$, $C' = C$, and $\delta = \mathrm{id}_C$. Further, the definition of $\bar{\otimes}$ here coincides with the definition of $\otimes'$ in Lemma A.14. Therefore Lemma A.14 implies that $(\bar{T}, C, \bar{\otimes})$ is an **NCP** preform, and that $[(T, C, \otimes), (\bar{T}, C, \bar{\otimes}), \bar{\tau}, \mathrm{id}_C]$ is an **NCP** isomorphism. Thus [Csq1] and [Csq2] remain to be shown.

For [Csq1], note that the definition of $\bar{T}$ implies that $\bar{T}$ is a collection of finite sequences. Further, since $t^o \in T$ by [P1], the definition of $\bar{T}$ implies that $(q\circ p^{k(t^o)-\ell}(t^o))_{\ell=1}^{k(t^o)} \in \bar{T}$. Thus, since $k(t^o) = 0$ by the definition of k, $(q\circ p^{0-\ell}(t^o))_{\ell=1}^{0} \in \bar{T}$. Thus $\{\} \in \bar{T}$.

For [Csq2], take $(\bar{t}, c, \bar{t}^\sharp) \in \bar{\otimes}$. Then by the definition of $\bar{\otimes}$, there are $t \in T$ and $t^\sharp \in T$ such that [a] $\bar{\tau}(t) = \bar{t}$, [b] $\bar{\tau}(t^\sharp) = \bar{t}^\sharp$, and [c] $(t, c, t^\sharp) \in \otimes^{\mathrm{gr}}$. [a], [b], and the definition of $\bar{\tau}$ imply [d] $\bar{t} = (q\circ p^{k(t)-\ell}(t))_{\ell=1}^{k(t)}$ and [e] $\bar{t}^\sharp = (q\circ p^{k(t^\sharp)-\ell}(t^\sharp))_{\ell=1}^{k(t^\sharp)}$. Also [c] and SP Proposition 3.1(b) imply [f] $t = p(t^\sharp)$ and [g] $c = q(t^\sharp)$. [f] and the definition of k imply [h] $k(t) = k(t^\sharp){-}1$. Finally, it is argued in steps that $\bar{t}\oplus(c)$ by [d] is $(q\circ p^{k(t)-\ell}(t))_{\ell=1}^{k(t)}\oplus(c)$, which by [f]–[h] is $(q\circ p^{k(t^\sharp)-1-\ell}\circ p(t^\sharp))_{\ell=1}^{k(t^\sharp)-1}\oplus(q(t^\sharp))$, which by rearrangement is $(q\circ p^{k(t^\sharp)-\ell}(t^\sharp))_{\ell=1}^{k(t^\sharp)-1}\oplus(q(t^\sharp))$, which by rearrangement is $(q\circ p^{k(t^\sharp)-\ell}(p(t^\sharp))_{\ell=1}^{k(t^\sharp)}$, which by [e] is $\bar{t}^\sharp$.

(b). By assumption, $(I, T, (C_i)_{i\in I}, \otimes)$ is an **NCF** form. Thus [F1] implies $(T, \cup_{i\in I}C_i, \otimes)$ is an **NCP** preform. Further, part (b) defines $\bar{T}$, $\bar{\tau}$, and $\bar{\otimes}$ as part (a) did. Thus part (a) implies [1] $(\bar{T}, \cup_{i\in I}C_i, \bar{\otimes})$ is a **CsqP** preform.

Meanwhile, Lemma B.2 implies $\bar{\tau}{:}T \to \bar{T}$ is a bijection. Thus the assumptions of Lemma A.15 are met at $I' = I$, $\iota = \mathrm{id}_I$, $T' = \bar{T}$, $C' = \cup_{i\in I}C_i$, and $\delta = \mathrm{id}_{\cup_{i\in I}C_i}$. Further, the definition of $\bar{\otimes}$ here coincides with the definition of $\otimes'$ in Lemma A.15. Also, the transparent definitions of ι and δ here, and the definition of $(C'_{i'})_{i'\in I'}$ in Lemma A.15, imply that $(C'_{i'})_{i'\in I'} = (C_i)_{i\in I}$. Hence Lemma A.15 implies that [2] $(I, \bar{T}, (C_i)_{i\in I}, \bar{\otimes})$ is an **NCF** form, and [3] $[(I, T, (C_i)_{i\in I}, \otimes), (I, \bar{T}, (C_i)_{i\in I}, \bar{\otimes}), \mathrm{id}_I, \bar{\tau}, \mathrm{id}_{\cup_{i\in I}C_i}]$ is an **NCF** isomorphism.

[1] and [2] imply that $(I, \bar{T}, (C_i)_{i\in I}, \bar{\otimes})$ is a **CsqF** form. This and [3] are part (b)'s conclusions. □

Lemma B.4 *Suppose $(T, C, \otimes)$ is a* **CsqP** *preform. Then the following are equivalent.*

(a) $(T, C, \otimes)$ has no absentmindedness.

(b) $(\nexists H{\in}\mathcal{H}, t{\in}H, \ell{<}|t|)\ {}_1t_\ell \in H$.
(c) $(\forall t{\in}T,\ H{\in}\mathcal{H})\ |\{\,\ell : 1{\leq}\ell{\leq}|t|,\ t_\ell{\in}F(H)\,\}| \leq 1$.[27]
(d) $(\forall t{\in}T)\ |R(t)| = |t|$.
(e) $R|_T$ *is injective.*

Proof The lemma follows from Claims 5–7.

Claim 1: *Not (a) ⇒ not (b).* Assume absentmindedness. Then there are [a] $H \in \mathcal{H}$, [b] $t \in H$, and [c] $s \in H$ such that [d] $s \prec t$. [d] and Proposition 3.1(i) imply [e] $|s| < |t|$ and [f] $s = {}_1t_{|s|}$. [f] and [c] imply [g] ${}_1t_{|s|} \in H$. [a], [b], [e], and [g] show property (b) is violated at $\ell = |s|$.

Claim 2: *Not (b) ⇒ not (c).* Assume not (b). Then there is $H \in \mathcal{H}$, [a] $t \in H$, and [b] $\ell < |t|$ such that [c] ${}_1t_\ell \in H$. [b] implies $t_{\ell+1}$ is well-defined; thus Proposition 3.1(d) implies $t_{\ell+1} \in F({}_1t_\ell)$; and thus [c] implies [d] $t_{\ell+1} \in F(H)$. [d] and [a] imply $t_{\ell+1} \in F(t)$; and thus $t^* = t{\oplus}(t_{\ell+1})$ is a member of T. Finally, it will be argued, in steps, that $|\{\ell'{:}1{\leq}\ell'{\leq}|t^*|,\ t^*_{\ell'}{\in}F(H)\}|$ by [d] is at least $|\{\ell'{:}1{\leq}\ell'{\leq}|t^*|, t^*_{\ell'}{=}t_{\ell+1}\}|$, which by the construction of t^* is at least $|\{\ell{+}1,\ |t|{+}1\}|$, which by [b] is 2.

Claim 3: *Not (c) ⇒ not (d).* Assume not (c). Then there are $t \in T$ and $H \in \mathcal{H}$ such that $|\{\ell{:}1{\leq}\ell{\leq}|t|, t_\ell{\in}F(H)\}| \geq 2$. Hence there are ℓ and ℓ' such that [a] $\ell < \ell'$, [b] $t_\ell \in F(H)$, and [c] $t_{\ell'} \in F(H)$. It will be argued in three steps that [c] by Proposition 3.1(b) implies $q({}_1t_{\ell'}) \in F(H)$, which by Lemma A.1(c) implies $p({}_1t_{\ell'}) \in H$, which by Proposition 3.1(b) implies [d] ${}_1t_{\ell'-1} \in H$. [b] and [d] imply $t_\ell \in F({}_1t_{\ell'-1})$. Thus $t^* = {}_1t_{\ell'-1}{\oplus}(t_\ell)$ is a member of T. Further, [a] implies t^*_ℓ is well-defined and equal to t_ℓ. Thus $t^*_\ell = t^*_{\ell'}$. This and [a] again imply $|R(t^*)| < |t^*|$.

Claim 4: *Not (d) ⇒ not (a).* Assume not (d). There there is $t \in T$ such that $|R(t)| \neq |t|$. Since $|R(t)| > |t|$ is inconceivable, $|R(t)| < |t|$. Thus there are ℓ and ℓ' in $\{1, 2, \ldots |t|\}$ such that [a] $\ell < \ell'$ and [b] $t_\ell = t_{\ell'}$. Proposition 3.1(d) implies [c] ${}_1t_{\ell-1} \in F^{-1}(t_\ell)$ and [d] ${}_1t_{\ell'-1} \in F^{-1}(t_{\ell'})$. [b] and [d] imply [e] ${}_1t_{\ell'-1} \in F^{-1}(t_\ell)$. [P3] implies [f] $F^{-1}(t_\ell) \in \mathcal{H}$. Finally, [a] implies $\ell{-}1 < \ell'{-}1$; thus Proposition 3.1(i) implies [g] ${}_1t_{\ell-1} \prec {}_1t_{\ell'-1}$. [f], [c], [e], and [g] imply absentmindedness.

Claim 5: *(a), (b), (c), and (d) are equivalent.* This follows from Claims 1–4.

Claim 6: *Not (d) ⇒ not (e).* Assume not (d). Then there is $t \in T$ such that $|R(t)| \neq |t|$. Thus since $|R(t)| > |t|$ is inconceivable, $|R(t)| < |t|$. Thus there are ℓ and ℓ' in $\{1, 2, \ldots |t|\}$ such that [a] $\ell < \ell'$ and [b] $t_\ell = t_{\ell'}$. [a] and [b] imply $R({}_1t_{\ell'-1}) = R({}_1t_{\ell'})$. Thus $R|_T$ is not injective.

Claim 7: *Not (e) ⇒ not (d).* Assume not (e). Then $R|_T$ is not injective. Then there are s and t in T such that [a] $s \neq t$ and [b] $R(s) = R(t)$.

On the one hand, suppose there is not an ℓ in $\{1, 2, \ldots \min\{|s|, |t|\}\}$ such that $s_\ell \neq t_\ell$. Then [c] ${}_1s_{\min\{|s|,|t|\}} = {}_1t_{\min\{|s|,|t|\}}$. Thus [a] implies

[27]The ":" replaces "|" for clarity.

$|s| \neq |t|$. Hence $|s| < |t|$ or $|t| < |s|$. Without loss of generality assume [d] $|s| < |t|$. Hence [c] implies [e] $s = {}_1t_{|s|}$. [d] implies $t_{|s|+1}$ exists. Thus [b] implies $s \neq \{\}$ and there is [f] $\ell \in \{1, 2, \dots |s|\}$ such that [g] $s_\ell = t_{|s|+1}$. But [e] implies $s_\ell = t_\ell$, and thus [g] implies $t_\ell = t_{|s|+1}$. This and [f] imply $|R(t)| < |t|$. In other words, property (d) is violated.

On the other hand, suppose there is an ℓ in $\{1, 2, \dots \min\{|s|, |t|\}\}$ such that $s_\ell \neq t_\ell$. Let j be the smallest such ℓ. Then [h] ${}_1s_{j-1} = {}_1t_{j-1}$ and [i] $s_j \neq t_j$. Proposition 3.1(d) implies $s_j \in F({}_1s_{j-1})$ and $t_j \in F({}_1t_{j-1})$, and thus, [h] implies [j] $\{s_j, t_j\} \subseteq F({}_1s_{j-1})$. A fortiori [j] and [P3] imply there is $H \in \mathcal{H}$ such that ${}_1s_{j-1} \in H$. Hence [j] also implies [k] $\{s_j, t_j\} \subseteq F(H)$. Further, [b] and [i] imply there is $j^* \in \{1, 2, \dots |s|\}$ such that [l] $j^* \neq j$ and [m] $s_{j^*} = t_j$. [m] and [k] imply [n] $\{s_j, s_{j^*}\} \subseteq F(H)$. Finally, it is argued, in steps, that $|\{\ell : 1 \leq \ell \leq |s|, s_\ell \in$
$F(H)\}|$ is at least as great as $|\{j, j^*\}|$ by [n], which is 2 by [l]. Thus the proposition's property (c) is violated. So Claim 5((c)⇔(d)) implies property (d) is violated.

□

Appendix C: CsetF

Lemma C.1 *Suppose C is a set, $t \subseteq C$, $c \in C$, and $t^\sharp \subseteq C$. Then the following are equivalent. (a) $c \notin t$ and $t \cup \{c\} = t^\sharp$. (b) $t \neq t^\sharp$ and $t \cup \{c\} = t^\sharp$. (c) $t \neq t^\sharp$ and $t = t^\sharp \smallsetminus \{c\}$. (d) $t \subseteq t^\sharp$ and $\{c\} = t^\sharp \smallsetminus t$.*

Proof *(a)⇔(b)*. It suffices to show that if $t \cup \{c\} = t^\sharp$, then $c \notin t$ and $t \neq t^\sharp$ are equivalent. Toward that end, assume $t \cup \{c\} = t^\sharp$. Then both directions of the equivalence hold by inspection.

(b)⇔(c). It suffices to show that if $t \neq t^\sharp$, then $t \cup \{c\} = t^\sharp$ and $t = t^\sharp \smallsetminus \{c\}$ are equivalent. Toward that end, assume [1] $t \neq t^\sharp$. For the forward direction, assume [2] $t \cup \{c\} = t^\sharp$. [1] and [2] imply [3] $c \notin t$. [2] implies $(t \cup \{c\}) \smallsetminus \{c\} = t^\sharp \smallsetminus \{c\}$, and [3] implies the left-hand side is t. For the reverse direction, assume [4] $t = t^\sharp \smallsetminus \{c\}$. [1] and [4] imply [5] $c \in t^\sharp$. [4] implies $t \cup \{c\} = (t^\sharp \smallsetminus \{c\}) \cup \{c\}$, and [5] implies the right-hand side is $t^\sharp$.

(a)⇔(d). Assume (a). That is, assume [a] $c \notin t$ and [b] $t \cup \{c\} = t^\sharp$. [b] implies $t \subseteq t^\sharp$. Further, [b] implies $(t \cup \{c\}) \smallsetminus t = t^\sharp \smallsetminus t$, and [a] implies that the left-hand side is $\{c\}$. Conversely, assume (d). That is, assume [c] $t \subseteq t^\sharp$ and [d] $\{c\} = t^\sharp \smallsetminus t$. [d] implies $c \notin t$. Further, [d] implies $t \cup \{c\} = t \cup (t^\sharp \smallsetminus t)$, and [c] implies the right-hand side is $t^\sharp$. □

Proof C.2 (for Proposition 4.1). The proposition is proved by Claims 1, 3, 5, 6, 9, 10, 11, 12, 14, 15, 16, 18, and 19.

Claim 1: *(a) holds.* Suppose [a] $t^o \neq \{\}$. [Cset1] states [b] $\{\} \in T$. [a] and [b] imply $\{\} \in T \smallsetminus \{t^o\}$. Thus by [P1], there is $t \in T$ and $c \in C$ such that $(t, c, \{\}) \in \otimes^{\mathsf{gr}}$. Thus by [Cset2], $t \cup \{c\} = \{\}$. This implies $c \in \{\}$, which is impossible.

Claim 2: $\otimes^{\mathsf{gr}} \subseteq \{\, (t, c, t^\sharp) \in T \times C \times T \mid c \notin t,\ t \cup \{c\} = t^\sharp \,\}$. Take $(t, c, t^\sharp) \in \otimes^{\mathsf{gr}}$. [P1] yields [a] $(t, c, t^\sharp) \in T \times C \times T$. [P2] yields [b] $t = p(t^\sharp)$. Remark [ii] in the paragraph following SP Eq. (1) yields [c] $p(t^\sharp) \neq t^\sharp$. [b] and [c] imply [d] $t \neq t^\sharp$. [Cset2] yields [e] $t \cup \{c\} = t^\sharp$. [d] and [e] yield [f] $c \notin t$. [a], [f], and [e] are the desired results.

Claim 3: *(b) holds.* Take $t^\sharp \in T \smallsetminus \{\{\}\}$. Claim 1 (a) implies that $t^\sharp \in T \smallsetminus \{t^o\}$. Thus SP Proposition 3.1(a) implies that $(p(t^\sharp), q(t^\sharp), t^\sharp) \in \otimes^{\mathsf{gr}}$. Thus Claim 2 implies that $q(t^\sharp) \notin p(t^\sharp)$ and $p(t^\sharp) \cup \{q(t^\sharp)\} = t^\sharp$.

Claim 4: $(\forall t \in T, \forall m \in \{0, 1, 2, \ldots k(t)\})\ |p^m(t)| = |t| - m$. Note by inspection, that Claim 3 (b) implies [a] $(\forall t^\sharp \in T \smallsetminus \{\{\}\})\ |p(t^\sharp)| = |t^\sharp| - 1$. To prove the present claim, take $t \in T$. Induction will show $(\forall m \in \{0, 1, 2, \ldots k(t)\})\ |p^m(t)| = |t| - m$. For the initial step ($m=0$), $|p^0(t)| = |t| = |t| - 0 = |t| - m$ by inspection. For the inductive step ($m \geq 1$), first note that by assumption $m \leq k(t)$, which trivially implies $m-1 < k(t)$, which by the definition of k implies $p^{m-1}(t) \neq t^o$, which by Claim 1 (a) implies [b] $p^{m-1}(t) \neq \{\}$. Then it is argued, in steps, that $|p^m(t)|$ by rearrangement equals $|p \circ p^{m-1}(t)|$, which by [b] and [a] at $t^\sharp = p^{m-1}(t)$ equals $|p^{m-1}(t)| - 1$, which by the inductive hypothesis equals $(|t| - (m-1)) - 1$, which by rearrangement equals $|t| - m$.

Claim 5: *(c) holds.* Take $t \in T$. Note [a] $|p^{k(t)}(t)| = |t| - k(t)$ by Claim 4 at $m = k(t)$. Also note [b] $|p^{k(t)}(t)| = |t^o| = |\{\}| = 0$ by the definition of $k(t)$ and by Claim 1 (a). [a] and [b] imply $|t| - k(t) = 0$. Hence $|t| = k(t)$.

Claim 6: *(d) holds.* Take $t \in T$. Consider an induction on $m \in \{0, 1, \ldots |t|\}$. For the initial step ($m=0$), $p^m(t) = p^0(t) = t$ and $t \smallsetminus p^m(t) = t \smallsetminus p^0(t) = t \smallsetminus t = \{\} = \{\, q \circ p^n(t) \mid 0 > n \geq 0 \,\} = \{\, q \circ p^n(t) \mid m > n \geq 0 \,\}$. For the inductive step ($m \geq 1$), note $m \leq |t|$ by assumption; which implies $m-1 < |t|$; which implies $m-1 < k(t)$ by Claim 5 (c); which implies $p^{m-1}(t) \neq t^o$ by the definition of k; which implies $p^{m-1}(t) \neq \{\}$ by Claim 1 (a). Hence, Claim 3 (b) at $t^\sharp = p^{m-1}(t)$ implies $q \circ p^{m-1}(t) \notin p \circ p^{m-1}(t)$ and $p \circ p^{m-1}(t) \cup \{q \circ p^{m-1}(t)\} = p^{m-1}(t)$. By Lemma C.1 (a)$\Leftrightarrow$(d), this is equivalent to $p \circ p^{m-1}(t) \subseteq p^{m-1}(t)$ and $p^{m-1}(t) \smallsetminus p \circ p^{m-1}(t) = \{q \circ p^{m-1}(t)\}$. By a small rearrangement, this is equivalent to [c] $p^m(t) \subseteq p^{m-1}(t)$ and [d] $p^{m-1}(t) \smallsetminus p^m(t) = \{q \circ p^{m-1}(t)\}$. Meanwhile, the inductive hypothesis is [e] $p^{m-1}(t) \subseteq t$ and [f] $t \smallsetminus p^{m-1}(t) = \{\, q \circ p^n(t) \mid m-1 > n \geq 0 \,\}$. [c] and [e] imply [g] $p^m(t) \subseteq t$. [c] and [e] also imply [h] $t \smallsetminus p^m(t) = (t \smallsetminus p^{m-1}(t)) \cup (p^{m-1}(t) \smallsetminus p^m(t))$. [h], [f], and [d] imply [i] $t \smallsetminus p^m(t) = \{\, q \circ p^n(t) \mid m-1 > n \geq 0 \,\} \cup \{q \circ p^{m-1}(t)\}$. The right-hand side of [i] is equal to $\{\, q \circ p^n(t) \mid m-1 \geq n \geq 0 \,\}$, which is equal to $\{\, q \circ p^n(t) \mid m > n \geq 0 \,\}$. Hence [i] is equivalent to [j] $t \smallsetminus p^m(t) = \{\, q \circ p^n(t) \mid m > n \geq 0 \,\}$. [g] and [j] are the desired results.

Claim 7: $(\forall t^A \in T, t^B \in T)\ t^A \prec t^B \Rightarrow t^A \subset t^B$. Suppose $t^A \prec t^B$. Then the definitions of $\prec$ and k imply there is [a] $m \in \{1, 2, ... k(t^B)\}$ such that [b] $t^A = p^m(t^B)$. [a] and Claim 5 (c) imply $m \in \{1, 2, ... |t^B|\}$. Thus Claim 6 (d) at $t = t^B$ implies [c] $p^m(t^B) \subseteq t^B$ and [d] $t^B \setminus p^m(t^B) = \{\, q \circ p^n(t^B) \mid m > n \geq 0 \,\}$. Since $m \geq 1$ by [a], $\{\, q \circ p^n(t^B) \mid m > n \geq 0 \,\}$ is nonempty. Thus [c] and [d] imply $p^m(t^B) \subset t^B$. Thus [b] implies $t^A \subset t^B$.

Claim 8: $(\forall t^A \in T, t^B \in T)\ t^A \preccurlyeq t^B \Rightarrow t^A \subseteq t^B$. Suppose $t^A \preccurlyeq t^B$. Then the definition of $\preccurlyeq$ implies $t^A \prec t^B$ or $t^A = t^B$. The first implies $t^A \subseteq t^B$ by Claim 7. The second implies $t^A \subseteq t^B$ trivially.

Claim 9: *(e) holds.* Take $t \in T$. It is argued, in steps, that t trivially equals $t \setminus \{\}$, which by Claim 1 (a) equals $t \setminus t^o$, which by the definition of k equals $t \setminus p^{k(t)}(t)$, which by Claim 5 (c) equals $t \setminus p^{|t|}(t)$, which by Claim 6 (d) at $m = |t|$ equals $\{\, q \circ p^n(t) \mid |t| > n \geq 0 \,\}$

Claim 10: *(f) holds.* Forward direction. Take $c \in C$. By [P3], $F^{-1}(c)$ is a member of a partition, and thus, it is nonempty. Take $t^* \in F^{-1}(c)$. By [P1], $t^* \otimes c \in T$. Thus by [Cset2], $t^* \cup \{c\} \in T$. Thus c belongs to an element of T. Reverse direction. Take any t. [Cset1] implies that t is a set. Take $b \in t$. By Claim 9 (e), there is n such that $b = q \circ p^n(t)$. Thus, since the codomain of q is C, $b \in C$.

Claim 11: *(g) holds.* Suppose there were $H \in \mathcal{H}$, [a] $t^A \in H$, and [b] $t^B \in H$ such that [c] $t^A \prec t^B$. [c] and the definition of $\prec$ imply there is [d] $m > 1$ such that [e] $t^A = p^m(t^B)$. [d] and [e] imply $t^A = p \circ p^{m-1}(t^B)$. Thus [P2]'s definition of p implies there is $c \in C$ such that [f] $(t^A, c, p^{m-1}(t^B)) \in \otimes^{\mathrm{gr}}$. Thus the definition of F implies $c \in F(t^A)$. This, [a], [b], and SP Proposition 3.2(16a) imply $c \in F(t^B)$. Thus the definition of F implies there is $t^\star \in T$ such that $(t^B, c, t^\star) \in \otimes^{\mathrm{gr}}$. This and Claim 2 implies [g] $c \notin t^B$. But, [f] and Claim 2 imply [h] $c \in p^{m-1}(t^B)$. And, the definition of $\preccurlyeq$ implies $p^{m-1}(t^B) \preccurlyeq t^B$, and thus Claim 8 implies [i] $p^{m-1}(t^B) \subseteq t^B$. [h] and [i] imply $c \in t^B$, which contradicts [g].

Claim 12: *(h) holds.* Suppose $|t \cap F(H)| \geq 2$. Then by Claim 9 (e), there exist distinct m' and m such that $\{\, q \circ p^{m'}(t),\ q \circ p^m(t) \,\} \subseteq F(H)$. Thus by Lemma A.1(c), [a] $\{\, p^{m'+1}(t),\ p^{m+1}(t) \,\} \subseteq H$. Without loss of generality assume $m' > m$. Then $p^{m'+1}(t) = p^{m'-m} \circ p^{m+1}(t)$. Hence [b] $p^{m'+1}(t) \prec p^{m+1}(t)$ by the definition of $\prec$. [a] and [b] show there is absentmindedness, which contradicts Claim 11 (g).

Claim 13: $(\forall t^A \in T, t^B \in T)\ \ t^A \subset t^B$ *implies* $(\forall m \in \{0, 1, ... |t^A|\})\ \ p^m(t^A) = p^{m+|t^B|-|t^A|}(t^B)$. Suppose [1] $t^A \subset t^B$. Consider the following induction on m.

For the initial step, assume [2] $m = |t^A|$. It is argued, in steps, that $p^m(t^A)$ by [2] equals $p^{|t^A|}(t^A)$, which by Claim 5 (c) equals $p^{k(t^A)}(t^A)$, which by the definition of k equals t^o, which by the definition of k again equals $p^{k(t^B)}(t^B)$, which by Claim 5 (c) again equals $p^{|t^B|}(t^B)$, which by manipulation equals $p^{|t^A|+|t^B|-|t^A|}(t^B)$, which by [2] again equals $p^{m+|t^B|-|t^A|}(t^B)$.

For the inductive step, assume [3] $m < |t^A|$. (The next two sentences concern $p^m(t^A)$ alone.) [3] and Claim 5 (c) imply $m < k(t^A)$, which by the definition of k implies $p^m(t^A) \neq t^o$. This and SP Proposition 3.1(a) at $t^\sharp = p^m(t^A)$ yield

$$[4]\ p^{m+1}(t^A) \otimes q \circ p^m(t^A) = p^m(t^A).$$

(The next three sentences concern $p^{m+|t^B|-|t^A|}(t^B)$ alone.) [3] and manipulation imply $m+|t^B|-|t^A| < |t^A|+|t^B|-|t^A| = |t^B|$, which by Claim 5 (c) implies $m+|t^B|-|t^A| < k(t^B)$, which by the definition of k implies $p^{m+|t^B|-|t^A|}(t^B) \neq t^o$. This and SP Proposition 3.1(a) at $t^\sharp = p^{m+|t^B|-|t^A|}(t^B)$ yield

$$[5]\ p^{m+1+|t^B|-|t^A|}(t^B) \otimes q \circ p^{m+|t^B|-|t^A|}(t^B) = p^{m+|t^B|-|t^A|}(t^B).$$

Since the inductive hypothesis is $p^{m+1}(t^A) = p^{m+1+|t^B|-|t^A|}(t^B)$, [5] yields

$$[6]\ p^{m+1}(t^A) \otimes q \circ p^{m+|t^B|-|t^A|}(t^B) = p^{m+|t^B|-|t^A|}(t^B).$$

[4], [6], and the definition of F yield [7] $\{\, q \circ p^m(t^A),\ q \circ p^{m+|t^B|-|t^A|}(t^B)\, \} \subseteq F(p^{m+1}(t^A))$. Also, Claim 9 (e) and [1] yield [8] $q \circ p^m(t^A) \in t^A \subseteq t^B$. Also, Claim 9 (e) yields [9] $q \circ p^{m+|t^B|-|t^A|}(t^B) \in t^B$. [7], [8], [9], and Claim 12 (h) imply [10] $q \circ p^m(t^A) = q \circ p^{m+|t^B|-|t^A|}(t^B)$. Finally, it is argued, in steps, that $p^m(t^A)$ by [4] equals $p^{m+1}(t^A) \otimes q \circ p^m(t^A)$, which by [10] equals $p^{m+1}(t^A) \otimes q \circ p^{m+|t^B|-|t^A|}(t^B)$, which by [6] equals $p^{m+|t^B|-|t^A|}(t^B)$.

Claim 14: *(i) holds.* This follows from Claim 13 at $m = 0$.

Claim 15: *(j) holds.* Because of Claim 7, it suffices to show the reverse direction. Toward that end, suppose [1] $t^A \subset t^B$. [1] and Claim 14 (i) imply $t^A = p^{|t^B|-|t^A|}(t^B)$. Further, [1] and [Cset1] imply $|t^B|-|t^A| > 0$. The last two sentences and the definition of $\prec$ imply $t^A \prec t^B$.

Claim 16: *(k) holds.* Because of Claim 8, it suffices to show the reverse direction. Toward that end, suppose $t^A \subseteq t^B$. Then either $t^A \subset t^B$ or $t^A = t^B$. In the first case, Claim 15 (j) implies $t^A \preccurlyeq t^B$. In the second case, $t^A \preccurlyeq t^B$ holds trivially by the definition of $\preccurlyeq$.

Claim 17: $(\forall t \in T, c \in C, t^\sharp \in T)\ (c \notin t \text{ and } t \cup \{c\} = t^\sharp) \Rightarrow (t, c) = (p(t^\sharp), q(t^\sharp))$. Suppose [a] $c \notin t$ and [b] $t \cup \{c\} = t^\sharp$. [a] and [b] imply [c] $t \subset t^\sharp$ and [d] $|t^\sharp|-|t| = 1$. [c] and Claim 14 (i) at $(t^A, t^B) = (t, t^\sharp)$ imply $t = p^{|t^\sharp|-|t|}(t^\sharp)$. This and [d] imply [e] $t = p(t^\sharp)$. Further, [c] implies [f] $t^\sharp \neq \{\}$. Next it is argued, in steps, that $\{c\}$ by [a]–[b] equals $t^\sharp \setminus t$, which by [e] equals $t^\sharp \setminus p(t^\sharp)$, which by [f] and Claim 3 (b) equals $\{q(t^\sharp)\}$. Thus [g] $c = q(t^\sharp)$. [e] and [g] are the required results.

Claim 18: *(l) holds.* By Claim 2, it suffices to show the reverse direction. Toward that end, suppose [a] $c \notin t$ and [b] $t \cup \{c\} = t^\sharp$. [b] implies $t^\sharp \neq \{\}$. Thus Claim 1 (a) implies $t^\sharp \neq t^o$. Thus SP Proposition 3.1(a) implies [c] $p(t^\sharp) \otimes q(t^\sharp) = t^\sharp$. Also, [a], [b], and Claim 17 imply [d] $(t, c) = (p(t^\sharp), q(t^\sharp))$. [c] and [d] imply $t \otimes c = t^\sharp$.

Claim 19: *(m) holds*. It is argued, in three steps, that $(t, c) \in F^{\mathrm{gr}}$ by [P1] is equivalent to [a] $(t, c) \in T \times C$ and [b] $(\exists t^\sharp \in T)\ (t, c, t^\sharp) \in \otimes^{\mathrm{gr}}$, which by Claim 18 (l) is equivalent to [a] and [b′] $(\exists t^\sharp \in T)\ c \notin t$ and $t \cup \{c\} = t^\sharp$, which by rearrangement is equivalent to [a] and [b″] $c \notin t$ and $t \cup \{c\} \in T$.

□

Proof C.3 (for Theorem 4.2). *(a)*. Lemma B.4[(a)⇒(e)] implies $R|_{\bar{T}}:\bar{T} \to T$ is a bijection. Thus the assumptions of Lemma A.14 are met at [1] its $(T, C, \otimes)$ equal to $(\bar{T}, \bar{C}, \bar{\otimes})$ here, [2] its $\tau:T \to T'$ equal to $R|_{\bar{T}}:\bar{T} \to T$ here, and [3] its $\delta:C \to C'$ equal to $\mathrm{id}_{\bar{C}}:\bar{C} \to \bar{C}$ here. Further, the definition of $\otimes'$ in the lemma coincides with the definition of $\otimes$ here. Therefore the lemma implies that $(T, \bar{C}, \otimes)$ is an **NCP** preform, and that $[(\bar{T}, \bar{C}, \bar{\otimes}), (T, \bar{C}, \otimes), R|_{\bar{T}}, \mathrm{id}_{\bar{C}}]$ is an **NCP** isomorphism. Thus it remains to show that $(T, \bar{C}, \otimes)$ is a **CsetP** preform. By definition, it suffices to show [Cset1] and [Cset2].

For [Cset1], first note that $\bar{T}$ is a collection of (finite) sequences by assumption. Hence T is a collection of finite sets by the definitions of T and R. Further, {} belongs to $\bar{T}$ by [Csq1]. Hence $R(\{\}) = \{\}$ belongs to T.

For [Cset2], take $(t, \bar{c}, t^\sharp) \in \otimes^{\mathrm{gr}}$. Then by the definition of $\otimes$, there are $\bar{t} \in \bar{T}$ and $\bar{t}^\sharp \in \bar{T}$ such that [a] $R(\bar{t}) = t$, [b] $R(\bar{t}^\sharp) = t^\sharp$, and [c] $(\bar{t}, \bar{c}, \bar{t}^\sharp) \in \bar{\otimes}$. [c] and [Csq2] imply [d] $\bar{t} \oplus (\bar{c}) = \bar{t}^\sharp$. Finally, in steps, $t \cup \{\bar{c}\}$ by [a] equals $R(\bar{t}) \cup \{\bar{c}\}$, which by inspection equals $R(\bar{t} \oplus (\bar{c}))$, which by ([d]) equals $R(\bar{t}^\sharp)$, which by [b] equals $t^\sharp$.

(b). Lemma B.4[(a)⇒(e)] implies $R|_{\bar{T}}:\bar{T} \to T$ is a bijection. Thus the assumptions of Lemma A.15 are met at [1] its $(I, T, (C_i)_{i \in I}, \otimes)$ equal to $(\bar{I}, \bar{T}, (\bar{C}_{\bar{i}})_{\bar{i} \in \bar{I}}, \bar{\otimes})$ here, [2] its $\iota:I \to I'$ equal to $\mathrm{id}_{\bar{I}}:\bar{I} \to \bar{I}$ here, [3] its $\tau:T \to T'$ equal to $R|_{\bar{T}}:\bar{T} \to T$ here, and [4] its $\delta:\cup_{i \in I} C_i \to C'$ equal to $\mathrm{id}_{\cup_{\bar{i} \in \bar{I}} \bar{C}_{\bar{i}}}:\cup_{\bar{i} \in \bar{I}} \bar{C}_{\bar{i}} \to \cup_{\bar{i} \in \bar{I}} \bar{C}_{\bar{i}}$ here. Also, the definition of $\otimes'$ in Lemma A.15 coincides with the definition of $\otimes$ here. Also, the transparent definitions of δ and ι here, and the definition of $(C'_{i'})_{i' \in I'}$ in Lemma A.15, imply $(C'_{i'})_{i' \in I'}$ there equals $(\bar{C}_{\bar{i}})_{\bar{i} \in \bar{I}}$ here. Hence Lemma A.15 implies that $(\bar{I}, T, (\bar{C}_{\bar{i}})_{\bar{i} \in \bar{I}}, \otimes)$ is an **NCF** form, and that $[(\bar{I}, \bar{T}, (\bar{C}_{\bar{i}})_{\bar{i} \in \bar{I}}, \bar{\otimes}), (\bar{I}, T, (\bar{C}_{\bar{i}})_{\bar{i} \in \bar{I}}, \otimes), \mathrm{id}_{\bar{I}}, R|_{\bar{T}}, \mathrm{id}_{\cup_{\bar{i} \in \bar{I}} C_{\bar{i}}}]$ is an **NCF** isomorphism.

It remains to show that $(\bar{I}, T, (\bar{C}_{\bar{i}})_{\bar{i} \in \bar{I}}, \otimes)$ is a **CsetF** form. Since the previous paragraph showed that it is an **NCF** form, it suffices to show that its preform $(T, \cup_{\bar{i} \in \bar{I}} \bar{C}_{\bar{i}}, \otimes)$ is an **CsetP** preform. By assumption, $(\bar{T}, \cup_{\bar{i} \in \bar{I}} \bar{C}_{\bar{i}}, \bar{\otimes})$ is a **CsqP**$_{\bar{a}}$ preform. Thus the assumption of part (a) is met at $(\bar{T}, \bar{C}, \bar{\otimes}) = (\bar{T}, \cup_{\bar{i} \in \bar{I}} \bar{C}_{\bar{i}}, \bar{\otimes})$. Further, part (b) defines T and $\otimes$ just as part (a) does. Hence part (a) implies $(T, \cup_{\bar{i} \in \bar{I}} \bar{C}_{\bar{i}}, \otimes)$ is a **CsetP** preform. □

References

Abramsky, S., R. Jagadeesan, and P. Malacaria. 2000. Full Abstraction for PCF. *Information and Computation* 163: 409–470.

Abramsky, S., and V. Winschel. 2017. Coalgebraic Analysis of Subgame-Perfect Equilibria in Infinite Games without Discounting. *Mathematical Structures in Computer Science* 27: 751–761.

Alós-Ferrer, C., and K. Ritzberger. 2016. *The Theory of Extensive Form Games*. Berlin: Springer.

Awodey, S. 2010. *Category Theory*, 2nd ed. Oxford.

Blackburn, P., M. de Rijke, and Y. Venema. 2001. *Modal Logic*. Cambridge.

Conradie, W., S. Ghilardi, and A. Palmigiano. 2014. Unified Correspondence. In *Johan van Benthem on Logic and Information Dynamics*, ed. A. Baltag and S. Smets, 933–975. Berlin: Springer.

Dockner, E., S. Jørgensen, N.V. Long, and G. Sorger. 2000. *Differential Games in Economics and Management Science*. Cambridge.

Ghani, N., C. Kupke, A. Lambert, and F.N. Forsberg. 2018. A Composition Treatment of Iterated Open Games. *Theoretical Computer Science* 741: 48–57.

Halmos, P.R. 1974. *Naive Set Theory*. Berlin: Springer.

Harris, C. 1985. Existence and Characterization of Perfect Equilibrium in Games of Perfect Information. *Econometrica* 53: 613–628.

Hedges, J. 2017. Morphisms of Open Games. ArXiv:1711.07059v2.

Hodges, W. 2013. In *Logic and Games, The Stanford Encyclopedia of Philosophy (Fall 2018 Edition)*, ed. Zalta, E.N. https://plato.stanford.edu/archives/fall2018/entries/logic-games/.

Honsell, F., M. Lenisa, and R. Redamalla. 2012. Categories of Coalgebraic Games. In *Mathematical Foundations of Computer Science 2012*, Lecture Notes in Computer Science, vol. 7464, ed. B. Rovan, V. Sassone, and P. Widmayer, 503–515. Berlin: Springer.

Hyland, J., and C.-H.L. Ong. 2000. On Full Abstraction for PCF: I, II, and III. *Information and Computation* 163: 285–408.

Jiménez, A. 2014. *Game Theory from the Category Theory Point of View, 20*. Colombia: Universidad Distrital Francisco José de Caldas.

Kline, J.J., and S. Luckraz. 2016. Equivalence between Graph-Based and Sequence-Based Extensive Form Games. *Economic Theory Bulletin* 4: 85–94. referred to as KL16.

Kuhn, H.W. 1953. Extensive Games and the Problem of Information. In *Classics in Game Theory*, ed. H.W. Kuhn, 46–68. Princeton (1997). Originally in *Contributions to the Theory of Games*, ed. H.W. Kuhn and A.W. Tucker, 193–216. Princeton (1953).

Lapitsky, V. 1999. On Some Categories of Games and Corresponding Equilibria. *International Game Theory Review* 1: 169–185.

Mac Lane, S. 1998. *Categories for the Working Mathematician*, 2nd ed. Berlin: Springer.

Machover, M., and S.D. Terrington. 2014. Mathematical Structures of Simple Voting Games. *Mathematical Social Sciences* 71: 61–68.

Mas-Colell, A., M.D. Whinston, and J.R. Green. 1995. *Microeconomic Theory*. Oxford.

McCusker, G. 2000. Games and Full Abstraction for FPC. *Information and Computation* 160: 1–61.

Mertens, J.-F. 2002. Stochastic Games. *Handbook of Game Theory*, vol. 3, 1811–1832. North-Holland.

Osborne, M.J., and A. Rubinstein. 1994. *A Course in Game Theory*. MIT.

Pacuit, E. 2007. Some Comments on History Based Structures. *Journal of Applied Logic* 5: 613–624.

Parikh, R., and R. Ramanujam. 1985. Distributed Processes and the Logic of Knowledge. In *Logic of Programs*, Lecture Notes in Computer Science, vol. 193, 256–268. Berlin: Springer.

Piccione, M., and A. Rubinstein. 1997. On the Interpretation of Decision Problems with Imperfect Recall. *Games and Economic Behavior* 20: 3–24.

Selten, R. 1975. Reexamination of the Perfectness Concept for Equilibrium Points in Extensive Games. *International Journal of Game Theory* 4: 25–55.

Shoham, Y., and K. Leyton-Brown. 2009. *Multiagent Systems: Algorithmic, Game-Theoretic, and Logical Foundations*. Cambridge

Simmons, H. 2011. *An Introduction to Category Theory*. Cambridge

Streufert, P.A. 2016. *The Category of Node-and-Choice Forms for Extensive-Form Games*. Western University, Department of Economics Research Report 2016–5.

Streufert, P.A. 2018. The Category of Node-and-Choice Preforms for Extensive-Form Games. *Studia Logica (Special Issue: Logics for Social Behaviour)* 106: 1001–1064. referred to as SP.
Streufert, P.A. 2019. Equivalences among Five Game Specifications, including a New Specification whose Nodes are Sets of Past Choices. *International Journal of Game Theory* 48: 1–32. referred to as SE.
Tutte, W.T. 1984. Graph Theory. Addison-Wesley.
van Benthem, J. 2001. Correspondence Theory. In *Handbook of Philosophical Logic*, vol. 3, 2nd ed, ed. D. Gabbay, and F. Guenthner, 325–408. Kluwer.
van Benthem, J. 2014. *Logic in Games*. MIT.
Vannucci, S. 2007. Game Formats as Chu Spaces. *International Game Theory Review* 9: 119–138.
von Neumann, J., and O. Morgenstern. 1944. *Theory of Games and Economic Behavior*. Princeton.

About the Temporal Logic of the Lexicographic Products of Unbounded Dense Linear Orders: A New Study of Its Computability

Philippe Balbiani

Abstract This article considers the temporal logic of the lexicographic products of unbounded dense linear orders and provides via mosaics a new proof of the membership in NP of the satisfiability problem it gives rise to.

Keywords Linear temporal logic · Lexicographic product · Satisfiability problem · Decidability · Complexity · Mosaic method · Decision procedure

1 Introduction

The mosaic method has been applied for proving completeness and decidability of temporal logics over multifarious linear flows of time (Caleiro et al. 2013; Marx et al. 2000; Marx and Local 2007; Reynolds 1997). The operation of lexicographic product of Kripke frames has been introduced as a variant of the more classical operation of Cartesian product (Gabbay et al. 2003; Gabbay and Shehtman 1998; Kurucz 2007; Reynolds and Zakharyaschev 2001). See Balbiani and Shapirovsky (2019) for details. It has been used for defining the semantical basis of different languages designed for time representation and temporal reasoning from the perspective of non-standard analysis (Balbiani 2008, 2009). In Balbiani (2010, 2013), the temporal logic of the lexicographic products of unbounded dense linear orders has been considered, its complete axiomatization has been given and its computability has been studied.

The purpose of this paper is give a new proof of the membership in NP of the satisfiability problem of the lexicographic products of linear temporal logics. Its section-by-section breakdown is as follows. Section 2 studies the elementary properties of the lexicographic products of unbounded dense linear orders. In Sect. 3, we present the syntax and the semantics of the temporal logic we will be working with. Sections 4 and 5 define mosaics and maps. In Sects. 6 and 7, we prove that the satisfiability problem in our temporal logic is in NP.

P. Balbiani (✉)
Institut de recherche en informatique de Toulouse, CNRS — Toulouse University, Toulouse, France
e-mail: Philippe.Balbiani@irit.fr

S. Ju et al. (eds.), *Nonclassical Logics and Their Applications*, Logic in Asia: Studia Logica Library,
https://doi.org/10.1007/978-981-15-1342-8_3

2 Products of Unbounded Dense Linear Orders

Let $\mathcal{F}_1 = (T_1, <_1)$ and $\mathcal{F}_2 = (T_2, <_2)$ be linear orders. Their *lexicographic product* is the structure $\mathcal{F} = (T, \prec_1, \prec_2)$ where

- $T = T_1 \times T_2$,
- $\prec_1$ and $\prec_2$ are the binary relations on T defined by putting $(s_1, s_2) \prec_1 (t_1, t_2)$ iff $s_1 <_1 t_1$ and $(s_1, s_2) \prec_2 (t_1, t_2)$ iff $s_1 = t_1$ and $s_2 <_2 t_2$.

We define the binary relation $\prec$ on T by putting $(s_1, s_2) \prec (t_1, t_2)$ iff $s_1 <_1 t_1$, or $s_1 = t_1$ and $s_2 <_2 t_2$. For all $(s_1, s_2), (t_1, t_2) \in T$, if $(s_1, s_2) \prec (t_1, t_2)$ then let

- $](s_1, s_2), (t_1, t_2)[$ be the set of all $(u_1, u_2) \in T$ such that $(s_1, s_2) \prec (u_1, u_2)$ and $(u_1, u_2) \prec (t_1, t_2)$,
- $](s_1, s_2), +\infty[$ be the set of all $(u_1, u_2) \in T$ such that $(s_1, s_2) \prec (u_1, u_2)$,
- $]-\infty, (t_1, t_2)[$ be the set of all $(u_1, u_2) \in T$ such that $(u_1, u_2) \prec (t_1, t_2)$.

The effect of the operation of lexicographic product may be described informally as follows: given 2 linear orders, their lexicographic product is the structure obtained by replacing each point of the first one by a copy of the second one. The global intuitions underlying such an operation is based upon the fact that, depending on the accuracy required or the available knowledge, one can describe a temporal situation at different levels of abstraction. See Van Benthem (1991, Sect. I.2.2), or Euzenat and Montanari (2005) for details. In Fig. 1, we have $s_1 <_1 t_1$ and $s_2 <_2 t_2$. As a result, we have $(s_1, s_2) \prec_2 (s_1, t_2)$, $(s_1, s_2) \prec_1 (t_1, s_2)$, $(s_1, s_2) \prec_1 (t_1, t_2)$, $(s_1, t_2) \prec_1 (t_1, s_2)$, $(s_1, t_2) \prec_1 (t_1, t_2)$ and $(t_1, s_2) \prec_2 (t_1, t_2)$.

In order to characterize the elementary properties of the operation of lexicographic product, we introduce a first-order language. Let Var denote a countable set of individual variables (with typical members denoted x, y, etc). The set $\mathcal{L}_{fo}$ of all *formulas* (with typical members denoted φ, ψ, etc) of the first-order language is given by the rule

- $\varphi ::= x <_1 y \mid x <_2 y \mid \bot \mid \neg\varphi \mid (\varphi \vee \psi) \mid \forall x\, \varphi \mid x = y$,

the formulas $x <_1 y$ and $x <_2 y$ being read "x precedes but is not infinitely close to y" and "x precedes and is infinitely close to y". We adopt the standard definitions for the remaining Boolean connectives. As usual, we define for all individual variables x,

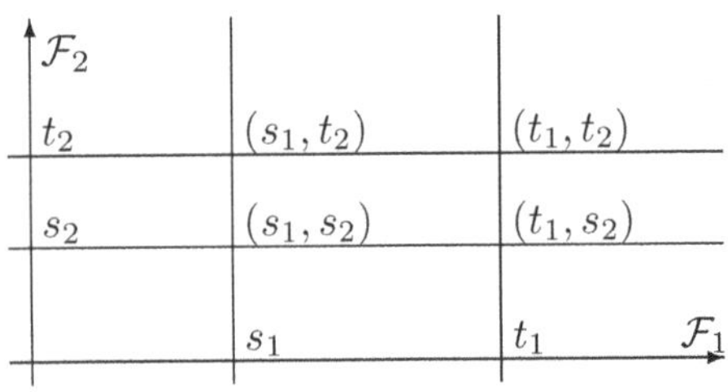

Fig. 1 Product of two linear orders

– $\exists x\ \varphi ::= \neg\forall x\ \neg\varphi$.

The notion of a *subformula* is standard. We adopt the standard rules for omission of the parentheses. The *size* of a formula φ, in symbols $\mid \varphi \mid$, is its length over the alphabet of $\mathcal{L}_{fo}$. Formulas in which every individual variable in an atomic subformula is in the scope of a corresponding quantifier are called *sentences*. *Models* for the first-order language are flows $\mathcal{F} = (T, \prec_1, \prec_2)$ where T is a nonempty set and $\prec_1$ and $\prec_2$ are binary relations on T. An *assignment* on $\mathcal{F}$ is a function $f\colon Var \longrightarrow T$. *Satisfaction* is a 3-place relation $\models$ between a flow $\mathcal{F} = (T, \prec_1, \prec_2)$, an assignment f on $\mathcal{F}$ and a formula φ. It is inductively defined as usual. In particular,

– $\mathcal{F} \models_f x <_1 y$ iff $f(x) \prec_1 f(y)$,
– $\mathcal{F} \models_f x <_2 y$ iff $f(x) \prec_2 f(y)$.

Obviously, if 2 linear orders are unbounded and dense then their lexicographic product satisfies the following sentences:

$IRRE$ $-\forall x\ x \not<_1 x$,
$-\forall x\ x \not<_2 x$,

$DISJ$ $-\forall x\ \forall y\ (x \not<_1 y \vee x \not<_2 y)$,

$TRAN$ $-\forall x\ \forall y\ (\exists z\ (x <_1 z \wedge z <_1 y) \rightarrow x <_1 y)$,
$-\forall x\ \forall y\ (\exists z\ (x <_1 z \wedge z <_2 y) \rightarrow x <_1 y)$,
$-\forall x\ \forall y\ (\exists z\ (x <_2 z \wedge z <_1 y) \rightarrow x <_1 y)$,
$-\forall x\ \forall y\ (\exists z\ (x <_2 z \wedge z <_2 y) \rightarrow x <_2 y)$,

$DENS$ $-\forall x\ \forall y\ (x <_1 y \rightarrow \exists z\ (x <_1 z \wedge z <_1 y))$,
$-\forall x\ \forall y\ (x <_1 y \rightarrow \exists z\ (x <_1 z \wedge z <_2 y))$,
$-\forall x\ \forall y\ (x <_1 y \rightarrow \exists z\ (x <_2 z \wedge z <_1 y))$,
$-\forall x\ \forall y\ (x <_2 y \rightarrow \exists z\ (x <_2 z \wedge z <_2 y))$,

$SERI$ $-\forall x\ \exists y\ x <_1 y$,
$-\forall x\ \exists y\ x <_2 y$,
$-\forall x\ \exists y\ y <_1 x$,
$-\forall x\ \exists y\ y <_2 x$,

$UNIV$ $-\forall x\ \forall y\ (x = y \vee x <_1 y \vee x <_2 y \vee y <_1 x \vee y <_2 x)$,

$IRRE, DISJ, TRAN, DENS, SERI$ and $UNIV$ standing for "irreflexive", "disjoint", "transitive", "dense", "serial" and "universal". Obviously, the sentences as above have not the finite model property. By Löwenheim–Skolem theorem, they have models in each infinite power. A flow $\mathcal{F} = (T, \prec_1, \prec_2)$ is said to be *standard* iff it satisfies the sentences as above. The *first-order theory* SF of standard flows has the following list of proper axioms: $IRRE, DISJ, TRAN, DENS, SERI$ and $UNIV$. The membership problem in the first-order theory SF is this:

– determine whether a given sentence is in the first-order theory SF.

There are several results about the first-order theory SF:

Proposition 1 (Balbiani 2008) *(i) SF is countably categorical; (ii) SF is not categorical in any uncountable power; (iii) SF is maximal consistent; (iv) SF is complete with respect to the lexicographic product of any unbounded dense linear orders.*

The results about the membership problem in the first-order theory SF are summarized in the following proposition:

Proposition 2 *(i) SF is decidable; (ii) SF is $PSPACE$-complete.*

See Balbiani (2008) for the proof of Propositions 1 and 2.

3 Syntax and Semantics

It is now time to meet the temporal language we will be working with. Let At be a countable set of atomic formulas (with typical members denoted p, q, etc). We define the set $\mathcal{L}_t$ of *formulas* of our temporal language (with typical members denoted φ, ψ, etc) as follows:

- $\varphi ::= p \mid \bot \mid \neg\varphi \mid (\varphi \vee \psi) \mid G_1\varphi \mid G_2\varphi \mid H_1\varphi \mid H_2\varphi$,

the formulas $G_1\varphi$, $G_2\varphi$, $H_1\varphi$ and $H_2\varphi$ being read "φ will be true at each point within the future of but not infinitely close to the present point", "φ will be true at each instant within the future of and infinitely close to the present instant", "φ has been true at each point within the past of but not infinitely close to the present point" and "φ has been true at each point within the past of and infinitely close to the present point". We adopt the standard definitions for the remaining Boolean connectives. As usual, we define for all $i \in \{1, 2\}$,

- $F_i\varphi ::= \neg G_i \neg\varphi$ and
- $P_i\varphi ::= \neg H_i \neg\varphi$.

The notion of a *subformula* is standard. It is usual to omit parentheses if this does not lead to any ambiguity. The *size* of a formula φ, in symbols $\mid \varphi \mid$, is the number of symbols of φ. A set Γ of formulas is said to be *full* iff there exists $i \in \{1, 2\}$ such that $G_i\bot \in \Gamma$, or $H_i\bot \in \Gamma$ and the following unique condition is satisfied:

- if there exists $i \in \{1, 2\}$ such that $G_i\varphi \in \Gamma$, or $H_i\varphi \in \Gamma$ then for all $i \in \{1, 2\}$, $G_i\varphi \in \Gamma$ and $H_i\varphi \in \Gamma$.

Remark that if Γ is a subformula closed full set of formulas then $\bot \in \Gamma$. Moreover, for all $i \in \{1, 2\}$, $G_i\bot \in \Gamma$ and $H_i\bot \in \Gamma$. In other respect,

Lemma 1 *Let φ be a formula. Let Γ be the least subformula closed full set of formulas containing φ. Then $card(\Gamma) \leq 4\times \mid \varphi \mid +2$.*

Proof By induction on φ.

A *model* is a structure $\mathcal{M} = (\mathcal{F}_1, \mathcal{F}_2, V)$ where $\mathcal{F}_1 = (T_1, <_1)$ and $\mathcal{F}_2 = (T_2, <_2)$ are linear orders and $V\colon At \longrightarrow 2^{T_1 \times T_2}$ is a function. *Satisfaction* is a 3-place relation $\models$ between a model $\mathcal{M} = (\mathcal{F}_1, \mathcal{F}_2, V)$, a pair $(s_1, s_2) \in T_1 \times T_2$ and a formula φ. It is defined by induction on φ as usual. In particular, for all $i \in \{1, 2\}$,

- $\mathcal{M}, (s_1, s_2) \models G_i\varphi$ iff $\mathcal{M}, (t_1, t_2) \models \varphi$ for each pair $(t_1, t_2) \in T_1 \times T_2$ such that $(s_1, s_2) \prec_i (t_1, t_2)$,
- $\mathcal{M}, (s_1, s_2) \models H_i\varphi$ iff $\mathcal{M}, (t_1, t_2) \models \varphi$ for each pair $(t_1, t_2) \in T_1 \times T_2$ such that $(t_1, t_2) \prec_i (s_1, s_2)$.

As a result, for all $i \in \{1, 2\}$,

- $\mathcal{M}, (s_1, s_2) \models F_i\varphi$ iff $\mathcal{M}, (t_1, t_2) \models \varphi$ for some pair $(t_1, t_2) \in T_1 \times T_2$ such that $(s_1, s_2) \prec_i (t_1, t_2)$,
- $\mathcal{M}, (s_1, s_2) \models P_i\varphi$ iff $\mathcal{M}, (t_1, t_2) \models \varphi$ for some pair $(t_1, t_2) \in T_1 \times T_2$ such that $(t_1, t_2) \prec_i (s_1, s_2)$.

$\mathcal{M}$ is said to be a *model* for φ iff there exists $(s_1, s_2) \in T_1 \times T_2$ such that $\mathcal{M}, (s_1, s_2) \models \varphi$. Let $\mathcal{C}_1$ and $\mathcal{C}_2$ be classes of linear orders. We shall say that a formula φ is *satisfiable* with respect to $(\mathcal{C}_1, \mathcal{C}_2)$ iff there exists a linear order $\mathcal{F}_1 = (T_1, <_1)$ in $\mathcal{C}_1$, there exists a linear order $\mathcal{F}_2 = (T_2, <_2)$ in $\mathcal{C}_2$ and there exists a function $V\colon At \longrightarrow 2^{T_1 \times T_2}$ such that $(\mathcal{F}_1, \mathcal{F}_2, V)$ is a model for φ. The *temporal logic* of $(\mathcal{C}_1, \mathcal{C}_2)$ is the set of all formulas φ such that $\neg\varphi$ is not satisfiable with respect to $(\mathcal{C}_1, \mathcal{C}_2)$. The class of all unbounded dense linear orders will be denoted $\mathcal{C}_{ud}$. Balbiani (2010) considers the temporal logic of $(\mathcal{C}_{ud}, \mathcal{C}_{ud})$ and gives its complete axiomatization based on the following 32 axioms:

- $\phi \to G_1 P_1 \phi$,
- $\phi \to H_1 F_1 \phi$,
- $\phi \to G_2 P_2 \phi$,
- $\phi \to H_2 F_2 \phi$,
- $F_1 F_1 \phi \to F_1 \phi$,
- $P_1 P_1 \phi \to P_1 \phi$,
- $F_1 F_2 \phi \to F_1 \phi$,
- $P_1 P_2 \phi \to P_1 \phi$,
- $F_2 F_1 \phi \to F_1 \phi$,
- $P_2 P_1 \phi \to P_1 \phi$,
- $F_2 F_2 \phi \to F_2 \phi$,
- $P_2 P_2 \phi \to P_2 \phi$,
- $F_1 \phi \to F_1 F_1 \phi$,
- $P_1 \phi \to P_1 P_1 \phi$,
- $F_1 \phi \to F_1 F_2 \phi$,
- $P_1 \phi \to P_1 P_2 \phi$,
- $F_1 \phi \to F_2 F_1 \phi$,
- $P_1 \phi \to P_2 P_1 \phi$,
- $F_2 \phi \to F_2 F_2 \phi$,

- $P_2\phi \to P_2P_2\phi$,
- $F_1\top$,
- $P_1\top$,
- $F_2\top$,
- $P_2\top$,
- $F_1\phi \wedge F_1\psi \to F_1(\phi \wedge \psi) \vee F_1(\phi \wedge F_1\psi) \vee F_1(\phi \wedge F_2\psi) \vee F_1(\psi \wedge F_1\phi) \vee F_1(\psi \wedge F_2\phi)$,
- $P_1\phi \wedge P_1\psi \to P_1(\phi \wedge \psi) \vee P_1(\phi \wedge P_1\psi) \vee P_1(\phi \wedge P_2\psi) \vee P_1(\psi \wedge P_1\phi) \vee P_1(\psi \wedge P_2\phi)$,
- $F_1\phi \wedge F_2\psi \to F_2(\psi \wedge F_1\phi)$,
- $P_1\phi \wedge P_2\psi \to P_2(\psi \wedge P_1\phi)$,
- $F_2\phi \wedge F_1\psi \to F_2(\phi \wedge F_1\psi)$,
- $P_2\phi \wedge P_1\psi \to P_2(\phi \wedge P_1\psi)$,
- $F_2\phi \wedge F_2\psi \to F_2(\phi \wedge \psi) \vee F_2(\phi \wedge F_2\psi) \vee F_2(\psi \wedge F_2\phi)$,
- $P_2\phi \wedge P_2\psi \to P_2(\phi \wedge \psi) \vee P_2(\phi \wedge P_2\psi) \vee P_2(\psi \wedge P_2\phi)$.

The *satisfiability problem* this temporal logic gives rise to is this:

- determine whether a given formula is satisfiable with respect to $(\mathcal{C}_{ud}, \mathcal{C}_{ud})$.

In order to provide a complete decision procedure in nondeterministic polynomial time for it, we use mosaics and maps.

4 Mosaics

Until the end of this paper, φ will denote a formula and Γ will denote the least subformula closed full set of formulas containing φ. A function $\sigma: \Gamma \longrightarrow \{0, 1\}$ is said to be *adequate* iff $\sigma(\bot) = 0$, for all $i \in \{1, 2\}$, $\sigma(G_i\bot) = 0$ and $\sigma(H_i\bot) = 0$ and the 2 following conditions are satisfied:

1. if $\neg\psi \in \Gamma$ then $\sigma(\neg\psi) = 1 - \sigma(\psi)$,
2. if $\psi \vee \chi \in \Gamma$ then $\sigma(\psi \vee \chi) = \max\{\sigma(\psi), \sigma(\chi)\}$.

A *mosaic* is a structure (σ, τ) where $\sigma: \Gamma \longrightarrow \{0, 1\}$ and $\tau: \Gamma \longrightarrow \{0, 1\}$ are adequate functions. We shall say that a mosaic (σ, τ) is 1-*temporal* iff the 2 following conditions are satisfied:

1. if $G_1\psi \in \Gamma$ then $\sigma(G_1\psi) \leq \min\{\tau(\psi), \tau(G_1\psi), \tau(G_2\psi), \tau(H_2\psi)\}$,
2. if $H_1\psi \in \Gamma$ then $\tau(H_1\psi) \leq \min\{\sigma(\psi), \sigma(G_2\psi), \sigma(H_1\psi), \sigma(H_2\psi)\}$.

A mosaic (σ, τ) is said to be 2-*temporal* iff the 6 following conditions are satisfied:

1. if $G_1\psi \in \Gamma$ then $\sigma(G_1\psi) \leq \tau(G_1\psi)$,
2. if $G_2\psi \in \Gamma$ then $\sigma(G_2\psi) \leq \min\{\tau(\psi), \tau(G_2\psi)\}$,
3. if $H_1\psi \in \Gamma$ then $\sigma(H_1\psi) \leq \tau(H_1\psi)$,
4. if $G_1\psi \in \Gamma$ then $\tau(G_1\psi) \leq \sigma(G_1\psi)$,
5. if $H_1\psi \in \Gamma$ then $\tau(H_1\psi) \leq \sigma(H_1\psi)$,
6. if $H_2\psi \in \Gamma$ then $\tau(H_2\psi) \leq \min\{\sigma(\psi), \sigma(H_2\psi)\}$.

A *premodel* is a structure (σ_0, M_1, M_2) where $\sigma_0 \colon \Gamma \longrightarrow \{0, 1\}$ is an adequate function, M_1 is a set of 1-temporal mosaics and M_2 is a set of 2-temporal mosaics. The *cardinality* of a premodel (σ_0, M_1, M_2), in symbols $\parallel (\sigma_0, M_1, M_2) \parallel$, is the number of mosaics in M_1 plus the number of mosaics in M_2. We shall say that a premodel (σ_0, M_1, M_2) is 0-*saturated* iff the 4 following conditions are satisfied:

1. if $G_1\psi \in \Gamma$ then $\sigma_0(G_1\psi) = 1$, or there exists an adequate function $\tau \colon \Gamma \longrightarrow \{0, 1\}$ such that $(\sigma_0, \tau) \in M_1$ and $\tau(\psi) = 0$,
2. if $G_2\psi \in \Gamma$ then $\sigma_0(G_2\psi) = 1$, or there exists an adequate function $\tau \colon \Gamma \longrightarrow \{0, 1\}$ such that $(\sigma_0, \tau) \in M_2$ and $\tau(\psi) = 0$,
3. if $H_1\psi \in \Gamma$ then $\sigma_0(H_1\psi) = 1$, or there exists an adequate function $\tau \colon \Gamma \longrightarrow \{0, 1\}$ such that $(\tau, \sigma_0) \in M_1$ and $\tau(\psi) = 0$,
4. if $H_2\psi \in \Gamma$ then $\sigma_0(H_2\psi) = 1$, or there exists an adequate function $\tau \colon \Gamma \longrightarrow \{0, 1\}$ such that $(\tau, \sigma_0) \in M_2$ and $\tau(\psi) = 0$.

A premodel (σ_0, M_1, M_2) is said to be 1-*saturated* iff for all mosaics $(\sigma, \tau) \in M_1$, the 8 following conditions are satisfied:

1. if $G_1\psi \in \Gamma$ then $\sigma(G_1\psi) = 1$, or there exists an adequate function $\mu \colon \Gamma \longrightarrow \{0, 1\}$ such that $(\sigma, \mu) \in M_1$, $(\mu, \tau) \in M_1 \cup M_2$ and $\mu(\psi) = 0$, or $\tau(\psi) = 0$, or $\tau(G_1\psi) = 0$, or $\tau(G_2\psi) = 0$,
2. if $G_2\psi \in \Gamma$ then $\sigma(G_2\psi) = 1$, or there exists an adequate function $\mu \colon \Gamma \longrightarrow \{0, 1\}$ such that $(\sigma, \mu) \in M_2$, $(\mu, \tau) \in M_1$ and $\mu(\psi) = 0$,
3. if $H_1\psi \in \Gamma$ then $\sigma(H_1\psi) = 1$, or there exists an adequate function $\mu \colon \Gamma \longrightarrow \{0, 1\}$ such that $(\mu, \sigma) \in M_1$ and $\mu(\psi) = 0$,
4. if $H_2\psi \in \Gamma$ then $\sigma(H_2\psi) = 1$, or there exists an adequate function $\mu \colon \Gamma \longrightarrow \{0, 1\}$ such that $(\mu, \sigma) \in M_2$ and $\mu(\psi) = 0$,
5. if $G_1\psi \in \Gamma$ then $\tau(G_1\psi) = 1$, or there exists an adequate function $\mu \colon \Gamma \longrightarrow \{0, 1\}$ such that $(\tau, \mu) \in M_1$ and $\mu(\psi) = 0$,
6. if $G_2\psi \in \Gamma$ then $\tau(G_2\psi) = 1$, or there exists an adequate function $\mu \colon \Gamma \longrightarrow \{0, 1\}$ such that $(\tau, \mu) \in M_2$ and $\mu(\psi) = 0$,
7. if $H_1\psi \in \Gamma$ then $\tau(H_1\psi) = 1$, or $\sigma(H_1\psi) = 0$, or $\sigma(H_2\psi) = 0$, or $\sigma(\psi) = 0$, or there exists an adequate function $\mu \colon \Gamma \longrightarrow \{0, 1\}$ such that $(\sigma, \mu) \in M_1 \cup M_2$, $(\mu, \tau) \in M_1$ and $\mu(\psi) = 0$,
8. if $H_2\psi \in \Gamma$ then $\tau(H_2\psi) = 1$, or there exists an adequate function $\mu \colon \Gamma \longrightarrow \{0, 1\}$ such that $(\sigma, \mu) \in M_1$, $(\mu, \tau) \in M_2$ and $\mu(\psi) = 0$.

We shall say that a premodel (σ_0, M_1, M_2) is 2-*saturated* iff for all mosaics $(\sigma, \tau) \in M_2$, the 8 following conditions are satisfied:

1. if $G_1\psi \in \Gamma$ then $\sigma(G_1\psi) = 1$, or there exists an adequate function $\mu \colon \Gamma \longrightarrow \{0, 1\}$ such that $(\tau, \mu) \in M_1$ and $\mu(\psi) = 0$,
2. if $G_2\psi \in \Gamma$ then $\sigma(G_2\psi) = 1$, or there exists an adequate function $\mu \colon \Gamma \longrightarrow \{0, 1\}$ such that $(\sigma, \mu) \in M_2$, $(\mu, \tau) \in M_2$ and $\mu(\psi) = 0$, or $\tau(\psi) = 0$, or $\tau(G_2\psi) = 0$,
3. if $H_1\psi \in \Gamma$ then $\sigma(H_1\psi) = 1$, or there exists an adequate function $\mu \colon \Gamma \longrightarrow \{0, 1\}$ such that $(\mu, \sigma) \in M_1$ and $\mu(\psi) = 0$,
4. if $H_2\psi \in \Gamma$ then $\sigma(H_2\psi) = 1$, or there exists an adequate function $\mu \colon \Gamma \longrightarrow \{0, 1\}$ such that $(\mu, \sigma) \in M_2$ and $\mu(\psi) = 0$,

5. if $G_1\psi \in \Gamma$ then $\tau(G_1\psi) = 1$, or there exists an adequate function $\mu: \Gamma \longrightarrow \{0, 1\}$ such that $(\tau, \mu) \in M_1$ and $\mu(\psi) = 0$,
6. if $G_2\psi \in \Gamma$ then $\tau(G_2\psi) = 1$, or there exists an adequate function $\mu: \Gamma \longrightarrow \{0, 1\}$ such that $(\tau, \mu) \in M_2$ and $\mu(\psi) = 0$,
7. if $H_1\psi \in \Gamma$ then $\tau(H_1\psi) = 1$, or there exists an adequate function $\mu: \Gamma \longrightarrow \{0, 1\}$ such that $(\mu, \sigma) \in M_1$ and $\mu(\psi) = 0$,
8. if $H_2\psi \in \Gamma$ then $\tau(H_2\psi) = 1$, or $\sigma(H_2\psi) = 0$, or $\sigma(\psi) = 0$, or there exists an adequate function $\mu: \Gamma \longrightarrow \{0, 1\}$ such that $(\sigma, \mu) \in M_2$, $(\mu, \tau) \in M_2$ and $\mu(\psi) = 0$.

A premodel is said to be *saturated* iff it is 0-, 1- and 2-saturated. We shall say that a premodel (σ_0, M_1, M_2) is 1-*dense* iff for all mosaics $(\sigma, \tau) \in M_1$, the following unique condition is satisfied:

- there exists an adequate function $\mu: \Gamma \longrightarrow \{0, 1\}$ such that $(\sigma, \mu) \in M_1$ and $(\mu, \tau) \in M_1$.

A premodel (σ_0, M_1, M_2) is said to be 2-*dense* iff for all mosaics $(\sigma, \tau) \in M_2$, the following unique condition is satisfied:

- there exists an adequate function $\mu: \Gamma \longrightarrow \{0, 1\}$ such that $(\sigma, \mu) \in M_2$ and $(\mu, \tau) \in M_2$.

We shall say that a premodel is *dense* iff it is 1- and 2-dense. A premodel (σ_0, M_1, M_2) is said to be *for* φ iff $\sigma_0(\varphi) = 1$.

5 Maps

Until the end of this section, (σ_0, M_1, M_2) will denote a saturated dense premodel. A *map* is a partial function $l: \mathbb{Q} \times \mathbb{Q} \longrightarrow (\Gamma \longrightarrow \{0, 1\})$ with finite domain and such that for all pairs $(s_1, s_2) \in dom(l)$, $l(s_1, s_2)$ is an adequate function. A map l is said to be *respectful* iff $(0, 0) \in dom(l)$, $l(0, 0) = \sigma_0$ and for all pairs $(s_1, s_2), (t_1, t_2) \in dom(l)$, the 2 following conditions are satisfied:

1. if $(s_1, s_2) \prec_1 (t_1, t_2)$ and $](s_1, s_2), (t_1, t_2)[\cap dom(l) = \emptyset$ then $(l(s_1, s_2), l(t_1, t_2)) \in M_1$,
2. if $(s_1, s_2) \prec_2 (t_1, t_2)$ and $](s_1, s_2), (t_1, t_2)[\cap dom(l) = \emptyset$ then $(l(s_1, s_2), l(t_1, t_2)) \in M_2$.

We first prove a simple lemma.

Lemma 2 *Let l_0 be the map defined by $dom(l_0) = \{(0, 0)\}$ and $l_0(0, 0) = \sigma_0$. Then l_0 is respectful.*

Proof By definition of l_0.

The map l_0 defined by Lemma 2 is called *init map* with respect to (σ_0, M_1, M_2). We shall say that a map l' *extends* a map l iff $dom(l') \supseteq dom(l)$ and $l'_{|dom(l)} = l$. Much more difficult than Lemma 2 are the following lemmas.

Lemma 3 *Let $i \in \{1, 2\}$ and ψ be a formula such that $G_i\psi \in \Gamma$. Let l be a respectful map and $(s_1, s_2) \in dom(l)$ such that $l(s_1, s_2)(G_i\psi) = 0$. There exists a respectful map l' and there exists $(t_1, t_2) \in \mathbb{Q} \times \mathbb{Q}$ such that l' extends l, $(t_1, t_2) \in dom(l')$, $(s_1, s_2) \prec_i (t_1, t_2)$ and $l'(t_1, t_2)(\psi) = 0$.*

Proof Since $dom(l)$ is finite, then there exists a nonnegative integer k and there exists $(u_1^1, u_2^1), \ldots, (u_1^k, u_2^k) \in \mathbb{Q} \times \mathbb{Q}$ such that $](s_1, s_2), +\infty[\cap dom(l) = \{(u_1^1, u_2^1), \ldots, (u_1^k, u_2^k)\}$. Without loss of generality, we may assume that $(s_1, s_2) \prec (u_1^1, u_2^1) \ldots \prec (u_1^k, u_2^k)$. Moreover, $(s_1, s_2) \prec_{j_1} (u_1^1, u_2^1) \ldots \prec_{j_k} (u_1^k, u_2^k)$ for exactly one k-tuple $(j_1, \ldots, j_k) \in \{1, 2\}^k$. Now, we proceed by induction on k.

Basis. Suppose $k = 0$. Now, consider the 2 following cases.

1. Suppose $]-\infty, (s_1, s_2)[\cap dom(l) = \emptyset$. Since l is respectful and $l(s_1, s_2)(G_i\psi) = 0$, then $(s_1, s_2) = (0, 0)$ and there exists an adequate function $\mu\colon \Gamma \longrightarrow \{0, 1\}$ such that $(\sigma_0, \mu) \in M_i$ and $\mu(\psi) = 0$. Let $(t_1, t_2) \in \mathbb{Q} \times \mathbb{Q}$ be such that $(0, 0) \prec_i (t_1, t_2)$ and l' be the least extension of l such that $(t_1, t_2) \in dom(l')$ and $l'(t_1, t_2) = \mu$. Obviously, l' is respectful.
2. Suppose $]-\infty, (s_1, s_2)[\cap dom(l) \neq \emptyset$. Since $dom(l)$ is finite, then there exists $(v_1, v_2) \in dom(l)$ such that $(v_1, v_2) \prec (s_1, s_2)$ and $](v_1, v_2), (s_1, s_2)[\cap dom(l) = \emptyset$. Moreover, $(v_1, v_2) \prec_j (s_1, s_2)$ for exactly one $j \in \{1, 2\}$. Since l is respectful and $l(s_1, s_2)(G_i\psi) = 0$, then $(l(v_1, v_2), l(s_1, s_2)) \in M_j$ and there exists an adequate function $\mu\colon \Gamma \longrightarrow \{0, 1\}$ such that $(l(s_1, s_2), \mu) \in M_i$ and $\mu(\psi) = 0$. Let $(t_1, t_2) \in \mathbb{Q} \times \mathbb{Q}$ be such that $(s_1, s_2) \prec_i (t_1, t_2)$ and l' be the least extension of l such that $(t_1, t_2) \in dom(l')$ and $l'(t_1, t_2) = \mu$. Obviously, l' is respectful.

Step. Suppose $k \geq 1$. Now, consider the 4 following cases.

1. Suppose $i = 1$ and $j_1 = 1$. Since l is respectful and $l(s_1, s_2)(G_1\psi) = 0$, then $(l(s_1, s_2), l(u_1^1, u_2^1)) \in M_1$ and **(i)** there exists an adequate function $\mu\colon \Gamma \longrightarrow \{0, 1\}$ such that $(l(s_1, s_2), \mu) \in M_1$, $(\mu, l(u_1^1, u_2^1)) \in M_1$ and $\mu(\psi) = 0$, or **(ii)** there exists an adequate function $\mu\colon \Gamma \longrightarrow \{0, 1\}$ such that $(l(s_1, s_2), \mu) \in M_1$, $(\mu, l(u_1^1, u_2^1)) \in M_2$ and $\mu(\psi) = 0$, or **(iii)** $l(u_1^1, u_2^1)(\psi) = 0$, or **(iv)** $l(u_1^1, u_2^1)(G_1\psi) = 0$, or **(v)** $l(u_1^1, u_2^1)(G_2\psi) = 0$. In case **(i)**, let $(t_1, t_2) \in \mathbb{Q} \times \mathbb{Q}$ be such that $(s_1, s_2) \prec_1 (t_1, t_2)$ and $(t_1, t_2) \prec_1 (u_1^1, u_2^1)$ and l' be the least extension of l such that $(t_1, t_2) \in dom(l')$ and $l'(t_1, t_2) = \mu$. In case **(ii)**, let $(t_1, t_2) \in \mathbb{Q} \times \mathbb{Q}$ be such that $(s_1, s_2) \prec_1 (t_1, t_2)$ and $(t_1, t_2) \prec_2 (u_1^1, u_2^1)$ and l' be the least extension of l such that $(t_1, t_2) \in dom(l')$ and $l'(t_1, t_2) = \mu$. In case **(iii)**, let $(t_1, t_2) \in \mathbb{Q} \times \mathbb{Q}$ be (u_1^1, u_2^1) and l' be l. In case **(iv)**, by induction hypothesis, there exists a respectful map l' and there exists $(t_1, t_2) \in \mathbb{Q} \times \mathbb{Q}$ such that l' extends l, $(t_1, t_2) \in dom(l')$, $(u_1^1, u_2^1) \prec_1 (t_1, t_2)$ and $l'(t_1, t_2)(\psi) = 0$. In case **(v)**, by induction hypothesis, there exists a respectful map l' and there exists $(t_1, t_2) \in \mathbb{Q} \times \mathbb{Q}$ such that l' extends l, $(t_1, t_2) \in dom(l')$, $(u_1^1, u_2^1) \prec_2 (t_1, t_2)$ and $l'(t_1, t_2)(\psi) = 0$. Obviously, in all cases, l' is respectful.

2. Suppose $i = 1$ and $j_1 = 2$. Since l is respectful and $l(s_1, s_2)(G_1\psi) = 0$, then $(l(s_1, s_2), l(u_1^1, u_2^1)) \in M_2$ and $l(u_1^1, u_2^1)(G_1\psi) = 0$. By induction hypothesis, there exists a respectful map l' and there exists $(t_1, t_2) \in \mathbb{Q} \times \mathbb{Q}$ such that l' extends l, $(t_1, t_2) \in dom(l')$, $(u_1^1, u_2^1) \prec_1 (t_1, t_2)$ and $l'(t_1, t_2)(\psi) = 0$. Obviously, l' is respectful.
3. Suppose $i = 2$ and $j_1 = 1$. Since l is respectful and $l(s_1, s_2)(G_2\psi) = 0$, then $(l(s_1, s_2), l(u_1^1, u_2^1)) \in M_1$ and there exists an adequate function $\mu: \Gamma \longrightarrow \{0, 1\}$ such that $(l(s_1, s_2), \mu) \in M_2$, $(\mu, l(u_1^1, u_2^1)) \in M_1$ and $\mu(\psi) = 0$. Let $(t_1, t_2) \in \mathbb{Q} \times \mathbb{Q}$ be such that $(s_1, s_2) \prec_2 (t_1, t_2)$ and $(t_1, t_2) \prec_1 (u_1^1, u_2^1)$ and l' be the least extension of l such that $(t_1, t_2) \in dom(l')$ and $l'(t_1, t_2) = \mu$. Obviously, l' is respectful.
4. Suppose $i = 2$ and $j_1 = 2$. Since l is respectful and $l(s_1, s_2)(G_2\psi) = 0$, then $(l(s_1, s_2), l(u_1^1, u_2^1)) \in M_2$ and **(i)** there exists an adequate function $\mu: \Gamma \longrightarrow \{0, 1\}$ such that $(l(s_1, s_2), \mu) \in M_2$, $(\mu, l(u_1^1, u_2^1)) \in M_2$ and $\mu(\psi) = 0$, or **(ii)** $l(u_1^1, u_2^1)(\psi) = 0$, or **(iii)** $l(u_1^1, u_2^1)(G_2\psi) = 0$. In case **(i)**, let $(t_1, t_2) \in \mathbb{Q} \times \mathbb{Q}$ be such that $(s_1, s_2) \prec_2 (t_1, t_2)$ and $(t_1, t_2) \prec_2 (u_1^1, u_2^1)$ and l' be the least extension of l such that $(t_1, t_2) \in dom(l')$ and $l'(t_1, t_2) = \mu$. In case **(ii)**, let $(t_1, t_2) \in \mathbb{Q} \times \mathbb{Q}$ be (u_1^1, u_2^1) and l' be l. In case **(iii)**, by induction hypothesis, there exists a respectful map l' and there exists $(t_1, t_2) \in \mathbb{Q} \times \mathbb{Q}$ such that l' extends l, $(t_1, t_2) \in dom(l')$, $(u_1^1, u_2^1) \prec_2 (t_1, t_2)$ and $l'(t_1, t_2)(\psi) = 0$. Obviously, in all cases, l' is respectful.

Lemma 4 *Let $i \in \{1, 2\}$ and ψ be a formula such that $H_i\psi \in \Gamma$. Let l be a respectful map and $(s_1, s_2) \in dom(l)$ such that $l(s_1, s_2)(H_i\psi) = 0$. There exists a respectful map l' and there exists $(t_1, t_2) \in \mathbb{Q} \times \mathbb{Q}$ such that l' extends l, $(t_1, t_2) \in dom(l')$, $(t_1, t_2) \prec_i (s_1, s_2)$ and $l'(t_1, t_2)(\psi) = 0$.*

Proof Similar to the proof of Lemma 3.

Lemma 5 *Let $i \in \{1, 2\}$. Let l be a respectful map and $(s_1, s_2), (t_1, t_2) \in dom(l)$ such that $(s_1, s_2) \prec_i (t_1, t_2)$ and $](s_1, s_2), (t_1, t_2)[\cap dom(l) = \emptyset$. There exists a respectful map l' and there exists $(u_1, u_2) \in \mathbb{Q} \times \mathbb{Q}$ such that l' extends l, $(u_1, u_2) \in dom(l')$, $(s_1, s_2) \prec_i (u_1, u_2)$ and $(u_1, u_2) \prec_i (t_1, t_2)$.*

Proof Since l is respectful and $(s_1, s_2) \prec_i (t_1, t_2)$, then $(l(s_1, s_2), l(t_1, t_2)) \in M_i$ and there exists an adequate function $\mu: \Gamma \longrightarrow \{0, 1\}$ such that $(l(s_1, s_2), \mu) \in M_i$ and $(\mu, l(t_1, t_2)) \in M_i$. Let $(u_1, u_2) \in \mathbb{Q} \times \mathbb{Q}$ be such that $(s_1, s_2) \prec_i (u_1, u_2)$ and $(u_1, u_2) \prec_i (t_1, t_2)$ and l' be the least extension of l such that $(u_1, u_2) \in dom(l')$ and $l'(t_1, t_2) = \mu$. Obviously, l' is respectful.

The maps l' defined by Lemmas 3–5 are respectively called *right completion* of l with respect to i, ψ and (s_1, s_2), *left completion* of l with respect to i, ψ and (s_1, s_2) and *dense completion* of l with respect to i, (s_1, s_2) and (t_1, t_2).

6 Correctness and Completeness of the Mosaic Method

Now, we are ready to formulate our main propositions.

6.1 Correctness

First, the correctness of the mosaic method.

Proposition 3 *If there exists a saturated dense premodel for φ then φ is satisfiable with respect to $(\mathcal{C}_{ud}, \mathcal{C}_{ud})$.*

Proof Suppose there exists a saturated dense premodel (σ_0, M_1, M_2) for φ. Following the line of reasoning suggested in Marx et al. (2000), we think of the construction of a model for φ as a process approaching a limit via a sequence l_0, l_1, ... of respectful maps. Lemma 2 is used to initiate the construction whereas Lemmas 3–5 are used to make improvements at each step of the construction. Consider an enumeration $(i^0, \psi^0, (s_1^0, s_2^0), (t_1^0, t_2^0)), (i^1, \psi^1, (s_1^1, s_2^1), (t_1^1, t_2^1)), \ldots$ of $\{1, 2\} \times \mathcal{L}_t \times (\mathbb{Q} \times \mathbb{Q}) \times (\mathbb{Q} \times \mathbb{Q})$ where each item appears infinitely often. We inductively define a sequence $l_0, l_1, \ldots$ of respectful maps as follows:

Basis. Let l_0 be the init map with respect to (σ_0, M_1, M_2).

Step. Let l_n^{right}, l_n^{left} and l_{n+1} be the respectful maps defined as follows:

(i) if i^n, ψ^n, l_n and (s_1^n, s_2^n) satisfy the conditions of Lemma 3 then let l_n^{right} be the right completion of l_n with respect to i^n, ψ^n and (s_1^n, s_2^n) else let l_n^{right} be l_n,

(ii) if i^n, ψ^n, l_n^{right} and (s_1^n, s_2^n) satisfy the conditions of Lemma 4 then let l_n^{left} be the left completion of l_n^{right} with respect to i^n, ψ^n and (s_1^n, s_2^n) else let l_n^{left} be l_n^{right},

(iii) if i^n, l_n^{left}, (s_1^n, s_2^n) and (t_1^n, t_2^n) satisfy the conditions of Lemma 5 then let l_{n+1} be the dense completion of l_n^{left} with respect to i^n, (s_1^n, s_2^n) and (t_1^n, t_2^n) else let l_{n+1} be l_n^{left}.

The reader may easily verify that the sequence $l_0, l_1, \ldots$ of respectful maps is such that $dom(l_0) \subseteq dom(l_1) \subseteq \ldots$ and for all nonnegative integers n, $l_{n+1|dom(l_n)} = l_n$. Let $l\colon \mathbb{Q} \times \mathbb{Q} \longrightarrow (\Gamma \longrightarrow \{0, 1\})$ be the partial function defined by $dom(l) = \bigcup\{dom(l_n)\colon n$ is a nonnegative integer$\}$ and $l(s_1, s_2) = l_n(s_1, s_2)$ for each $(s_1, s_2) \in dom(l)$, n being a nonnegative integer such that $(s_1, s_2) \in dom(l_n)$. Obviously, $(0, 0) \in dom(l)$ and $l(0, 0) = \sigma_0$. Let $\mathcal{F} = (T, \lessdot_1, \lessdot_2)$ be the structure defined by

- $T = dom(l)$,
- $\lessdot_1$ and $\lessdot_2$ are the binary relations on T defined by putting $(s_1, s_2) \lessdot_1 (t_1, t_2)$ iff $s_1 < t_1$ and $(s_1, s_2) \lessdot_2 (t_1, t_2)$ iff $s_1 = t_1$ and $s_2 < t_2$.

We define the binary relation $\lessdot$ on T by putting $(s_1, s_2) \lessdot (t_1, t_2)$ iff $s_1 < t_1$, or $s_1 = t_1$ and $s_2 < t_2$. By Lemmas 3–5, $\mathcal{F}$ satisfies $SERI$ and $DENS$. Since $\mathcal{F}$ is a substructure of $\mathbb{Q} \times \mathbb{Q}$, then it satisfies $IRRE$, $DISJ$, $TRAN$ and $UNIV$. Hence, $\mathcal{F}$ is standard. Thus, by item 1 of Proposition 1, $\mathcal{F}$ is isomorphic with $\mathbb{Q} \times \mathbb{Q}$. Without loss of generality, we may assume that $\mathcal{F}$ is equal to $\mathbb{Q} \times \mathbb{Q}$. Let $V\colon At \longrightarrow 2^T$ be the function defined by $(s_1, s_2) \in V(p)$ iff $p \in \Gamma$ and $l(s_1, s_2)(p) = 1$—i.e. V makes p true at (s_1, s_2) iff l says so—and $\mathcal{M} = (T, \lessdot_1, \lessdot_2, V)$. By induction on ψ, let us demonstrate that if $\psi \in \Gamma$ then

- if $l(s_1, s_2)(\psi) = 1$ then $\mathcal{M}, (s_1, s_2) \models \psi$,
- if $l(s_1, s_2)(\psi) = 0$ then $\mathcal{M}, (s_1, s_2) \not\models \psi$.

Basis. Suppose $\psi = p$. Now, if $\psi \in \Gamma$ then consider the 2 following cases.

1. Suppose $l(s_1, s_2)(\psi) = 1$. Therefore, $(s_1, s_2) \in V(p)$. Consequently, $\mathcal{M}, (s_1, s_2) \models \psi$.
2. Suppose $l(s_1, s_2)(\psi) = 0$. Hence, $(s_1, s_2) \notin V(p)$. Thus, $\mathcal{M}, (s_1, s_2) \not\models \psi$.

Step. Suppose $\psi = \bot$, or $\psi = \neg\chi$, or $\psi = \chi \vee \rho$, or $\psi = G_1\chi$, or $\psi = G_2\chi$, or $\psi = H_1\chi$, or $\psi = H_2\chi$. Leaving the cases $\psi = \bot$, $\psi = \neg\chi$ and $\psi = \chi \vee \rho$ to the reader, we only consider the cases $\psi = G_1\chi$ and $\psi = G_2\chi$, the cases $\psi = H_1\chi$ and $\psi = H_2\chi$ being similar to them. Now, if $G_i\chi \in \Gamma$ then consider the 2 following cases.

1. Suppose $l(s_1, s_2)(G_i\chi) = 1$. Therefore, $l_m(s_1, s_2)(G_i\chi) = 1$, m being a nonnegative integer such that $(s_1, s_2) \in dom(l_m)$. Let $(t_1, t_2) \in T$ such that $(s_1, s_2) \lessdot_i (t_1, t_2)$, we demonstrate $\mathcal{M}, (t_1, t_2) \models \chi$. Since $(t_1, t_2) \in T$, then there exists a nonnegative integer n such that $(t_1, t_2) \in dom(l_n)$. Let $o = \max\{m, n\}$. Since $l_m(s_1, s_2)(G_i\chi) = 1$, then $l_o(s_1, s_2)(G_i\chi) = 1$. Moreover, since $dom(l_o)$ is finite, then there exists a nonnegative integer k and there exists $(u_1^1, u_2^1), \ldots, (u_1^k, u_2^k) \in T$ such that $](s_1, s_2), (t_1, t_2)[\cap dom(l_o) = \{(u_1^1, u_2^1), \ldots, (u_1^k, u_2^k)\}$. Without loss of generality, we may assume that $(s_1, s_2) \lessdot (u_1^1, u_2^1) \ldots \lessdot (u_1^k, u_2^k)$ and $(u_1^k, u_2^k) \lessdot (t_1, t_2)$. Moreover, $(s_1, s_2) \lessdot_{j_1} (u_1^1, u_2^1) \ldots \lessdot_{j_k} (u_1^k, u_2^k)$ for exactly one k-tuple $(j_1, \ldots, j_k) \in \{1, 2\}^k$ and $(u_1^k, u_2^k) \lessdot_j (t_1, t_2)$ for exactly one $j \in \{1, 2\}$. Since l_o is respectful, then $(l_o(s_1, s_2), l_o(u_1^1, u_2^1)) \in M_{j_1}, \ldots, (l_o(u_1^{k-1}, u_2^{k-1}), l_o(u_1^k, u_2^k)) \in M_{j_k}$ and $(l_o(u_1^k, u_2^k), l_o(t_1, t_2)) \in M_j$. Remark that $\min\{\min\{j_1, \ldots, j_k\}, j\} = i$. Since $l_o(s_1, s_2)(G_i\chi) = 1$, mosaics in M_1 are 1-temporal and mosaics in M_2 are 2-temporal, then $l_o(t_1, t_2)(\chi) = 1$. Consequently, $l(t_1, t_2)(\chi) = 1$. By induction hypothesis, $\mathcal{M}, (t_1, t_2) \models \chi$.
2. Suppose $l(s_1, s_2)(G_i\chi) = 0$. Hence, $l_m(s_1, s_2)(G_i\chi) = 0$, m being a nonnegative integer such that $(s_1, s_2) \in dom(l_m)$. By definition of the sequence $l_0, l_1, \ldots$ of respectful maps, there exists a nonnegative integer n and there exists $(t_1, t_2) \in T$ such that $(t_1, t_2) \in dom(l_n)$, $(s_1, s_2) \lessdot_i (t_1, t_2)$ and $l_n(t_1, t_2)(\chi) = 0$. Consequently, $l(t_1, t_2)(\chi) = 0$. By induction hypothesis, $\mathcal{M}, (t_1, t_2) \not\models \chi$. Since $(s_1, s_2) \lessdot_i (t_1, t_2)$, then $\mathcal{M}, (s_1, s_2) \not\models G_i\chi$.

Since (σ_0, M_1, M_2) is for φ, then $\sigma_0(\varphi) = 1$. Since $(0, 0) \in dom(l)$ and $l(0, 0) = \sigma_0$, then $l(0, 0)(\varphi) = 1$ and $\mathcal{M}, (0, 0) \models \varphi$. Since $\mathcal{F}$ is isomorphic with $\mathbb{Q} \times \mathbb{Q}$, then φ is satisfiable with respect to $(\mathcal{C}_{ud}, \mathcal{C}_{ud})$.

6.2 Completeness

Second, the completeness of the mosaic method.

Proposition 4 *If φ is satisfiable with respect to $(\mathcal{C}_{ud}, \mathcal{C}_{ud})$ then there exists a saturated dense premodel for φ of cardinality bounded by $2 \times (8\times \mid \varphi \mid +6)^8$.*

Proof Suppose φ is satisfiable with respect to $(\mathcal{C}_{ud}, \mathcal{C}_{ud})$. Hence, there exists a linear order $\mathcal{F}_1 = (T_1, <_1)$ in $\mathcal{C}_{ud}$, there exists a linear order $\mathcal{F}_2 = (T_2, <_2)$ in $\mathcal{C}_{ud}$ and there exists a function $V: At \longrightarrow 2^{T_1 \times T_2}$ such that $(\mathcal{F}_1, \mathcal{F}_2, V)$ is a model for φ. For all $(s_1, s_2) \in T_1 \times T_2$, let $\sigma(s_1, s_2): \Gamma \longrightarrow \{0, 1\}$ be the function such that if $\psi \in \Gamma$ then $\sigma(s_1, s_2)(\psi) =$ if $(\mathcal{F}_1, \mathcal{F}_2, V), (s_1, s_2) \models \psi$ then 1 else 0. Obviously, $\sigma(s_1, s_2)$ is an adequate function. Let $s_1 \in T_1$. Let $\equiv_{s_1}$ be the binary relation on T_2 defined as follows: $t_2 \equiv_{s_1} u_2$ iff $\sigma(s_1, t_2) = \sigma(s_1, u_2)$. Obviously, $\equiv_{s_1}$ is an equivalence relation on T_2. Moreover, since Γ is finite, then there exists finitely many equivalence classes modulo $\equiv_{s_1}$. We define

- $\Delta_2^{s_1} = \{G_2\psi: G_2\psi \in \Gamma$ is such that $\sigma(s_1, t_2)(\psi) = 0$ for some $t_2 \in T_2\}$,
- $\Lambda_2^{s_1} = \{H_2\psi: H_2\psi \in \Gamma$ is such that $\sigma(s_1, t_2)(\psi) = 0$ for some $t_2 \in T_2\}$.

Obviously, $\Delta_2^{s_1} \subseteq \Gamma$ and $\Lambda_2^{s_1} \subseteq \Gamma$. Therefore, by Lemma 1, $card(\Delta_2^{s_1}) \leq 4\times \mid \varphi \mid +2$ and $card(\Lambda_2^{s_1}) \leq 4\times \mid \varphi \mid +2$. Let

- for all $G_2\psi \in \Delta_2^{s_1}$, $f_2^{s_1}(G_2\psi) = \{t_2: t_2 \in T_2$ is such that $\sigma(s_1, t_2)(\psi) = 0\}$,
- for all $H_2\psi \in \Lambda_2^{s_1}$, $g_2^{s_1}(H_2\psi) = \{t_2: t_2 \in T_2$ is such that $\sigma(s_1, t_2)(\psi) = 0\}$.

We shall say that

- $t_2 \in f_2^{s_1}(G_2\psi)$ is maximal iff for all $u_2 \in f_2^{s_1}(G_2\psi)$, $t_2 \equiv_{s_1} u_2$, or there exists $v_2 \in f_2^{s_1}(G_2\psi)$ such that $u_2 <_2 v_2$ and $t_2 \equiv_{s_1} v_2$,
- $t_2 \in g_2^{s_1}(H_2\psi)$ is minimal iff for all $u_2 \in g_2^{s_1}(H_2\psi)$, $t_2 \equiv_{s_1} u_2$, or there exists $v_2 \in g_2^{s_1}(H_2\psi)$ such that $v_2 <_2 u_2$ and $t_2 \equiv_{s_1} v_2$.

Since there exists finitely many equivalence classes modulo $\equiv_{s_1}$, then for all $G_2\psi \in \Delta_2^{s_1}$, there exists $t_2(G_2\psi) \in f_2^{s_1}(G_2\psi)$ such that $t_2(G_2\psi)$ is maximal and for all $H_2\psi \in \Lambda_2^{s_1}$, there exists $t_2(H_2\psi) \in g_2^{s_1}(H_2\psi)$ such that $t_2(H_2\psi)$ is minimal. For all $s_2 \in T_2$, let

- $\Pi_2^{s_1}(s_2) = \{s_2\} \cup \{t_2(G_2\psi): G_2\psi \in \Delta_2^{s_1}\} \cup \{t_2(H_2\psi): H_2\psi \in \Lambda_2^{s_1}\}$.

Since $card(\Delta_2^{s_1}) \leq 4\times \mid \varphi \mid +2$ and $card(\Lambda_2^{s_1}) \leq 4\times \mid \varphi \mid +2$, then $card(\Pi_2^{s_1}(s_2)) \leq 8\times \mid \varphi \mid +5$. Let $t_2, u_2 \in \Pi_2^{s_1}(s_2)$ be such that $t_2 \equiv_{s_1} \circ <_2 \circ \equiv_{s_1} u_2$. A witness of 2-density for t_2 and u_2 is $v_2 \in T_2$ such that $t_2 \equiv_{s_1} \circ <_2 \circ \equiv_{s_1} v_2$, $v_2 \equiv_{s_1} \circ <_2 \circ \equiv_{s_1} u_2$ and $v_2 \equiv_{s_1} \circ <_2 \circ \equiv_{s_1} v_2$. Since there exists finitely many equivalence classes modulo $\equiv_{s_1}$ and $t_2 \equiv_{s_1} \circ <_2 \circ \equiv_{s_1} u_2$, then such a witness exists. Now, let

- $\Pi_{2,witness}^{s_1}(s_2)$ be the least subset of T_2 containing $\Pi_2^{s_1}(s_2)$ and such that for all $t_2, u_2 \in \Pi_2^{s_1}(s_2)$, if $t_2 \equiv \circ <_2 \circ \equiv u_2$ then $\Pi_{2,witness}^{s_1}(s_2)$ contains a witness of 2-density for t_2 and u_2.

Remark that the cardinality of $\Pi_{2,witness}^{s_1}(s_2)$ is bounded by $(8\times \mid \varphi \mid +6)^2$. Let $\cong$ be the binary relation on $T_1 \times T_2$ defined as follows: $(s_1, s_2) \cong (t_1, t_2)$ iff $\{\sigma(s_1, u_2)$: $u_2 \in \Pi_2^{s_1}(s_2)\} = \{\sigma(t_1, v_2)$: $v_2 \in \Pi_2^{t_1}(t_2)\}$. Obviously, $\cong$ is an equivalence relation on $T_1 \times T_2$. Moreover, since Γ is finite, then there exists finitely many equivalence classes modulo $\cong$. We define

- $\Delta_1 = \{G_1\psi$: $G_1\psi \in \Gamma$ is such that $\sigma(s_1, s_2)(\psi) = 0$ for some $(s_1, s_2) \in T_1 \times T_2\}$,
- $\Lambda_1 = \{H_1\psi$: $H_1\psi \in \Gamma$ is such that $\sigma(s_1, s_2)(\psi) = 0$ for some $(s_1, s_2) \in T_1 \times T_2\}$.

Obviously, $\Delta_1 \subseteq \Gamma$ and $\Lambda_1 \subseteq \Gamma$. Therefore, by Lemma 1, $card(\Delta_1) \leq 4\times \mid \varphi \mid +2$ and $card(\Lambda_1) \leq 4\times \mid \varphi \mid +2$. Let

- for all $G_1\psi \in \Delta_1$, $f_1(G_1\psi) = \{(s_1, s_2)$: $(s_1, s_2) \in T_1 \times T_2$ is such that $\sigma(s_1, s_2)(\psi) = 0\}$,
- for all $H_1\psi \in \Lambda_1$, $g_1(H_1\psi) = \{(s_1, s_2)$: $(s_1, s_2) \in T_1 \times T_2$ is such that $\sigma(s_1, s_2)(\psi) = 0\}$.

We shall say that

- $(s_1, s_2) \in f_1(G_1\psi)$ is maximal iff for all $(t_1, t_2) \in f_1(G_1\psi)$, $(s_1, s_2) \cong (t_1, t_2)$, or there exists $(u_1, u_2) \in f_1(G_1\psi)$ such that $(t_1, t_2) \prec (u_1, u_2)$ and $(s_1, s_2) \cong (u_1, u_2)$,
- $(s_1, s_2) \in g_1(H_1\psi)$ is minimal iff for all $(t_1, t_2) \in g_1(H_1\psi)$, $(s_1, s_2) \cong (t_1, t_2)$, or there exists $(u_1, u_2) \in g_1(H_1\psi)$ such that $(u_1, u_2) \prec (t_1, t_2)$ and $(s_1, s_2) \cong (u_1, u_2)$.

Since there exists finitely many equivalence classes modulo $\cong$, then for all $G_1\psi \in \Delta_1$, there exists $(s_1(G_1\psi), s_2(G_1\psi)) \in f_1(G_1\psi)$ such that $(s_1(G_1\psi), s_2(G_1\psi))$ is maximal and for all $H_1\psi \in \Lambda_1$, there exists $(s_1(H_1\psi), s_2(H_1\psi)) \in g_1(H_1\psi)$ such that $(s_1(H_1\psi), s_2(H_1\psi))$ is minimal. Since $(\mathcal{F}_1, \mathcal{F}_2, V)$ is a model for φ, there exists $(s_1^0, s_2^0) \in T_1 \times T_2$ such that $(\mathcal{F}_1, \mathcal{F}_2, V), (s_1^0, s_2^0) \models \varphi$. Let

- $\Pi_1 = \{(s_1^0, s_2^0)\} \cup \{(s_1(G_1\psi), s_2(G_1\psi))$: $G_1\psi \in \Delta_1\} \cup \{(s_1(H_1\psi), s_2(H_1\psi))$: $H_1\psi \in \Lambda_1\}$.

Since $card(\Delta_1) \leq 4\times \mid \varphi \mid +2$ and $card(\Lambda_1) \leq 4\times \mid \varphi \mid +2$, then $card(\Pi_1) \leq 8\times \mid \varphi \mid +5$. Let $(s_1, s_2), (t_1, t_2) \in \Pi_1$ be such that $(s_1, s_2) \cong \circ \prec_1 \circ \cong (t_1, t_2)$. A witness of 1-density for (s_1, s_2) and (t_1, t_2) is $(u_1, u_2) \in T_1 \times T_2$ such that $(s_1, s_2) \cong \circ \prec_1 \circ \cong (u_1, u_2)$, $(u_1, u_2) \cong \circ \prec_1 \circ \cong (t_1, t_2)$ and $(u_1, u_2) \cong \circ \prec_1 \circ \cong (u_1, u_2)$. Since there exists finitely many equivalence classes modulo $\cong$ and $(s_1, s_2) \cong \circ \prec_1 \circ \cong (t_1, t_2)$, then such a witness exists. Now, let

- $\Pi_{1,witness}$ be the least subset of $T_1 \times T_2$ containing Π_1 and such that for all $(s_1, s_2), (t_1, t_2) \in \Pi_1$, if $(s_1, s_2) \cong \circ \prec_1 \circ \cong (t_1, t_2)$ then $\Pi_{1,witness}$ contains a witness of 1-density for (s_1, s_2) and (t_1, t_2).

Remark that the cardinality of $\Pi_{1,witness}$ is bounded by $(8\times \mid \varphi \mid +6)^2$. Now, let (σ_0, M_1, M_2) be the structure defined by

- $\sigma_0 = \sigma(s_1^0, s_2^0)$,
- M_1 is the set of all mosaics of the form (σ, τ) such that for some $(s_1, s_2), (t_1, t_2) \in \Pi_{1,witness}$, there exists $u_2 \in \Pi_{2,witness}^{s_1}(s_2)$ and there exists $v_2 \in \Pi_{2,witness}^{t_1}(t_2)$ such that $\sigma = \sigma(s_1, u_2)$, $\tau = \sigma(t_1, v_2)$ and $(s_1, s_2) \cong \circ \prec_1 \circ \cong (t_1, t_2)$,

- M_2 is the set of all mosaics of the form (σ, τ) such that for some $(s_1, s_2) \in \Pi_{1,witness}$, there exists $t_2, u_2 \in \Pi^{s_1}_{2,witness}(s_2)$ such that $\sigma = \sigma(s_1, t_2)$, $\tau = \sigma(s_1, u_2)$ and $t_2 \equiv_{s_1} \circ <_2 \circ \equiv_{s_1} u_2$.

Remark that the cardinality of (σ_0, M_1, M_2) is bounded by $2 \times (8\times \mid \varphi \mid +6)^8$. Moreover, $\sigma_0(\varphi) = 1$. Hence, it suffices to prove the following properties:

- (σ_0, M_1, M_2) is a premodel,
- (σ_0, M_1, M_2) is saturated,
- (σ_0, M_1, M_2) is dense.

The proofs of these properties are left to the reader.

7 Decidability and Complexity

Now, we are ready to provide a complete decision procedure in nondeterministic polynomial time for the satisfiability problem the temporal logic of the lexicographic products of unbounded dense linear orders gives rise to.

Theorem 1 *The satisfiability problem with respect to* $(\mathcal{C}_{ud}, \mathcal{C}_{ud})$ *is in* NP.

Proof Given a formula φ, let us proceed as follows:

1. compute the least subformula closed full set Γ of formulas containing φ,
2. nondeterministically choose a premodel (σ_0, M_1, M_2) for φ of cardinality bounded by $2 \times (8\times \mid \varphi \mid +6)^8$,
3. check whether (σ_0, M_1, M_2) is saturated and dense.

By Lemma 1, $card(\Gamma) \leq 4\times \mid \varphi \mid +2$. Hence, (σ_0, M_1, M_2) is polysize in $\mid \varphi \mid$. Thus, we can execute steps (i), (ii) and (iii) in time polynomial in $\mid \varphi \mid$. Obviously, the execution of steps (i), (ii) and (iii) leads to an "accept" iff there exists a saturated dense premodel for φ of cardinality bounded by $2 \times (8\times \mid \varphi \mid +6)^8$. Therefore, by Propositions 3 and 4, the execution of steps (i), (ii) and (iii) leads to an "accept" iff φ is satisfiable with respect to $(\mathcal{C}_{ud}, \mathcal{C}_{ud})$. Consequently, the satisfiability problem with respect to $(\mathcal{C}_{ud}, \mathcal{C}_{ud})$ is in NP.

8 Conclusion

Temporal logics in which one can assign a proper meaning to the association of statements about different grained temporal domains have been considered. See Euzenat and Montanari (2005), Gagné and Plaice (1996), Nakamura and Fusaoka (2007) for details. Nevertheless, it seems that the results concerning the issues of axiomatization/completeness and decidability/complexity presented in Balbiani (2010, 2013) and in this paper constitute the first steps towards a temporal logic based on different levels of abstraction. Much remains to be done.

Let us, for instance, consider the lexicographic products of special linear flows of time like $\mathbb{Z}$, $\mathbb{Q}$ and $\mathbb{R}$. For each of these products, what is the corresponding first-order theory in terms of the binary predicates $<_1$ and $<_2$? Is this first-order theory decidable? What is the corresponding modal logic in terms of the temporal connectives G_1, G_2, H_1 and H_2? Is this modal logic decidable?

Concerning the problems of decidability and complexity, all normal extensions of $S4.3$, as proved in Bull (1966), Fine (1971), possess the finite model property and all finitely axiomatizable normal extensions of $K4.3$, as proved in Zakharyaschev and Alekseev (1995), are decidable. Is it possible to demonstrate similar results in our lexicographic setting? Or could undecidability results similar to the ones obtained by Reynolds and Zakharyaschev (2001) within the context of the products of the modal logics determined by arbitrarily long linear orders be demonstrated in our lexicographic setting?

In other respect, we can associate with $<_1$ and $<_2$ the until-like connectives U_1 and U_2 and the since-like connectives S_1 and S_2, the formulas $\varphi U_1 \psi$, $\varphi U_2 \psi$, $\varphi S_1 \psi$ and $\varphi S_2 \psi$ being read as one reads the formulas $\varphi U \psi$ and $\varphi S \psi$ in classical temporal logic, this time with $<_1$ and $<_2$. What can be done concerning the problems of axiomatization/completeness and decidability/complexity these new temporal connectives give rise to?

An important result in temporal logic is Kamp's Theorem. It concerns the functional completeness of the until operator (Gabbay et al. 1980). Kamp's Theorem says that for all monadic formulas of the first-order theory of linear orders, there exists a temporal formula using only the until and since temporal operators such that both formulas—the monadic one and the temporal one—are equally interpreted in any temporal model. What can be done about Kamp's Theorem in terms of the binary predicates $<_1$ and $<_2$ and the temporal connectives U_1, U_2, S_1 and S_2?

Acknowledgements The author makes a point of thanking his colleagues of the *Institut de recherche en informatique de Toulouse* as well as the participants of the *8th International Workshop on Logic and Cognition* who, by their comments and their suggestions, contributed to the development of the present paper. We also make a point of thanking the referees for their feedback.

References

Balbiani, P. 2008. Time representation and temporal reasoning from the perspective of non-standard analysis. In *Eleventh International Conference on Principles of Knowledge Representation and Reasoning. AAAI* ed. G. Brewka, and J. Lang, 695–704.

Balbiani, P. 2009. Axiomatization and completeness of lexicographic products of modal logics. In *Frontiers of Combining Systems*, ed. S. Ghilardi, and R. Sebastiani, 165–180. Berlin: Springer.

Balbiani, P. 2010. Axiomatizing the temporal logic defined over the class of all lexicographic products of dense linear orders without endpoints. In *Temporal Representation and Reasoning*, ed. N. Markey, and J. Wijsen, 19–26. IEEE.

Balbiani, P., and S. Mikulás. 2013. Decidability and complexity via mosaics of the temporal logic of the lexicographic products of unbounded dense linear orders. In *Frontiers of Combining Systems*, ed. P. Fontaine, C. Ringeissen, and R. Schmidt, 151–164. Berlin: Springer.

Balbiani, P., and I. Shapirovsky. 2019. Complete axiomatizations of lexicographic sums and products of modal logics. In preparation.

Bull, R. 1966. That all normal extensions of $S4.3$ have the finite model property. *Zeitschrift für mathematische Logik und Grundlagen der Mathematik* 12: 314–344.

Caleiro, C., L. Viganò, and M. Volpe. 2013. On the mosaic method for many-dimensional modal logics: a case study combining tense and modal operators. *Logica Universalis* 7: 33–69.

Euzenat, J., and A. Montanari. 2005. Time granularity. In *Handbook of Temporal Reasoning in Artificial Intelligence*, ed. M. Fisher, D. Gabbay, and L. Vila, 59–118. Amsterdam: Elsevier.

Fine, K. 1971. The logics containing $S4.3$. *Zeitschrift für mathematische Logik und Grundlagen der Mathematik* 17: 371–376.

Gabbay, D., and V. Shehtman. 1998. Products of modal logics, part 1. *Logic Journal of the IGPL* 6: 73–146.

Gabbay, D., A. Pnueli, S. Shelah, and J. Stavi. 1980. On the temporal analysis of fairness. In *Proceedings of the 7th ACM SIGPLAN-SIGACT Symposium on Principles of Programming Languages*, 163–173. New York: ACM.

Gabbay, D., A. Kurucz, F. Wolter, and M. Zakharyaschev. 2003. *Many-Dimensional Modal Logics: Theory and Applications*. Amsterdam: Elsevier.

Gagné, J.-R., and J. Plaice. 1996. A nonstandard temporal deductive database system. *Journal of Symbolic Computation* 22: 649–664.

Kurucz, A. 2007. Combining modal logics. In *Handbook of Modal Logic*, ed. P. Blackburn, J. van Benthem, and F. Wolter, 869–924. Amsterdam: Elsevier.

Marx, M., and Y. Venema. 2007. Local variations on a loose theme: Modal logic and decidability. In *Finite Model Theory and Its Applications*, ed. E. Grädel, P. Kolaitis, L. Libkin, M. Marx, J. Spencer, M. Vardi, Y. Venema, and S. Weinstein, 371–429. Berlin: Springer.

Marx, M., S. Mikulás, and M. Reynolds. 2000. The mosaic method for temporal logics. In *Automated Reasoning with Analytic Tableaux and Related Methods*, ed. R. Dyckhoff, 324–340. Berlin: Springer.

Nakamura, K., and A. Fusaoka. 2007. Reasoning about hybrid systems based on a nonstandard model. In *AI 2007: Advances in Artificial Intelligence*, ed. M. Orgun, and J. Thornton, 749–754. Berlin: Springer.

Reynolds, M. 1997. A decidable temporal logic of parallelism. *Notre Dame Journal of Formal Logic* 38: 419–436.

Reynolds, M., and M. Zakharyaschev. 2001. On the products of linear modal logics. *Journal of Logic and Compution* 11: 909–931.

Van Benthem, J. 1991. *The Logic of Time*. Dordrecht: Kluwer.

Zakharyaschev, M., and A. Alekseev. 1995. All finitely axiomatizable normal extensions of $K4.3$ are decidable. *Mathematical Logic Quarterly* 41: 15–23.

Contact Logic is Finitary for Unification with Constants

Philippe Balbiani and Çiğdem Gencer

Abstract Contact Logic is a formalism for reasoning about the relation of contact between regions. We introduce a new inference problem for Contact Logic, the unification problem, which extends the validity problem by allowing one to replace variables by terms before testing for validity. Our main result is the proof that unification with constants in Contact Logic is finitary.

Keywords Contact Logic · Unification with constants · Unification type

1 Introduction

Contact Logic is a formalism for reasoning about the relation of contact between regions represented by Boolean terms. Its language is obtained from the language of Boolean algebras by the addition of a binary predicate representing this relation. Its role in artificial intelligence and computer science stems from the importance of spatial information in systems for natural language understanding, robotic navigation, etc. (Cohn and Renz 2008; Renz 2002). Different variants of Contact Logic have been proposed (Kontchakov et al. 2010a, b, 2013; Tinchev and Vakarelov 2010; Wolter and Zakharyaschev 2000). Some of them are based on the predicate of boundedness or the predicate of connectedness. Their semantics can be given by interpreting terms

---

P. Balbiani (✉)
Institut de recherche en informatique de Toulouse, CNRS—Toulouse University, Toulouse, France
e-mail: Philippe.Balbiani@irit.fr

Ç. Gencer
Faculty of Engineering and Natural Sciences, Sabancı University, Istanbul, Turkey

Faculty of Arts and Sciences, Aydın University, Istanbul, Turkey

S. Ju et al. (eds.), *Nonclassical Logics and Their Applications*,
Logic in Asia: Studia Logica Library,
https://doi.org/10.1007/978-981-15-1342-8_4

and formulas either in topological spaces (Dimov and Vakarelov 2006a, b; Düntsch and Winter 2005), or in frames (Balbiani et al. 2007; Vakarelov 2007).

The above-mentioned applications of Contact Logic require the resolution of unification problems. There is a wide variety of situations where unification problems arise. Suppose the formula $\varphi(p_1, \ldots, p_m)$ describes a given geographic configuration of constant regions $p_1, \ldots, p_m$ and the formula $\psi(x_1, \ldots, x_n)$ represents a desirable geographic property of variable regions $x_1, \ldots, x_n$. It may happen that $\varphi(p_1, \ldots, p_m) \to \psi(x_1, \ldots, x_n)$ is not valid in the considered geographic environment. Hence, one may ask whether there are n-tuples $(a_1, \ldots, a_n)$ of terms such that $\varphi(p_1, \ldots, p_m) \to \psi(a_1, \ldots, a_n)$ is valid in this environment. Moreover, one may be interested to obtain, if possible, the most general n-tuples $(a_1, \ldots, a_n)$ of terms such that $\varphi(p_1, \ldots, p_m) \to \psi(a_1, \ldots, a_n)$ is valid.

Little is known about the unification problem in Contact Logic. For example, without the unary predicate of connectedness, elementary unification is NP-complete whereas unification with constants is decidable, its exact complexity being still unknown. See Balbiani and Gencer (2015) for details about the computability of the unification problem in Contact Logic without the unary predicate of connectedness. A logic is said to be unitary when the solutions of a unification problem can always be represented by a most general one whereas it is said to be finitary when the solutions of a unification problem can always be represented by finitely many maximal ones. In this paper, we prove that unification with constants in Contact Logic either with, or without the unary predicate of connectedness is finitary.

2 Syntax

We adopt the standard rules for omission of the parentheses.

Atoms Let **CON** be a countable set of *propositional constants* (with typical members denoted p, q, etc) and **VAR** be a countable set of *propositional variables* (with typical members denoted x, y, etc). Let $(p_1, p_2, \ldots)$ be an enumeration of **CON** without repetitions and $(x_1, x_2, \ldots)$ be an enumeration of **VAR** without repetitions. A *propositional atom* is either a constant, or a variable. Let **ATO** be the set of all atoms (with typical members denoted α, β, etc).

Terms The *Boolean terms* (denoted a, b, etc) are defined as follows:

- $a ::= \alpha \mid 0 \mid a^\star \mid (a \cup b)$.

The other Boolean constructs for terms (for instance, 1 and $\cap$) are defined as usual. We use the following notations for terms: a^0 for $a^\star$ and a^1 for a. Reading terms as regions, the constructs 0, * and $\cup$ should be regarded as the empty region, the complement operation and the union operation. As a result, the constructs 1 and $\cap$ should be regarded as the full region and the intersection operation. For all nonnegative integers m, n, let $\mathbf{TER}_{m,n}$ be the set of all terms whose constants form a subset of $\{p_1, \ldots, p_m\}$ and whose variables form a subset of $\{x_1, \ldots, x_n\}$. Let **TER** be the set of all terms.

Formulas The *formulas* (denoted φ, ψ, etc) are defined as follows:

- $\varphi ::= C(a,b) \mid a \equiv b \mid con(a) \mid \bot \mid \neg\varphi \mid (\varphi \vee \psi)$.

Here, a and b are terms whereas C is the binary predicate of contact, $\equiv$ is the binary predicate of equality and con is the unary predicate of connectedness. For C, $\equiv$ and con, we propose the following readings:

- $C(a,b)$: "the regions denoted a and b are in contact",
- $a \equiv b$: "the regions denoted a and b are equal",
- $con(a)$: "the region denoted a is connected".

The other connectives for formulas (for instance, $\top$ and $\wedge$) are defined as usual. We use the following notation for formulas: $a \leq b$ for $a \cup b \equiv b$. A formula φ is *equational* if $\equiv$ is the only predicate possibly occurring in φ. For all nonnegative integers m, n, let $\mathbf{FOR}_{m,n}$ be the set of all formulas whose constants form a subset of $\{p_1, \ldots, p_m\}$ and whose variables form a subset of $\{x_1, \ldots, x_n\}$. Let **FOR** be the set of all formulas. Note that **FOR** is denoted $\mathcal{C}c$ in Kontchakov et al. (2010a, b, 2013) and L^c in Tinchev and Vakarelov (2010).

Example 1 Within the context of the unification problem, we will interest with the following formulas:

- $\varphi_{01} \;=\; x \equiv 0 \vee x \equiv 1$,
- $\varphi_{pq} \;=\; C(p,q) \rightarrow x \not\equiv 0 \wedge x \leq p \cup q$.

The first formula is equational. It says that the region denoted x is either the empty region, or the full region. As for the second formula, it says that if the regions denoted p and q are in contact then the region denoted x is a nonempty region included in the region denoted $p \cup q$.

Substitutions A *substitution* is a function σ : **VAR** $\longrightarrow$ **TER** which moves at most finitely many variables. Given a substitution σ, let $\bar{\sigma}$: **TER** $\cup$ **FOR** $\longrightarrow$ **TER** $\cup$ **FOR** be the endomorphism such that for all variables x, $\bar{\sigma}(x) = \sigma(x)$. The *composition* of the substitutions σ and τ is the substitution $\sigma \circ \tau$ such that for all $x \in$ **VAR**, $(\sigma \circ \tau)(x) = \bar{\tau}(\sigma(x))$.

Example 2 Within the context of the unification problem, we will interest with the following substitutions:

- the substitution σ_0 such that $\sigma_0(x) = 0$ and for all variables y, if $x \neq y$ then $\sigma_0(y) = y$,
- the substitution σ_1 such that $\sigma_1(x) = 1$ and for all variables y, if $x \neq y$ then $\sigma_1(y) = y$,
- the substitution σ_p such that $\sigma_p(x) = p \cup (q \cap x)$ and for all variables y, if $x \neq y$ then $\sigma_p(y) = y$,
- the substitution σ_q such that $\sigma_q(x) = q \cup (p \cap x)$ and for all variables y, if $x \neq y$ then $\sigma_q(y) = y$.

3 Semantics

The semantics of Contact Logic can be given by interpreting terms and formulas either in topological spaces (Balbiani et al. 2007; Wolter and Zakharyaschev 2000), or in frames (Balbiani et al. 2007; Vakarelov 2007).

Topological semantics The best way to understand the meaning of the predicates of contact and connectedness is by interpreting them in *topological spaces,* i.e. structures of the form (X, τ) where X is a nonempty set (with typical members denoted A, B, etc) and τ is a set of subsets of X such that

- $\emptyset \in \tau$,
- $X \in \tau$,
- if $\{A_i : i \in I\}$ is a finite subset of τ then $\bigcap\{A_i : i \in I\} \in \tau$,
- if $\{A_i : i \in I\}$ is a subset of τ then $\bigcup\{A_i : i \in I\} \in \tau$.

The subsets of X in τ are called *open sets* whereas their complements are called *closed sets.* The most important topological spaces considered within the context of Contact Logic are for all $n \geq 1$, the topological space $(\mathbb{R}^n, \tau_{\mathbb{R}^n})$, i.e. the real space $\mathbb{R}^n$ together with its ordinary topology $\tau_{\mathbb{R}^n}$. See Kontchakov et al. (2010a, b, 2013), Wolter and Zakharyaschev (2000) for details. Let (X, τ) be a topological space. Let Int_τ and Cl_τ denote the *interior operator* and the *closure operator* in (X, τ). A subset A of X is *regular closed* if $Cl_\tau(Int_\tau(A)) = A$. Regular closed subsets of X will also be called *regions.* It is well-known that the set $RC(X, \tau)$ of all regular closed subsets of X forms a Boolean algebra $(RC(X, \tau), 0_X, \star_X, \cup_X)$ where for all $A, B \in RC(X, \tau)$:

- $0_X = \emptyset$,
- $A^{\star_X} = Cl_\tau(X \setminus A)$,
- $A \cup_X B = A \cup B$.

As a result, for all $A, B \in RC(X, \tau)$, $1_X = X$ and $A \cap_X B = Cl_\tau(Int_\tau(A \cap B))$. Regions being regular closed subsets of X, two regions are in contact if they have a nonempty intersection whereas a region is connected if it cannot be represented as the union of two disjoint nonempty regions. For this reason, we associate to the predicates C and con the relations $C^{(X,\tau)}$ and $con^{(X,\tau)}$ on the set of all regular closed subsets of X as follows:

- $C^{(X,\tau)}(A, B)$ if $A \cap B \neq \emptyset$,
- $con^{(X,\tau)}(A)$ if for all nonempty regular closed subsets B, B' of X, if $B \cup B' = A$ then $B \cap B' \neq \emptyset$.

The following conditions hold for all regular closed subsets A, B of X:

- if $A \neq \emptyset$ then $C^{(X,\tau)}(A, A)$,
- if $C^{(X,\tau)}(A, B)$ then $C^{(X,\tau)}(B, A)$.

If τ only lets $\emptyset$ and X being open then the following condition holds for all regular closed subsets A, B of X:

– if $A \neq \emptyset$ and $B \neq \emptyset$ then $C^{(X,\tau)}(A, B)$.

If X cannot be represented as the union of two disjoint nonempty open sets then the following conditions hold for all regular closed subsets A of X:

– if $A \neq \emptyset$ and $A \neq X$ then $C^{(X,\tau)}(A, Cl_\tau(X \setminus A))$,
– $con^{(X,\tau)}(X)$.

A *valuation* on X is a map $\mathcal{V}$ associating with every atom α a regular closed subset $\mathcal{V}(\alpha)$ of X. Given a valuation $\mathcal{V}$ on X, we define

– $\bar{\mathcal{V}}(\alpha) = \mathcal{V}(\alpha)$,
– $\bar{\mathcal{V}}(0) = \emptyset$,
– $\bar{\mathcal{V}}(a^\star) = Cl_\tau(X \setminus \bar{\mathcal{V}}(a))$,
– $\bar{\mathcal{V}}(a \cup b) = \bar{\mathcal{V}}(a) \cup \bar{\mathcal{V}}(b)$.

Thus, every term is interpreted as a regular closed subset of X. A *model* on (X, τ) is a structure $\mathcal{M} = (X, \tau, \mathcal{V})$ where $\mathcal{V}$ is a valuation on X. The connectives $\bot$, $\neg$ and $\vee$ being classically interpreted, the *satisfiability* of a formula φ in $\mathcal{M}$ (in symbols $\mathcal{M} \models \varphi$) is defined as follows:

– $\mathcal{M} \models C(a, b)$ if $C^{(X,\tau)}(\bar{\mathcal{V}}(a), \bar{\mathcal{V}}(b))$,
– $\mathcal{M} \models a \equiv b$ if $\bar{\mathcal{V}}(a) = \bar{\mathcal{V}}(b)$,
– $\mathcal{M} \models con(a)$ if $con^{(X,\tau)}(\bar{\mathcal{V}}(a))$.

A formula φ is *valid* in (X, τ) if for all valuations $\mathcal{V}$ on X, $(X, \tau, \mathcal{V}) \models \varphi$. A formula φ is *satisfiable* in (X, τ) if there exists a valuation $\mathcal{V}$ on X such that $(X, \tau, \mathcal{V}) \models \varphi$.

Galton's perspective A relational perspective is suggested by Galton (2000) who introduces the notion of adjacency space. Galton's spaces are frames (W, R) where W is a nonempty set of cells and R is an adjacency relation between cells. Galton defines regions to be sets of cells. He also defines two regions A and B to be in contact if some point in A is adjacent to some point in B. This definition relates Galton's adjacency spaces to the relational semantics of modal logic which makes it possible to use methods from modal logic for studying region-based theories of space. The truth is that the above-mentioned topological semantics and the relational perspective suggested by Galton are equivalent (Tinchev and Vakarelov 2010; Vakarelov 2007). In this paper, we adopt Galton's perspective.

Relational semantics A *frame* is a structure of the form (W, R) where W is a nonempty set (with typical members denoted s, t, etc) and R is a binary relation on W. Let (W, R) be a frame. We will denote by R^+ the transitive closure of R, i.e. the least transitive relation on W containing R. Subsets of W will also be called *regions*. Naturally, two regions are in contact if they have R-related elements whereas a region is connected if there is an R-path between any pair of elements in that region. For this reason, we associate to the predicates C and con the relations $C^{(W,R)}$ and $con^{(W,R)}$ on the set of all subsets of W as follows:

- $C^{(W,R)}(A, B)$ if $R \cap (A \times B) \neq \emptyset$,
- $con^{(W,R)}(A)$ if $A \times A \subseteq R^+$.

If R is reflexive and symmetric then the following conditions hold for all subsets A, B of W:

- if $A \neq \emptyset$ then $C^{(W,R)}(A, A)$,
- if $C^{(W,R)}(A, B)$ then $C^{(W,R)}(B, A)$.

If $R = W \times W$ then the following condition holds for all subsets A, B of W:

- if $A \neq \emptyset$ and $B \neq \emptyset$ then $C^{(W,R)}(A, B)$.

If $R^+ = W \times W$ then the following conditions hold for all subsets A of W:

- if $A \neq \emptyset$ and $A \neq W$ then $C^{(W,R)}(A, W \setminus A)$,
- $con^{(W,R)}(W)$.

A *valuation* on W is a map $\mathcal{V}$ associating with every atom α a subset $\mathcal{V}(\alpha)$ of W. Given a valuation $\mathcal{V}$ on W, we define

- $\bar{\mathcal{V}}(\alpha) = \mathcal{V}(\alpha)$,
- $\bar{\mathcal{V}}(0) = \emptyset$,
- $\bar{\mathcal{V}}(a^\star) = W \setminus \bar{\mathcal{V}}(a)$,
- $\bar{\mathcal{V}}(a \cup b) = \bar{\mathcal{V}}(a) \cup \bar{\mathcal{V}}(b)$.

Thus, every term is interpreted as a subset of W. A *model* on (W, R) is a structure $\mathcal{M} = (W, R, \mathcal{V})$ where $\mathcal{V}$ is a valuation on W. The connectives $\bot$, $\neg$ and $\vee$ being classically interpreted, the *satisfiability* of a formula φ in $\mathcal{M}$ (in symbols $\mathcal{M} \models \varphi$) is defined as follows:

- $\mathcal{M} \models C(a, b)$ if $C^{(W,R)}(\bar{\mathcal{V}}(a), \bar{\mathcal{V}}(b))$,
- $\mathcal{M} \models a \equiv b$ if $\bar{\mathcal{V}}(a) = \bar{\mathcal{V}}(b)$,
- $\mathcal{M} \models con(a)$ if $con^{(W,R)}(\bar{\mathcal{V}}(a))$.

A formula φ is *valid* in (W, R) if for all valuations $\mathcal{V}$ on W, $(W, R, \mathcal{V}) \models \varphi$. A formula φ is *satisfiable* in (W, R) if there exists a valuation $\mathcal{V}$ on W such that $(W, R, \mathcal{V}) \models \varphi$.
Validity and satisfiability Let $\mathcal{C}$ be a class of frames. We say that a formula φ is $\mathcal{C}$-*valid* if for all frames (W, R) in $\mathcal{C}$, φ is valid in (W, R). We say that a formula φ is $\mathcal{C}$-*satisfiable* if there exists a frame (W, R) in $\mathcal{C}$ such that φ is satisfiable in (W, R). The $\mathcal{C}$-*satisfiability problem* consists in determining whether a given formula is $\mathcal{C}$-satisfiable. A frame (W, R) is *indiscrete* if $R = W \times W$. A frame (W, R) is *connected* if $R^+ = W \times W$. Let $\mathcal{C}_{all}$ denote the class of all reflexive and symmetric frames, $\mathcal{C}_{ind}$ denote the class of all reflexive, symmetric and indiscrete frames and $\mathcal{C}_{con}$ denote the class of all reflexive, symmetric and connected frames. The topological counterparts of these classes of frames are the class of all topological spaces, the class of all indiscrete spaces (those spaces that only let $\emptyset$ and the full set being open) and the class of all connected spaces (those spaces that cannot be represented as the

union of two disjoint nonempty open sets). For con-free formulas, the satisfiability problem is known to be NP-complete in $\mathcal{C}_{all}$ and $\mathcal{C}_{ind}$ and $PSPACE$-complete in $\mathcal{C}_{con}$, the situation being generally more complex for arbitrary formulas. See Balbiani et al. (2007), Kontchakov et al. (2010a, b, 2013), Wolter and Zakharyaschev (2000) for details. The following formulas are $\mathcal{C}_{all}$-valid:

- $x \not\equiv 0 \rightarrow C(x, x)$,
- $C(x, y) \rightarrow C(y, x)$.

The following formula is $\mathcal{C}_{ind}$-valid:

- $x \not\equiv 0 \wedge y \not\equiv 0 \rightarrow C(x, y)$.

The following formulas are $\mathcal{C}_{con}$-valid:

- $x \not\equiv 0 \wedge x \not\equiv 1 \rightarrow C(x, x^{\star})$,
- $con(1)$.

4 About Classes of Frames

Let $\mathcal{C}$ be a class of frames. We say that $\mathcal{C}$ *agrees with unions* if for all disjoint frames (W, R), (W', R') in $\mathcal{C}$, there exists a frame (W'', R'') in $\mathcal{C}$ such that $W \cup W' = W''$ and $R'' \not\subseteq Id_{W''}$. We should stress here that many natural classes of frames agree with unions. In particular,

Proposition 1 *$\mathcal{C}_{all}$, $\mathcal{C}_{ind}$ and $\mathcal{C}_{con}$ agree with unions.*

Proof By the definition of what it means for classes of frames to agree with unions.

We say that $\mathcal{C}$ is *determined* by a formula φ if $\mathcal{C}$ is the class of all frames validating φ.

Proposition 2 *$\mathcal{C}_{all}$ is determined by $x \not\equiv 0 \rightarrow C(x, x)$ and $C(x, y) \rightarrow C(y, x)$, $\mathcal{C}_{ind}$ is determined by $x \not\equiv 0 \wedge y \not\equiv 0 \rightarrow C(x, y)$ and $\mathcal{C}_{con}$ is both determined by $x \not\equiv 0 \wedge x \not\equiv 1 \rightarrow C(x, x^{\star})$ and $con(1)$.*

Proof Left to the reader.

We say that $\mathcal{C}$ is *large* if there exists a frame (W, R) in $\mathcal{C}$ such that $\mathbf{Card}(W) \geq 3$ and $R \not\subseteq Id_W$. We should stress here that many natural classes of frames are large. In particular, obviously, if a nonempty class of frames agrees with unions then it is large. A valuation $\mathcal{V}$ on a nonempty set W is *balanced* if for all terms a, either $\bar{\mathcal{V}}(a) = \emptyset$, or $\bar{\mathcal{V}}(a) = W$, or $\bar{\mathcal{V}}(a)$ is infinite and coinfinite. We say that $\mathcal{C}$ is *balanced* if for all formulas φ, if φ is $\mathcal{C}$-satisfiable then there exists a countable frame (W, R) in $\mathcal{C}$ and there exists a balanced valuation $\mathcal{V}$ on W such that $(W, R, \mathcal{V}) \models \varphi$. We should stress here that many natural classes of frames are balanced. In particular,

Proposition 3 *$\mathcal{C}_{all}$, $\mathcal{C}_{ind}$ and $\mathcal{C}_{con}$ are balanced.*

Proof Let $\mathcal{C}$ be one of the above-mentioned classes of frames. In order to demonstrate that $\mathcal{C}$ is balanced, let φ be a $\mathcal{C}$-satisfiable formula. By Balbiani et al. (2007, Theorem 4.2), φ is satisfiable in a finite frame of $\mathcal{C}$. Let (W, R) be a finite frame of $\mathcal{C}$ and $\mathcal{V}$ be a valuation on it such that $(W, R, \mathcal{V}) \models \varphi$. Let $W' = W \times \mathbb{N}$ and R' be the binary relation on W' defined by $(x, i)R'(y, j)$ if xRy. Since the frame (W, R) is finite, therefore the frame (W', R') is countable. Let $\mathcal{V}'$ be the valuation on (W', R') defined by $\mathcal{V}'(\alpha) = \mathcal{V}(\alpha) \times \mathbb{N}$ for each atom α. As the reader may easily verify by induction, $\bar{\mathcal{V}}'(a) = \mathcal{V}(a) \times \mathbb{N}$ for each term a. It follows that the model $(W', R', \mathcal{V}')$ is balanced. As the reader may easily verify by induction, $(W, R, \mathcal{V}) \models \psi$ iff $(W', R', \mathcal{V}') \models \psi$ for each formula ψ. Since $(W, R, \mathcal{V}) \models \varphi$, therefore $(W', R', \mathcal{V}') \models \varphi$. Finally, it suffices now to remark that if (W, R) is in $\mathcal{C}_{all}$ (respectively, in $\mathcal{C}_{ind}$, in $\mathcal{C}_{con}$) then (W', R') is in $\mathcal{C}_{all}$ (respectively, in $\mathcal{C}_{ind}$, in $\mathcal{C}_{con}$) too.

5 Unifiability

Now, following the standard notations and definitions about the unifiability problem in modal and description logics (Baader et al. 2012; Baader and Ghilardi 2011; Baader and Morawska 2009; Gencer 2002; Gencer and de Jongh 2009; Ghilardi 2000; Ghilardi and Sacchetti 2004; Rybakov 1997), we introduce the unifiability problem in Contact Logic. Let $\mathcal{C}$ be a class of frames.
Unifiable formulas We say that a substitution σ is *$\mathcal{C}$-equivalent* to a substitution τ (in symbols $\sigma \simeq_{\mathcal{C}} \tau$) if for all variables x, $\sigma(x) \equiv \tau(x)$ is $\mathcal{C}$-valid.

Example 3 As the reader can easily show, the substitutions σ_0 and σ_1 introduced in Example 2 are respectively $\mathcal{C}$-equivalent to the substitutions σ_0' and σ_1' such that $\sigma_0'(x) = x \cap x^\star$, $\sigma_1'(x) = x \cup x^\star$ and for all variables y, if $x \neq y$ then $\sigma_0'(y) = y$ and $\sigma_1'(y) = y$.

We say that a substitution σ is more *$\mathcal{C}$-general* than a substitution τ (in symbols $\sigma \preceq_{\mathcal{C}} \tau$) if there exists a substitution υ such that $\sigma \circ \upsilon \simeq_{\mathcal{C}} \tau$. In that case, we also say that τ is *less $\mathcal{C}$-general* than σ.

Example 4 As the reader can easily show, the substitutions σ_p and σ_q introduced in Example 2 are respectively less $\mathcal{C}$-general than the substitutions σ_p' and σ_q' such that $\sigma_p'(x) = x \cup p$, $\sigma_q'(x) = x \cup q$ and for all variables y, if $x \neq y$ then $\sigma_p'(y) = y$ and $\sigma_q'(y) = y$.

We say that a formula φ is *$\mathcal{C}$-unifiable* if there exists a substitution σ such that $\bar{\sigma}(\varphi)$ is $\mathcal{C}$-valid. In that case, we say that σ is a *$\mathcal{C}$-unifier* of φ.

Example 5 As the reader can easily show, the substitutions σ_0 and σ_1 introduced in Example 2 are $\mathcal{C}$-unifiers of the formula φ_{01} introduced in Example 1. In other respect, the substitutions σ_p and σ_q introduced in Example 2 are $\mathcal{C}$-unifiers of the formula φ_{pq} introduced in Example 1.

The $\mathcal{C}$-*unifiability problem with constants* consists in determining whether a given formula is $\mathcal{C}$-unifiable. We say that a set of $\mathcal{C}$-unifiers of a formula φ is *complete* if for all $\mathcal{C}$-unifiers σ of φ, there exists a $\mathcal{C}$-unifier τ of φ in that set such that $\tau \preceq_{\mathcal{C}} \sigma$. An important question is: when a formula is $\mathcal{C}$-unifiable, has it a minimal complete set of $\mathcal{C}$-unifiers? When the answer is "yes", how large is this set?
Unification types We say that a $\mathcal{C}$-unifiable formula φ is $\mathcal{C}$-*finitary* if there exists a finite minimal complete set of $\mathcal{C}$-unifiers of φ but there exists no with cardinality 1. We say that a $\mathcal{C}$-unifiable formula φ is $\mathcal{C}$-*unitary* if there exists a minimal complete set of $\mathcal{C}$-unifiers of φ with cardinality 1. We say that unification in $\mathcal{C}$ is *finitary* if every $\mathcal{C}$-unifiable formula is either $\mathcal{C}$-finitary, or $\mathcal{C}$-unitary and there exists a $\mathcal{C}$-finitary formula. We say that unification in $\mathcal{C}$ is *unitary* if every $\mathcal{C}$-unifiable formula is $\mathcal{C}$-unitary. See Balbiani and Gencer (2017), Dzik (2007), Jerabek (2015) for an introduction to the unification types in logics.

6 Examples

From now on, when we write "**CPL**", we mean "Classical Propositional Logic". Let $\mathcal{C}$ be a class of frames.

As remarked above, the substitutions σ_0 and σ_1 introduced in Example 2 are $\mathcal{C}$-unifiers of the formula φ_{01} introduced in Example 1. Moreover,

- $\sigma_0 \not\preceq_{\mathcal{C}} \sigma_1$ and $\sigma_1 \not\preceq_{\mathcal{C}} \sigma_0$.

To see why, suppose $\sigma_0 \preceq_{\mathcal{C}} \sigma_1$, or $\sigma_1 \preceq_{\mathcal{C}} \sigma_0$. Without loss of generality, suppose $\sigma_0 \preceq_{\mathcal{C}} \sigma_1$. Let τ be a substitution such that $\sigma_0 \circ \tau \simeq_{\mathcal{C}} \sigma_1$. Hence, $\bar{\tau}(\sigma_0(x)) \equiv \sigma_1(x)$ is $\mathcal{C}$-valid. Since $\sigma_0(x) = 0$ and $\sigma_1(x) = 1$, therefore $0 \equiv 1$ is $\mathcal{C}$-valid: a contradiction. In other respect,

- if $\mathcal{C}$ agrees with unions then σ_0 and σ_1 form a complete set of $\mathcal{C}$-unifiers of φ_{01}.

To see why, suppose $\mathcal{C}$ agrees with unions and σ_0 and σ_1 do not form a complete set of $\mathcal{C}$-unifiers of φ_{01}. Let τ be a $\mathcal{C}$-unifier of φ_{01} such that neither $\sigma_0 \preceq_{\mathcal{C}} \tau$, nor $\sigma_1 \preceq_{\mathcal{C}} \tau$. Thus, neither $0 \equiv \tau(x)$ is $\mathcal{C}$-valid, nor $1 \equiv \tau(x)$ is $\mathcal{C}$-valid. Let (W, R) and (W', R') be disjoint frames in $\mathcal{C}$, $\mathcal{V}$ be a valuation on W and $\mathcal{V}'$ be a valuation on W' such that neither $\bar{\mathcal{V}}(\tau(x)) = \emptyset$, nor $\bar{\mathcal{V}}'(\tau(x)) = W'$. Since $\mathcal{C}$ agrees with unions, therefore let (W'', R'') be a frame in $\mathcal{C}$ such that $W \cup W' = W''$. Let $\mathcal{V}''$ be a valuation on W'' such that for all atoms α, $\mathcal{V}''(\alpha) = \mathcal{V}(\alpha) \cup \mathcal{V}'(\alpha)$. Obviously, for all terms a, $\bar{\mathcal{V}}''(a) = \bar{\mathcal{V}}(a) \cup \bar{\mathcal{V}}'(a)$. Since neither $\bar{\mathcal{V}}(\tau(x)) = \emptyset$, nor $\bar{\mathcal{V}}'(\tau(x)) = W'$, therefore neither $\bar{\mathcal{V}}''(\tau(x)) = \emptyset$, nor $\bar{\mathcal{V}}''(\tau(x)) = W''$. Consequently, τ is not a $\mathcal{C}$-unifier of φ_{01}: a contradiction. Remark that the above argument shows that if $\mathcal{C}$ agrees with unions then φ_{01} is either $\mathcal{C}$-unitary, or $\mathcal{C}$-finitary. Finally,

- if $\mathcal{C}$ agrees with unions then φ_{01} is $\mathcal{C}$-finitary.

To see why, suppose $\mathcal{C}$ agrees with unions and φ_{01} is not $\mathcal{C}$-finitary. Since if $\mathcal{C}$ agrees with unions then φ_{01} is either $\mathcal{C}$-unitary, or $\mathcal{C}$-finitary, therefore φ_{01} is $\mathcal{C}$-unitary. Let τ be a $\mathcal{C}$-unifier of φ_{01} such that $\tau \preceq_{\mathcal{C}} \sigma_0$ and $\tau \preceq_{\mathcal{C}} \sigma_1$. Since σ_0 and σ_1 form a complete set of $\mathcal{C}$-unifiers of φ_{01}, therefore $\sigma_0 \preceq_{\mathcal{C}} \tau$, or $\sigma_1 \preceq_{\mathcal{C}} \tau$. Without loss of generality, suppose $\sigma_0 \preceq_{\mathcal{C}} \tau$. Since $\tau \preceq_{\mathcal{C}} \sigma_1$, therefore $\sigma_0 \preceq_{\mathcal{C}} \sigma_1$: a contradiction.

As remarked above, the substitutions σ_p and σ_q introduced in Example 2 are $\mathcal{C}$-unifiers of the formula φ_{pq} introduced in Example 1. Moreover,

- if $p \neq q$ then $\sigma_p \not\preceq_{\mathcal{C}} \sigma_q$ and $\sigma_q \not\preceq_{\mathcal{C}} \sigma_p$.

To see why, suppose $p \neq q$ and $\sigma_p \preceq_{\mathcal{C}} \sigma_q$, or $\sigma_q \preceq_{\mathcal{C}} \sigma_p$. Without loss of generality, suppose $\sigma_p \preceq_{\mathcal{C}} \sigma_q$. Let τ be a substitution such that $\sigma_p \circ \tau \simeq_{\mathcal{C}} \sigma_q$. Hence, $\bar{\tau}(\sigma_p(x)) \equiv \sigma_q(x)$ is $\mathcal{C}$-valid. Since $\sigma_p(x) = p \cup (q \cap x)$ and $\sigma_q(x) = q \cup (p \cap x)$, therefore $p \cup (q \cap \tau(x)) \equiv q \cup (p \cap x)$ is $\mathcal{C}$-valid. Thus, after replacing in $p \cup (q \cap \tau(x)) \equiv q \cup (p \cap x)$ each occurrence of q and x by 0, $p \equiv 0$ is $\mathcal{C}$-valid: a contradiction. In other respect,

- if $\mathcal{C}$ agrees with unions and $p \neq q$ then σ_p and σ_q form a complete set of $\mathcal{C}$-unifiers of φ_{pq}.

To see why, suppose $\mathcal{C}$ agrees with unions, $p \neq q$ and σ_p and σ_q do not form a complete set of $\mathcal{C}$-unifiers of φ_{pq}. As a result, remark that $\mathcal{C}$ is large. Let τ be a $\mathcal{C}$-unifier of φ_{pq} such that neither $\sigma_p \preceq_{\mathcal{C}} \tau$, nor $\sigma_q \preceq_{\mathcal{C}} \tau$. Thus, $C(p, q) \rightarrow \tau(x) \not\equiv 0 \wedge \tau(x) \leq p \cup q$ is $\mathcal{C}$-valid. Without loss of generality, we can assume that $\tau(x) = (p^0 \cap q^0 \cap a_{00}) \cup (p^0 \cap q^1 \cap a_{01}) \cup (p^1 \cap q^0 \cap a_{10}) \cup (p^1 \cap q^1 \cap a_{11})$ for some $\{p, q\}$-free terms $a_{00}, a_{01}, a_{10}, a_{11}$. Since $\mathcal{C}$ is large, therefore let (W, R) be a frame in $\mathcal{C}$ such that $\mathbf{Card}(W) \geq 3$ and $R \not\subseteq Id_W$. As a result, remark that $R \neq \emptyset$. Suppose we succeed in proving the following:

(1) $a_{00} \equiv 0$ is $\mathcal{C}$-valid,
(2) $a_{11} \equiv 1$ is $\mathcal{C}$-valid,
(3) $a_{01} \cup a_{10} \equiv 1$ is $\mathcal{C}$-valid,
(4) $a_{01} \leq a_{10}$ is $\mathcal{C}$-valid, or $a_{10} \leq a_{01}$ is $\mathcal{C}$-valid.

Consequently, by (1) and (2), considered as formulas in **CPL**, $\tau(x)$ and $(p^0 \cap q^1 \cap a_{01}) \cup (p^1 \cap q^0 \cap a_{10}) \cup (p^1 \cap q^1)$ are equivalent. Moreover, by (3) and (4), $a_{10} \equiv 1$ is $\mathcal{C}$-valid, or $a_{01} \equiv 1$ is $\mathcal{C}$-valid. In the former case, considered as formulas in **CPL**, $\tau(x)$ and $p \cup (q \cap a_{01})$ are equivalent. Hence, $\sigma_p \preceq_{\mathcal{C}} \tau$: a contradiction. In the latter case, considered as formulas in **CPL**, $\tau(x)$ and $q \cup (p \cap a_{10})$ are equivalent. Thus, $\sigma_q \preceq_{\mathcal{C}} \tau$: a contradiction. Consequently, it remains to prove (1), (2), (3) and (4).

(1) Suppose $a_{00} \equiv 0$ is not $\mathcal{C}$-valid. Let $\mathcal{V}$ be a valuation on (W, R) such that $\bar{\mathcal{V}}(a_{00}) = W$. Since a_{00} is $\{p, q\}$-free, $\mathbf{Card}(W) \geq 3$ and $R \neq \emptyset$, therefore we can assume that $\mathcal{V}(p) \cup \mathcal{V}(q) \neq W$ and $R \cap (\mathcal{V}(p) \times \mathcal{V}(q)) \neq \emptyset$. Hence, $(W, R, \mathcal{V}) \models C(p, q)$ and $\bar{\mathcal{V}}(\tau(x)) \not\subseteq \bar{\mathcal{V}}(p \cup q)$: a contradiction with the $\mathcal{C}$-validity of $C(p, q) \rightarrow \tau(x) \not\equiv 0 \wedge \tau(x) \leq p \cup q$. Thus, $a_{00} \equiv 0$ is $\mathcal{C}$-valid and (1) is proved. Moreover, considered as formulas in **CPL**, $\tau(x)$ and $(p^0 \cap q^1 \cap a_{01}) \cup (p^1 \cap q^0 \cap a_{10}) \cup (p^1 \cap q^1 \cap a_{11})$ are equivalent.

(2) Suppose $a_{11} \equiv 1$ is not $\mathcal{C}$-valid. Let $\mathcal{V}$ be a valuation on (W, R) such that $\bar{\mathcal{V}}(a_{11}) = \emptyset$. Since a_{11} is $\{p, q\}$-free and $R \neq \emptyset$, therefore we can assume that $\mathcal{V}(p) = \mathcal{V}(q)$ and $R \cap (\mathcal{V}(p) \times \mathcal{V}(q)) \neq \emptyset$. Consequently, $(W, R, \mathcal{V}) \models C(p, q)$ and $\bar{\mathcal{V}}(\tau(x)) = \emptyset$: a contradiction with the $\mathcal{C}$-validity of $C(p, q) \to \tau(x) \not\equiv 0 \wedge \tau(x) \leq p \cup q$. Hence, $a_{11} \equiv 1$ is $\mathcal{C}$-valid and (2) is proved. Moreover, considered as formulas in **CPL**, $\tau(x)$ and $(p^0 \cap q^1 \cap a_{01}) \cup (p^1 \cap q^0 \cap a_{10}) \cup (p^1 \cap q^1)$ are equivalent.

(3) Suppose $a_{01} \cup a_{10} \equiv 1$ is not $\mathcal{C}$-valid. Let $\mathcal{V}$ be a valuation on (W, R) such that $\bar{\mathcal{V}}(a_{01}) = \emptyset$ and $\bar{\mathcal{V}}(a_{10}) = \emptyset$. Since $p \neq q$, a_{01} and a_{10} are $\{p, q\}$-free and $R \not\subseteq Id_W$, therefore we can assume that $\mathcal{V}(p) \cap \mathcal{V}(q) = \emptyset$ and $R \cap (\mathcal{V}(p) \times \mathcal{V}(q)) \neq \emptyset$. Thus, $(W, R, \mathcal{V}) \models C(p, q)$ and $\bar{\mathcal{V}}(\tau(x)) = \emptyset$: a contradiction with the $\mathcal{C}$-validity of $C(p, q) \to \tau(x) \not\equiv 0 \wedge \tau(x) \leq p \cup q$. Consequently, $a_{01} \cup a_{10} \equiv 1$ is $\mathcal{C}$-valid and (3) is proved.

(4) Suppose neither $a_{01} \leq a_{10}$ is $\mathcal{C}$-valid, nor $a_{10} \leq a_{01}$ is $\mathcal{C}$-valid. By using an argument similar to the one showing above that σ_0 and σ_1 form a complete set of unifiers of φ_{01}, we obtain a frame (W'', R'') in $\mathcal{C}$ such that $R'' \not\subseteq Id_W$ and a valuation $\mathcal{V}''$ on W'' such that $\bar{\mathcal{V}}''(a_{01}) \not\subseteq \bar{\mathcal{V}}''(a_{10})$ and $\bar{\mathcal{V}}''(a_{10}) \not\subseteq \bar{\mathcal{V}}''(a_{01})$. Since $p \neq q$ and a_{01} and a_{10} are $\{p, q\}$-free, therefore we can assume that $\mathcal{V}''(p)$ and $\mathcal{V}''(q)$ are such that $\mathcal{V}''(p) \subseteq \bar{\mathcal{V}}''(a_{01})$, $\mathcal{V}''(p) \cap \bar{\mathcal{V}}''(a_{10}) = \emptyset$, $\mathcal{V}''(q) \cap \bar{\mathcal{V}}''(a_{01}) = \emptyset$, $\mathcal{V}''(q) \subseteq \bar{\mathcal{V}}''(a_{10})$ and $R \cap (\mathcal{V}''(p) \times \mathcal{V}''(q)) \neq \emptyset$. Thus, $\bar{\mathcal{V}}''(\tau(x)) = \emptyset$ and $(W'', R'', \mathcal{V}'') \models C(p, q)$: a contradiction with the $\mathcal{C}$-validity of $C(p, q) \to \tau(x) \not\equiv 0 \wedge \tau(x) \leq p \cup q$. Thus, $a_{01} \leq a_{10}$ is $\mathcal{C}$-valid, or $a_{10} \leq a_{01}$ is $\mathcal{C}$-valid and (4) is proved.

Remark that the above argument shows that if $\mathcal{C}$ agrees with unions and $p \neq q$ then φ_{pq} is either $\mathcal{C}$-unitary, or $\mathcal{C}$-finitary. Finally,

- if $\mathcal{C}$ agrees with unions and $p \neq q$ then φ_{pq} is $\mathcal{C}$-finitary.

To see why, suppose $\mathcal{C}$ agrees with unions, $p \neq q$ and φ_{pq} is not $\mathcal{C}$-finitary. Since if $\mathcal{C}$ agrees with unions and $p \neq q$ then φ_{pq} is either $\mathcal{C}$-unitary, or $\mathcal{C}$-finitary, therefore φ_{pq} is either $\mathcal{C}$-unitary, or $\mathcal{C}$-finitary. Since φ_{pq} is not $\mathcal{C}$-finitary, therefore φ_{pq} is $\mathcal{C}$-unitary. Let τ be a $\mathcal{C}$-unifier of φ_{pq} such that $\tau \preceq_{\mathcal{C}} \sigma_p$ and $\tau \preceq_{\mathcal{C}} \sigma_q$. Since σ_p and σ_q form a complete set of $\mathcal{C}$-unifiers of φ_{pq}, therefore $\sigma_p \preceq_{\mathcal{C}} \tau$, or $\sigma_q \preceq_{\mathcal{C}} \tau$. Without loss of generality, suppose $\sigma_p \preceq_{\mathcal{C}} \tau$. Since $\tau \preceq_{\mathcal{C}} \sigma_q$, therefore $\sigma_p \preceq_{\mathcal{C}} \sigma_q$: a contradiction.

7 Monomials and Polynomials

The purpose of this section is to introduce definitions and properties about terms. These definitions and properties are purely Boolean. Let k, m, n be nonnegative integers. An *m-vector* is a map $\mathbf{s}$ associating with every positive integer $i \leq m$ an element $\mathbf{s}(i)$ of $\{0, 1\}$. A *(k, m, n)-correspondence* is a map f associating with every m-vector $\mathbf{s}$ a function $f_{\mathbf{s}} : \{0, 1\}^k \longrightarrow \{0, 1\}^n$. Let f be a (k, m, n)-correspondence. An *n-monomial* is a term of the form

- $x_1^{\beta_1} \cap \cdots \cap x_n^{\beta_n}$

where $(\beta_1, \ldots, \beta_n) \in \{0, 1\}^n$. For all m-vectors $\mathbf{s}$, considering a term a in $\mathbf{TER}_{m,n}$ as a formula in Classical Propositional Logic (**CPL**), we define:

- $mon_{\mathbf{s}}(n, a) = \{x_1^{\beta_1} \cap \cdots \cap x_n^{\beta_n}$: a is a tautological consequence of $p_1^{\mathbf{s}(1)} \cap \cdots \cap p_m^{\mathbf{s}(m)} \cap x_1^{\beta_1} \cap \cdots \cap x_n^{\beta_n}\}$.

Hence, for all m-vectors $\mathbf{s}$, considering a as a formula in **CPL**, $mon_{\mathbf{s}}(n, a)$ is nothing but the set of all conjunctions of the form $x_1^{\beta_1} \cap \cdots \cap x_n^{\beta_n}$ that appear in the Disjunctive Normal Form of a in conjunction with the conjunction $p_1^{\mathbf{s}(1)} \cap \cdots \cap p_m^{\mathbf{s}(m)}$. An *n-polynomial* is a term of the form

- $(x_1^{\beta_{11}} \cap \cdots \cap x_n^{\beta_{1n}}) \cup \cdots \cup (x_1^{\beta_{j1}} \cap \cdots \cap x_n^{\beta_{jn}})$

where j is a nonnegative integer and $(\beta_{11}, \ldots, \beta_{1n}), \ldots, (\beta_{j1}, \ldots, \beta_{jn}) \in \{0, 1\}^n$. Remark that for all m-vectors $\mathbf{s}$ and for all terms a in $\mathbf{TER}_{m,n}$, $\bigcup mon_{\mathbf{s}}(n, a)$ is an n-polynomial.

Proposition 4 *Let* $a \in \mathbf{TER}_{m,n}$. *Considered as formulas in* **CPL**, *the following terms are equivalent:*

- a,
- $\bigcup\{p_1^{\mathbf{s}(1)} \cap \cdots \cap p_m^{\mathbf{s}(m)} \cap \bigcup mon_{\mathbf{s}}(n, a)$: $\mathbf{s}$ *is an* m*-vector*$\}$.

Proof By the definition of $mon_{\mathbf{s}}$, $\mathbf{s}$ being an arbitrary m-vector.

For all positive integers $i \leq n$, let $\pi_i : \{0, 1\}^n \longrightarrow \{0, 1\}$ be the function such that for all $(\beta_1, \ldots, \beta_n) \in \{0, 1\}^n$,

- $\pi_i(\beta_1, \ldots, \beta_n) = \beta_i$.

For all m-vectors $\mathbf{s}$ and for all $(\beta_1, \ldots, \beta_n) \in \{0, 1\}^n$, we define:

- $f_{\mathbf{s}}^{-1}(\beta_1, \ldots, \beta_n) = \{(\alpha_1, \ldots, \alpha_k) \in \{0, 1\}^k : f_{\mathbf{s}}(\alpha_1, \ldots, \alpha_k) = (\beta_1, \ldots, \beta_n)\}$.

Obviously, for all m-vectors $\mathbf{s}$ and for all $(\beta_1, \ldots, \beta_n) \in \{0, 1\}^n$, $f_{\mathbf{s}}^{-1}(\beta_1, \ldots, \beta_n) \subseteq \{0, 1\}^k$. For all m-vectors $\mathbf{s}$ and for all positive integers $i \leq n$, we define:

- $\Delta_{\mathbf{s},i} = \{(\alpha_1, \ldots, \alpha_k) \in \{0, 1\}^k : \pi_i(f_{\mathbf{s}}(\alpha_1, \ldots, \alpha_k)) = 1\}$,
- $c_{\mathbf{s},i} = \bigcup\{x_1^{\alpha_1} \cap \cdots \cap x_k^{\alpha_k} : (\alpha_1, \ldots, \alpha_k) \in \Delta_{\mathbf{s},i}\}$.

Obviously, for all positive integers $i \leq n$, $\Delta_{\mathbf{s},i} \subseteq \{0, 1\}^k$ and $c_{\mathbf{s},i}$ is a k-polynomial. Remark that $\Delta_{\mathbf{s},i}$ and $c_{\mathbf{s},i}$ depend on f—more precisely, on $f_{\mathbf{s}}$—too.

Proposition 5 *For all* m*-vectors* $\mathbf{s}$ *and for all* $(\beta_1, \ldots, \beta_n) \in \{0, 1\}^n$, *considered as formulas in* **CPL**, *the following terms are equivalent:*

- $\bigcup\{x_1^{\alpha_1} \cap \cdots \cap x_k^{\alpha_k} : (\alpha_1, \ldots, \alpha_k) \in f_{\mathbf{s}}^{-1}(\beta_1, \ldots, \beta_n)\}$,
- $c_{\mathbf{s},1}^{\beta_1} \cap \cdots \cap c_{\mathbf{s},n}^{\beta_n}$.

Proof Let $\mathbf{s}$ be an m-vector and $(\beta_1, \dots, \beta_n) \in \{0, 1\}^n$. It suffices to show that considered as formulas in **CPL**, for all $\theta_1, \dots, \theta_k \in \{0, 1\}$, if $x_1, \dots, x_k$ are respectively interpreted by $\theta_1, \dots, \theta_k$ then $\bigcup\{x_1^{\alpha_1} \cap \dots \cap x_k^{\alpha_k} : (\alpha_1, \dots, \alpha_k) \in f_{\mathbf{s}}^{-1}(\beta_1, \dots, \beta_n)\}$ is equivalent to 1 iff $c_{\mathbf{s},1}^{\beta_1} \cap \dots \cap c_{\mathbf{s},n}^{\beta_n}$ is equivalent to 1. Let $\theta_1, \dots, \theta_k \in \{0, 1\}$. Let $x_1, \dots, x_k$ be respectively interpreted by $\theta_1, \dots, \theta_k$.

Suppose $\bigcup\{x_1^{\alpha_1} \cap \dots \cap x_k^{\alpha_k} : (\alpha_1, \dots, \alpha_k) \in f_{\mathbf{s}}^{-1}(\beta_1, \dots, \beta_n)\}$ is equivalent to 1. Hence, $(\theta_1, \dots, \theta_k) \in f_{\mathbf{s}}^{-1}(\beta_1, \dots, \beta_n)$. Thus, $f_{\mathbf{s}}(\theta_1, \dots, \theta_k) = (\beta_1, \dots, \beta_n)$. For the sake of the contradiction, suppose $c_{\mathbf{s},1}^{\beta_1} \cap \dots \cap c_{\mathbf{s},n}^{\beta_n}$ is equivalent to 0. Let $i \leq n$ be a positive integer such that $c_{\mathbf{s},i}^{\beta_i}$ is equivalent to 0. Since either $\beta_i = 0$, or $\beta_i = 1$, therefore we have to consider two cases. In the former case, $\beta_i = 0$ and therefore $\bigcup\{x_1^{\alpha_1} \cap \dots \cap x_k^{\alpha_k} : (\alpha_1, \dots, \alpha_k) \in \Delta_{\mathbf{s},i}\}$ is equivalent to 1. Consequently, $(\theta_1, \dots, \theta_k) \in \Delta_{\mathbf{s},i}$. Hence, $\pi_i(f_{\mathbf{s}}(\theta_1, \dots, \theta_k)) = 1$. Since $f_{\mathbf{s}}(\theta_1, \dots, \theta_k) = (\beta_1, \dots, \beta_n)$, therefore $\beta_i = 1$: a contradiction. In the latter case, $\beta_i = 1$ and therefore $\bigcup\{x_1^{\alpha_1} \cap \dots \cap x_k^{\alpha_k} : (\alpha_1, \dots, \alpha_k) \in \Delta_{\mathbf{s},i}\}$ is equivalent to 0. Thus, $(\theta_1, \dots, \theta_k) \notin \Delta_{\mathbf{s},i}$. Hence, $\pi_i(f_{\mathbf{s}}(\theta_1, \dots, \theta_k)) = 0$. Since $f_{\mathbf{s}}(\theta_1, \dots, \theta_k) = (\beta_1, \dots, \beta_n)$, therefore $\beta_i = 0$: a contradiction.

Suppose $c_{\mathbf{s},1}^{\beta_1} \cap \dots \cap c_{\mathbf{s},n}^{\beta_n}$ is equivalent to 1. Let $i \leq n$ be an arbitrary positive integer. Since $c_{\mathbf{s},1}^{\beta_1} \cap \dots \cap c_{\mathbf{s},n}^{\beta_n}$ is equivalent to 1, therefore $c_{\mathbf{s},i}^{\beta_i}$ is equivalent to 1. Since either $\beta_i = 0$, or $\beta_i = 1$, therefore we have to consider two cases. In the former case, $\beta_i = 0$ and therefore $c_{\mathbf{s},i}$ is equivalent to 0. Hence, $(\theta_1, \dots, \theta_k) \notin \Delta_{\mathbf{s},i}$. Thus, $\pi_i(f_{\mathbf{s}}(\theta_1, \dots, \theta_k)) = 0$. Since $\beta_i = 0$, therefore $\pi_i(f_{\mathbf{s}}(\theta_1, \dots, \theta_k)) = \beta_i$. In the latter case, $\beta_i = 1$ and therefore $c_{\mathbf{s},i}$ is equivalent to 1. Consequently, $(\theta_1, \dots, \theta_k) \in \Delta_{\mathbf{s},i}$. Hence, $\pi_i(f_{\mathbf{s}}(\theta_1, \dots, \theta_k)) = 1$. Since $\beta_i = 1$, therefore $\pi_i(f_{\mathbf{s}}(\theta_1, \dots, \theta_k)) = \beta_i$. In both cases, $\pi_i(f_{\mathbf{s}}(\theta_1, \dots, \theta_k)) = \beta_i$. Since i was arbitrary, therefore $f_{\mathbf{s}}(\theta_1, \dots, \theta_k) = (\beta_1, \dots, \beta_n)$. Thus, $(\theta_1, \dots, \theta_k) \in f_{\mathbf{s}}^{-1}(\beta_1, \dots, \beta_n)$. Consequently, $\bigcup\{x_1^{\alpha_1} \cap \dots \cap x_k^{\alpha_k} : (\alpha_1, \dots, \alpha_k) \in f_{\mathbf{s}}^{-1}(\beta_1, \dots, \beta_n)\}$ is equivalent to 1.

The property about m-vector $\mathbf{s}$ and $(\beta_1, \dots, \beta_n) \in \{0, 1\}^n$ described in Proposition 5 is purely Boolean. It will be used in the proof of Proposition 6 which will itself be used in the proof of Proposition 12, our main result.

8 Equivalence Relations Between Tuples of Bits

Let k, m, n be nonnegative integers. Let $(a_1, \dots, a_n) \in \mathbf{TER}_{m,k}^n$. For all m-vectors $\mathbf{s}$, we define on $\{0, 1\}^k$ the equivalence relation $\sim_{(a_1,\dots,a_n)}^{k,\mathbf{s}}$ as follows:

- $(\alpha_1, \dots, \alpha_k) \sim_{(a_1,\dots,a_n)}^{k,\mathbf{s}} (\alpha_1', \dots, \alpha_k')$ if for all positive integers $i \leq n$, $x_1^{\alpha_1} \cap \dots \cap x_k^{\alpha_k} \in mon_{\mathbf{s}}(k, a_i)$ iff $x_1^{\alpha_1'} \cap \dots \cap x_k^{\alpha_k'} \in mon_{\mathbf{s}}(k, a_i)$.

Obviously, for all m-vectors $\mathbf{s}$, $\sim_{(a_1,\dots,a_n)}^{k,\mathbf{s}}$ has at most 2^n equivalence classes on $\{0, 1\}^k$. Let f be a (k, m, n)-correspondence such that for all m-vectors $\mathbf{s}$ and for all $(\alpha_1, \dots, \alpha_k), (\alpha_1', \dots, \alpha_k') \in \{0, 1\}^k$, if $f_{\mathbf{s}}(\alpha_1, \dots, \alpha_k) = f_{\mathbf{s}}(\alpha_1', \dots, \alpha_k')$ then

$(\alpha_1, \ldots, \alpha_k) \sim^{k,\mathbf{s}}_{(a_1,\ldots,a_n)} (\alpha'_1, \ldots, \alpha'_k)$. For all m-vectors $\mathbf{s}$, by means of f—more precisely, of $f_{\mathbf{s}}$—, we define the n-tuple $(b_{\mathbf{s},1}, \ldots, b_{\mathbf{s},n})$ of n-polynomials as follows:

- $b_{\mathbf{s},i} = \bigcup\{x_1^{\beta_1} \cap \cdots \cap x_n^{\beta_n} : x_1^{\alpha_1} \cap \cdots \cap x_k^{\alpha_k} \in mon_{\mathbf{s}}(k, a_i)$ and $f_{\mathbf{s}}(\alpha_1, \ldots, \alpha_k) = (\beta_1, \ldots, \beta_n)\}$.

An n-tuple $(b_1, \ldots, b_n) \in \mathbf{TER}^n_{m,n}$ of terms is *properly obtained* from $(a_1, \ldots, a_n)$ if for all positive integers $i \leq n$, considered as formulas in **CPL**, the following terms are equivalent:

- b_i,
- $\bigcup\{p_1^{\mathbf{s}(1)} \cap \cdots \cap p_m^{\mathbf{s}(m)} \cap b_{\mathbf{s},i} :$ $\mathbf{s}$ is an m-vector$\}$.

For all m-vectors $\mathbf{s}$ and for all $(\beta_1, \ldots, \beta_n) \in \{0, 1\}^n$, let $f_{\mathbf{s}}^{-1}(\beta_1, \ldots, \beta_n)$ be as in Sect. 7. For all m-vectors $\mathbf{s}$ and for all positive integers $i \leq n$, let $\Delta_{\mathbf{s},i}$ and $c_{\mathbf{s},i}$ be as in Sect. 7. A substitution υ is *properly obtained* from $(a_1, \ldots, a_n)$ if for all variables y, if $y \notin \{x_1, \ldots, x_n\}$ then $\upsilon(y) = y$ and for all positive integers $i \leq n$, considered as formulas in **CPL**, the following terms are equivalent:

- $\upsilon(x_i)$,
- $\bigcup\{p_1^{\mathbf{s}(1)} \cap \cdots \cap p_m^{\mathbf{s}(m)} \cap c_{\mathbf{s},i} :$ $\mathbf{s}$ is an m-vector$\}$.

The proof of the following result uses the purely Boolean property described in Proposition 5.

Proposition 6 *Let $(a_1, \ldots, a_n) \in \mathbf{TER}^n_{m,k}$, $(b_1, \ldots, b_n) \in \mathbf{TER}^n_{m,n}$ be an n-tuple of terms and υ be a substitution. If $(b_1, \ldots, b_n)$ and υ are properly obtained from $(a_1, \ldots, a_n)$ then for all positive integers $i \leq n$, considered as formulas in* **CPL**, *the following terms are equivalent:*

- a_i,
- $\bar{\upsilon}(b_i)$.

Proof Suppose $(b_1, \ldots, b_n)$ and υ are properly obtained from $(a_1, \ldots, a_n)$. Let $i \leq n$ be a positive integer. Considered as formulas in **CPL**, the following terms are equivalent:

1. $\bar{\upsilon}(b_i)$.
2. $\bigcup\{p_1^{\mathbf{s}(1)} \cap \cdots \cap p_m^{\mathbf{s}(m)} \cap \bar{\upsilon}(b_{\mathbf{s},i}) :$ $\mathbf{s}$ is an m-vector$\}$.
3. $\bigcup\{p_1^{\mathbf{s}(1)} \cap \cdots \cap p_m^{\mathbf{s}(m)} \cap \upsilon(x_1)^{\beta_1} \cap \cdots \cap \upsilon(x_n)^{\beta_n} :$ $\mathbf{s}$ is an m-vector, $x_1^{\alpha_1} \cap \cdots \cap x_k^{\alpha_k} \in mon_{\mathbf{s}}(k, a_i)$ and $f_{\mathbf{s}}(\alpha_1, \ldots, \alpha_k) = (\beta_1, \ldots, \beta_n)\}$.
4. $\bigcup\{p_1^{\mathbf{s}(1)} \cap \cdots \cap p_m^{\mathbf{s}(m)} \cap c_{\mathbf{s},1}^{\beta_1} \cap \cdots \cap c_{\mathbf{s},n}^{\beta_n} :$ $\mathbf{s}$ is an m-vector, $x_1^{\alpha_1} \cap \cdots \cap x_k^{\alpha_k} \in mon_{\mathbf{s}}(k, a_i)$ and $f_{\mathbf{s}}(\alpha_1, \ldots, \alpha_k) = (\beta_1, \ldots, \beta_n)\}$.
5. $\bigcup\{p_1^{\mathbf{s}(1)} \cap \cdots \cap p_m^{\mathbf{s}(m)} \cap x_1^{\alpha'_1} \cap \cdots \cap x_k^{\alpha'_k} :$ $\mathbf{s}$ is an m-vector, $x_1^{\alpha_1} \cap \cdots \cap x_k^{\alpha_k} \in mon_{\mathbf{s}}(k, a_i)$, $f_{\mathbf{s}}(\alpha_1, \ldots, \alpha_k) = (\beta_1, \ldots, \beta_n)$ and $(\alpha'_1, \ldots, \alpha'_k) \in f_{\mathbf{s}}^{-1}(\beta_1, \ldots, \beta_n)\}$.
6. $\bigcup\{p_1^{\mathbf{s}(1)} \cap \cdots \cap p_m^{\mathbf{s}(m)} \cap \bigcup mon_{\mathbf{s}}(k, a_i) :$ $\mathbf{s}$ is an m-vector$\}$.
7. a_i.

The equivalence between 1 and 2 is a consequence of the definition of $(b_1, \ldots, b_n)$. The equivalence between 2 and 3 is a consequence of the definition of $(b_{\mathbf{s},1}, \ldots, b_{\mathbf{s},n})$, $\mathbf{s}$ being an arbitrary m-vector. The equivalence between 3 and 4 is a consequence of the definition of v. The equivalence between 4 and 5 is a consequence of Proposition 5. The equivalence between 5 and 6 is a consequence of the definitions of $mon_{\mathbf{s}}$ and $\sim^{k,\mathbf{s}}_{(a_1,\ldots,a_n)}$ and the fact that for all $(\alpha_1, \ldots, \alpha_k), (\alpha'_1, \ldots, \alpha'_k) \in \{0, 1\}^k$, if $f_{\mathbf{s}}(\alpha_1, \ldots, \alpha_k) = f_{\mathbf{s}}(\alpha'_1, \ldots, \alpha'_k)$ then $(\alpha_1, \ldots, \alpha_k) \sim^{k,\mathbf{s}}_{(a_1,\ldots,a_n)} (\alpha'_1, \ldots, \alpha'_k)$, $\mathbf{s}$ being an arbitrary m-vector. The equivalence between 6 and 7 is a consequence of Proposition 4.

Proposition 7 *Let $(a_1, \ldots, a_n) \in \mathbf{TER}^n_{m,k}$ and $(b_1, \ldots, b_n) \in \mathbf{TER}^n_{m,n}$ be an n-tuple of terms. Let W be a nonempty set. If $(b_1, \ldots, b_n)$ is properly obtained from $(a_1, \ldots, a_n)$ then:*

- *for all valuations $\mathcal{V}$ on W, there exists a valuation $\mathcal{V}'$ on W such that for all positive integers $i \leq n$, $\bar{\mathcal{V}}(a_i) = \bar{\mathcal{V}}'(b_i)$,*
- *for all valuations $\mathcal{V}$ on W, there exists a valuation $\mathcal{V}'$ on W such that for all positive integers $i \leq n$, $\bar{\mathcal{V}}(b_i) = \bar{\mathcal{V}}'(a_i)$.*

Proof Let $\mathcal{V}$ be a valuation on W. Let $\mathcal{V}'$ be a valuation on W such that for all positive integers $i \leq m$, $\mathcal{V}'(p_i) = \mathcal{V}(p_i)$ and for all positive integers $i \leq n$, $\mathcal{V}'(x_i) = \bar{\mathcal{V}}(\bigcup\{p_1^{\mathbf{s}(1)} \cap \cdots \cap p_m^{\mathbf{s}(m)} \cap c_{\mathbf{s},i} : \mathbf{s}$ is an m-vector$\})$. For all positive integers $i \leq n$, the following subsets of W are equal:

1. $\bar{\mathcal{V}}'(b_i)$.
2. $\bigcup\{\bar{\mathcal{V}}'(p_1^{\mathbf{s}(1)} \cap \cdots \cap p_m^{\mathbf{s}(m)}) \cap \bar{\mathcal{V}}'(b_{\mathbf{s},i}) : \mathbf{s}$ is an m-vector$\}$.
3. $\bigcup\{\bar{\mathcal{V}}'(p_1^{\mathbf{s}(1)} \cap \cdots \cap p_m^{\mathbf{s}(m)}) \cap \bar{\mathcal{V}}'(x_1^{\beta_1} \cap \cdots \cap x_n^{\beta_n}) : \mathbf{s}$ is an m-vector, $x_1^{\alpha_1} \cap \cdots \cap x_k^{\alpha_k} \in mon_{\mathbf{s}}(k, a_i)$ and $f_{\mathbf{s}}(\alpha_1, \ldots, \alpha_k) = (\beta_1, \ldots, \beta_n)\}$.
4. $\bar{\mathcal{V}}(\bigcup\{p_1^{\mathbf{s}(1)} \cap \cdots \cap p_m^{\mathbf{s}(m)} \cap c_{\mathbf{s},1}^{\beta_1} \cap \cdots \cap c_{\mathbf{s},n}^{\beta_n} : \mathbf{s}$ is an m-vector, $x_1^{\alpha_1} \cap \cdots \cap x_k^{\alpha_k} \in mon_{\mathbf{s}}(k, a_i)$ and $f_{\mathbf{s}}(\alpha_1, \ldots, \alpha_k) = (\beta_1, \ldots, \beta_n)\})$.
5. $\bar{\mathcal{V}}(\bigcup\{p_1^{\mathbf{s}(1)} \cap \cdots \cap p_m^{\mathbf{s}(m)} \cap x_1^{\alpha'_1} \cap \cdots \cap x_k^{\alpha'_k} : \mathbf{s}$ is an m-vector, $x_1^{\alpha_1} \cap \cdots \cap x_k^{\alpha_k} \in mon_{\mathbf{s}}(k, a_i)$, $f_{\mathbf{s}}(\alpha_1, \ldots, \alpha_k) = (\beta_1, \ldots, \beta_n)$ and $(\alpha'_1, \ldots, \alpha'_k) \in f_{\mathbf{s}}^{-1}(\beta_1, \ldots, \beta_n)\})$.
6. $\bar{\mathcal{V}}(\bigcup\{p_1^{\mathbf{s}(1)} \cap \cdots \cap p_m^{\mathbf{s}(m)} \cap \bigcup mon_{\mathbf{s}}(k, a_i) : \mathbf{s}$ is an m-vector$\})$.
7. $\bar{\mathcal{V}}(a_i)$.

The equivalence between 1 and 2 is a consequence of the definition of $(b_1, \ldots, b_n)$. The equivalence between 2 and 3 is a consequence of the definition of $(b_{\mathbf{s},1}, \ldots, b_{\mathbf{s},n})$, $\mathbf{s}$ being an arbitrary m-vector. The equivalence between 3 and 4 is a consequence of the definition of $\mathcal{V}'$. The equivalence between 4 and 5 is a consequence of Proposition 5. The equivalence between 5 and 6 is a consequence of the definitions of $mon_{\mathbf{s}}$ and $\sim^{k,\mathbf{s}}_{(a_1,\ldots,a_n)}$ and the fact that for all $(\alpha_1, \ldots, \alpha_k), (\alpha'_1, \ldots, \alpha'_k) \in \{0, 1\}^k$, if $f_{\mathbf{s}}(\alpha_1, \ldots, \alpha_k) = f_{\mathbf{s}}(\alpha'_1, \ldots, \alpha'_k)$ then $(\alpha_1, \ldots, \alpha_k) \sim^{k,\mathbf{s}}_{(a_1,\ldots,a_n)} (\alpha'_1, \ldots, \alpha'_k)$, $\mathbf{s}$ being an arbitrary m-vector. The equivalence between 6 and 7 is a consequence of Proposition 4.

Let $\mathcal{V}$ be a valuation on W. Let $\mathcal{V}'$ be a valuation on W such that for all positive integers $i \leq m$, $\mathcal{V}'(p_i) = \mathcal{V}(p_i)$ and for all m-vectors $\mathbf{s}$ and for all $(\alpha_1, \ldots, \alpha_k) \in \{0, 1\}^k$,

$\bar{\mathcal{V}}'(p_1^{\mathbf{s}(1)} \cap \cdots \cap p_m^{\mathbf{s}(m)} \cap x_1^{\alpha_1} \cap \cdots \cap x_k^{\alpha_k}) = \bar{\mathcal{V}}(p_1^{\mathbf{s}(1)} \cap \cdots \cap p_m^{\mathbf{s}(m)} \cap x_1^{\beta_1} \cap \cdots \cap x_n^{\beta_n})$ where $(\beta_1, \ldots, \beta_n) \in \{0, 1\}^n$ is such that $f_{\mathbf{s}}(\alpha_1, \ldots, \alpha_k) = (\beta_1, \ldots, \beta_n)$. For all positive integers $i \leq n$, the following subsets of W are equal:

1. $\bar{\mathcal{V}}'(a_i)$.
2. $\bigcup\{\bar{\mathcal{V}}'(p_1^{\mathbf{s}(1)} \cap \cdots \cap p_m^{\mathbf{s}(m)}) \cap \bar{\mathcal{V}}'(\bigcup mon_{\mathbf{s}}(k, a_i)) : \mathbf{s}$ is an m-vector$\}$.
3. $\bar{\mathcal{V}}(\bigcup\{p_1^{\mathbf{s}(1)} \cap \cdots \cap p_m^{\mathbf{s}(m)} \cap x_1^{\beta_1} \cap \cdots \cap x_n^{\beta_n} : \mathbf{s}$ is an m-vector, $x_1^{\alpha_1} \cap \cdots \cap x_k^{\alpha_k} \in mon_{\mathbf{s}}(k, a_i)$ and $f_{\mathbf{s}}(\alpha_1, \ldots, \alpha_k) = (\beta_1, \ldots, \beta_n)\})$.
4. $\bar{\mathcal{V}}(\bigcup\{p_1^{\mathbf{s}(1)} \cap \cdots \cap p_m^{\mathbf{s}(m)} \cap b_{\mathbf{s},i} : \mathbf{s}$ is an m-vector$\})$.
5. $\bar{\mathcal{V}}(b_i)$.

The equivalence between 1 and 2 is a consequence of Proposition 4. The equivalence between 2 and 3 is a consequence of the definitions of $mon_{\mathbf{s}}$ and $\mathcal{V}'$, $\mathbf{s}$ being an arbitrary m-vector. The equivalence between 3 and 4 is a consequence of the definition of $(b_{\mathbf{s},1}, \ldots, b_{\mathbf{s},n})$, $\mathbf{s}$ being an arbitrary m-vector. The equivalence between 4 and 5 is a consequence of the definition of $(b_1, \ldots, b_n)$.

9 Equivalence Relations Between Formulas

Let k, m, n be nonnegative integers and $\mathcal{C}$ be a class of frames. We define on $\mathbf{FOR}_{m,n}$ the equivalence relation $\equiv_{\mathcal{C}}^{m,n}$ as follows:

- $\varphi \equiv_{\mathcal{C}}^{m,n} \psi$ if $\varphi \leftrightarrow \psi$ is $\mathcal{C}$-valid.

Since each formula φ in $\mathbf{FOR}_{m,n}$ is a combination of formulas of the form $C(a, b)$, $a \equiv b$ and $con(a)$ where a and b are terms in $\mathbf{TER}_{m,n}$, therefore $\equiv_{\mathcal{C}}^{m,n}$ has finitely many equivalence classes on $\mathbf{FOR}_{m,n}$. Let A_n be the set of all n-tuples of terms. Note that n-tuples of terms in A_n may contain occurrences of constants outside $\{p_1, \ldots, p_m\}$ and occurrences of variables outside $\{x_1, \ldots, x_n\}$. Given a model $(W, R, \mathcal{V})$ on a frame in $\mathcal{C}$ and $(a_1, \ldots, a_n) \in A_n$, let $\Phi_{(a_1,\ldots,a_n)}^{(W,R,\mathcal{V})}$ be the set of all equational formulas $\varphi(x_1, \ldots, x_n)$ in $\mathbf{FOR}_{m,n}$ such that $(W, R, \mathcal{V}) \models \varphi(a_1, \ldots, a_n)$. Consider a complete list of representatives for each equivalence class on $\Phi_{(a_1,\ldots,a_n)}^{(W,R,\mathcal{V})}$ modulo $\equiv_{\mathcal{C}}^{m,n}$ and let $\varphi_{(a_1,\ldots,a_n)}^{(W,R,\mathcal{V})}(x_1, \ldots, x_n)$ be their conjunction. We define on A_n the equivalence relation $\cong_{\mathcal{C}}^{m,n}$ as follows:

- $(a_1, \ldots, a_n) \cong_{\mathcal{C}}^{m,n} (b_1, \ldots, b_n)$ if for all formulas $\varphi(x_1, \ldots, x_n)$ in $\mathbf{FOR}_{m,n}$, $\varphi(a_1, \ldots, a_n)$ is $\mathcal{C}$-valid iff $\varphi(b_1, \ldots, b_n)$ is $\mathcal{C}$-valid.

Since $\equiv_{\mathcal{C}}^{m,n}$ has finitely many equivalence classes on $\mathbf{FOR}_{m,n}$, therefore $\cong_{\mathcal{C}}^{m,n}$ has finitely many equivalence classes on A_n. Now, we define on A_n the equivalence relation $\simeq_{\mathcal{C}}^{m,n}$ as follows:

- $(a_1, \ldots, a_n) \simeq_{\mathcal{C}}^{m,n} (b_1, \ldots, b_n)$ if for all equational formulas $\varphi(x_1, \ldots, x_n)$ in $\mathbf{FOR}_{m,n}$, $\varphi(a_1, \ldots, a_n)$ is $\mathcal{C}$-valid iff $\varphi(b_1, \ldots, b_n)$ is $\mathcal{C}$-valid.

Obviously, $\cong_{\mathcal{C}}^{m,n}$ is finer than $\simeq_{\mathcal{C}}^{m,n}$. Since $\cong_{\mathcal{C}}^{m,n}$ has finitely many equivalence classes on A_n, therefore $\simeq_{\mathcal{C}}^{m,n}$ has finitely many equivalence classes on A_n.

Proposition 8 *Let* $(a_1, \ldots, a_n) \in \mathbf{TER}^n_{m,k}$ *and* $(b_1, \ldots, b_n) \in \mathbf{TER}^n_{m,n}$ *be an n-tuple of terms. If* $(b_1, \ldots, b_n)$ *is properly obtained from* $(a_1, \ldots, a_n)$ *then* $(a_1, \ldots, a_n) \simeq_{\mathcal{C}}^{m,n} (b_1, \ldots, b_n)$.

Proof By Proposition 7.

Proposition 9 $\mathbf{TER}^n_{m,n}$ *constitutes a complete set of representatives for each equivalence class on* A_n *modulo* $\simeq_{\mathcal{C}}^{m,n}$.

Proof By Proposition 8.

Proposition 10 *Let* $(W, R, \mathcal{V})$ *be a model on a frame in* $\mathcal{C}$ *and* $(a_1, \ldots, a_n) \in A_n$. $(W, R, \mathcal{V}) \models \varphi^{(W,R,\mathcal{V})}_{(a_1,\ldots,a_n)}(a_1, \ldots, a_n)$.

Proof By the definition of $\varphi^{(W,R,\mathcal{V})}_{(a_1,\ldots,a_n)}(x_1, \ldots, x_n)$.

Proposition 11 *If* $\mathcal{C}$ *is balanced then for all* $(a_1, \ldots, a_n), (b_1, \ldots, b_n) \in A_n$, *if* $(a_1, \ldots, a_n) \simeq_{\mathcal{C}}^{m,n} (b_1, \ldots, b_n)$ *then* $(a_1, \ldots, a_n) \cong_{\mathcal{C}}^{m,n} (b_1, \ldots, b_n)$.

Proof Suppose $\mathcal{C}$ is balanced. Let $(a_1, \ldots, a_n), (b_1, \ldots, b_n) \in A_n$ be such that $(a_1, \ldots, a_n) \simeq_{\mathcal{C}}^{m,n} (b_1, \ldots, b_n)$ and $(a_1, \ldots, a_n) \not\cong_{\mathcal{C}}^{m,n} (b_1, \ldots, b_n)$. Let $\varphi(x_1, \ldots, x_n)$ be a formula in $\mathbf{FOR}_{m,n}$ such that $\varphi(a_1, \ldots, a_n)$ is $\mathcal{C}$-valid not-iff $\varphi(b_1, \ldots, b_n)$ is $\mathcal{C}$-valid. Without loss of generality, let us assume that $\varphi(a_1, \ldots, a_n)$ is $\mathcal{C}$-valid and $\varphi(b_1, \ldots, b_n)$ is not $\mathcal{C}$-valid. Remind that $\equiv$ may not be the only predicate occurring in $\varphi(x_1, \ldots, x_n)$. Since $\mathcal{C}$ is balanced, therefore let $(W, R, \mathcal{V})$ be a balanced model on a countable frame in $\mathcal{C}$ such that $(W, R, \mathcal{V}) \not\models \varphi(b_1, \ldots, b_n)$. By Proposition 10, $(W, R, \mathcal{V}) \models \varphi^{(W,R,\mathcal{V})}_{(b_1,\ldots,b_n)}(b_1, \ldots, b_n)$. Hence, $\neg\varphi^{(W,R,\mathcal{V})}_{(b_1,\ldots,b_n)}(b_1, \ldots, b_n)$ is not $\mathcal{C}$-valid. Remind that $\varphi^{(W,R,\mathcal{V})}_{(b_1,\ldots,b_n)}(x_1, \ldots, x_n)$ is equational. Since $(a_1, \ldots, a_n) \simeq_{\mathcal{C}}^{m,n} (b_1, \ldots, b_n)$, therefore $\neg\varphi^{(W,R,\mathcal{V})}_{(b_1,\ldots,b_n)}(a_1, \ldots, a_n)$ is not $\mathcal{C}$-valid. Since $\mathcal{C}$ is balanced, therefore let $(W', R', \mathcal{V}')$ be a balanced model on a countable frame in $\mathcal{C}$ such that $(W', R', \mathcal{V}') \models \varphi^{(W,R,\mathcal{V})}_{(b_1,\ldots,b_n)}(a_1, \ldots, a_n)$. Now, consider $(\beta_1, \ldots, \beta_n) \in \{0, 1\}^n$. If $\bar{\mathcal{V}}(b_1^{\beta_1} \cap \cdots \cap b_n^{\beta_n}) = \emptyset$ then $(W, R, \mathcal{V}) \models b_1^{\beta_1} \cap \cdots \cap b_n^{\beta_n} \equiv 0$. Thus, $\varphi^{(W,R,\mathcal{V})}_{(b_1,\ldots,b_n)}(x_1, \ldots, x_n) \to x_1^{\beta_1} \cap \cdots \cap x_n^{\beta_n} \equiv 0$ is $\mathcal{C}$-valid. Since $(W', R', \mathcal{V}') \models \varphi^{(W,R,\mathcal{V})}_{(b_1,\ldots,b_n)}(a_1, \ldots, a_n)$, therefore $(W', R', \mathcal{V}') \models a_1^{\beta_1} \cap \cdots \cap a_n^{\beta_n} \equiv 0$. Consequently, $\bar{\mathcal{V}}'(a_1^{\beta_1} \cap \cdots \cap a_n^{\beta_n}) = \emptyset$. Similarly, the reader may easily verify that if $\bar{\mathcal{V}}(b_1^{\beta_1} \cap \cdots \cap b_n^{\beta_n}) = W$ then $\bar{\mathcal{V}}'(a_1^{\beta_1} \cap \cdots \cap a_n^{\beta_n}) = W'$ and if $\bar{\mathcal{V}}(b_1^{\beta_1} \cap \cdots \cap b_n^{\beta_n})$ is infinite and coinfinite then $\bar{\mathcal{V}}'(a_1^{\beta_1} \cap \cdots \cap a_n^{\beta_n})$ is infinite and coinfinite. In all cases, there exists a bijection $g_{(\beta_1,\ldots,\beta_n)}$ from $\bar{\mathcal{V}}(b_1^{\beta_1} \cap \cdots \cap b_n^{\beta_n})$ to $\bar{\mathcal{V}}'(a_1^{\beta_1} \cap \cdots \cap a_n^{\beta_n})$. Let g be the union of all $g_{(\beta_1,\ldots,\beta_n)}$ when $(\beta_1, \ldots, \beta_n)$ describes $\{0, 1\}^n$. The reader may easily verify that g is a bijection from W to W' such that for all $u \in W$ and for all $(\beta_1, \ldots, \beta_n) \in \{0, 1\}^n$, $u \in \bar{\mathcal{V}}(b_1^{\beta_1} \cap \cdots \cap b_n^{\beta_n})$ iff $g(u) \in \bar{\mathcal{V}}'(a_1^{\beta_1} \cap \cdots \cap a_n^{\beta_n})$. Let R'_g be the binary relation on W' defined by $u' R'_g v'$ if $g^{-1}(u') R g^{-1}(v')$. Obviously, g is an isomorphism from (W, R) to (W', R'_g). Since $\varphi(a_1, \ldots, a_n)$ is $\mathcal{C}$-valid, therefore $(W', R'_g, \mathcal{V}') \models \varphi(a_1, \ldots, a_n)$. Hence, $(W, R, \mathcal{V}) \models \varphi(b_1, \ldots, b_n)$: a contradiction.

10 Unification Type

Let $\mathcal{C}$ be a class of frames.

Proposition 12 *If $\mathcal{C}$ is balanced then unification in $\mathcal{C}$ is either finitary, or unitary. Moreover, if $\mathcal{C}$ agrees with unions then unification in $\mathcal{C}$ is finitary.*

Proof Suppose $\mathcal{C}$ is balanced. Let φ be an arbitrary $\mathcal{C}$-unifiable formula. Let σ be an arbitrary substitution such that $\bar{\sigma}(\varphi)$ is $\mathcal{C}$-valid. Let $m, n \geq 0$ be such that φ's constants form a subset of $\{p_1, \ldots, p_m\}$ and φ's variables form a subset of $\{x_1, \ldots, x_n\}$. Let σ' be the substitution defined by $\sigma'(x_i) = \sigma(x_i)$ for all $i = 1 \ldots n$ and $\sigma'(y) = y$ for all y not in $\{x_1, \ldots, x_n\}$. Obviously, σ' is a unifier of φ too. Now, it may happen that for some $i \in \{1, \ldots, n\}$, $\sigma'(x_i)$ contains extra constants outside $\{p_1, \ldots, p_m\}$. If it is, then let $q_1, \ldots, q_l$ be the list of these extra constants. Take new variables $z_1, \ldots, z_l$ and define σ'' by uniformly replacing in $\sigma'(x_1), \ldots, \sigma'(x_n)$ each occurrence of $q_1, \ldots, q_l$ by, respectively, $z_1, \ldots, z_l$. Obviously, σ'' is a unifier of φ too. As a result, for all constants q, if $q \notin \{p_1, \ldots, p_m\}$ then for all positive integers $i \leq n$, q does not occur in $\sigma''(x_i)$ and for all variables y, if $y \notin \{x_1, \ldots, x_n\}$ then $\sigma''(y) = y$. Let k be a nonnegative integer and $(a_1, \ldots, a_n) \in \mathbf{TER}^n_{m,k}$ be such that for all positive integers $i \leq n$, $\sigma''(x_i) = a_i$. Since $\bar{\sigma''}(\varphi)$ is $\mathcal{C}$-valid, therefore $\varphi(a_1, \ldots, a_n)$ is $\mathcal{C}$-valid. For all m-vectors $\mathbf{s}$, let $\sim^{k,\mathbf{s}}_{(a_1,\ldots,a_n)}$ be as in Sect. 8. Let f be a (k, m, n)-correspondence such that for all m-vectors $\mathbf{s}$ and for all $(\alpha_1, \ldots, \alpha_k), (\alpha'_1, \ldots, \alpha'_k) \in \{0, 1\}^k$, if $f_{\mathbf{s}}(\alpha_1, \ldots, \alpha_k) = f_{\mathbf{s}}(\alpha'_1, \ldots, \alpha'_k)$ then $(\alpha_1, \ldots, \alpha_k) \sim^{k,\mathbf{s}}_{(a_1,\ldots,a_n)} (\alpha'_1, \ldots, \alpha'_k)$. For all m-vectors $\mathbf{s}$ and for all $(\beta_1, \ldots, \beta_n) \in \{0, 1\}^n$, let $f^{-1}_{\mathbf{s}}(\beta_1, \ldots, \beta_n)$ be as in Sect. 7. For all m-vectors $\mathbf{s}$ and for all positive integers $i \leq n$, let $\Delta_{\mathbf{s},i}$ and $c_{\mathbf{s},i}$ be as in Sect. 7. Let $(b_1, \ldots, b_n) \in \mathbf{TER}^n_{m,n}$ be an n-tuple of terms properly obtained from $(a_1, \ldots, a_n)$. By Proposition 8, $(a_1, \ldots, a_n) \simeq^{m,n}_{\mathcal{C}} (b_1, \ldots, b_n)$. Since $\mathcal{C}$ is balanced, therefore by Proposition 11, $(a_1, \ldots, a_n) \cong^{m,n}_{\mathcal{C}} (b_1, \ldots, b_n)$. Let τ be the substitution such that for all variables y, if $y \notin \{x_1, \ldots, x_n\}$ then $\tau(y) = y$ and for all positive integers $i \leq n$, $\tau(x_i) = b_i$. Remark that $(\tau(x_1), \ldots, \tau(x_n)) \in \mathbf{TER}^n_{m,n}$. Moreover, since $\varphi(a_1, \ldots, a_n)$ is $\mathcal{C}$-valid and $(a_1, \ldots, a_n) \cong^{m,n}_{\mathcal{C}} (b_1, \ldots, b_n)$, therefore $\varphi(b_1, \ldots, b_n)$ is $\mathcal{C}$-valid. Hence, τ is a $\mathcal{C}$-unifier of φ. Let υ be a substitution properly obtained from $(a_1, \ldots, a_n)$. By Proposition 6, for all positive integers $i \leq n$, considered as formulas in **CPL**, the terms a_i and $\bar{\upsilon}(b_i)$ are equivalent. Thus, for all positive integers $i \leq n$, $\bar{\upsilon}(\tau(x_i)) \equiv \sigma''(x_i)$ is $\mathcal{C}$-valid. Consequently, $\tau \circ \upsilon \simeq_{\mathcal{C}} \sigma''$. Hence, $\tau \preceq_{\mathcal{C}} \sigma''$. By the construction of τ, one can deduce that $\tau \preceq_{\mathcal{C}} \sigma$. Since σ was arbitrary and $(\tau(x_1), \ldots, \tau(x_n)) \in \mathbf{TER}^n_{m,n}$, therefore φ is either $\mathcal{C}$-finitary, or $\mathcal{C}$-unitary. Since φ was arbitrary, therefore unification in $\mathcal{C}$ is either finitary, or unitary. Now, suppose $\mathcal{C}$ agrees with unions. By the examples considered in Sect. 6, unification in $\mathcal{C}$ is not unitary. Since unification in $\mathcal{C}$ is either finitary, or unitary, therefore unification in $\mathcal{C}$ is finitary.

It follows from the above discussion that unification in $\mathcal{C}_{all}$, $\mathcal{C}_{ind}$ and $\mathcal{C}_{con}$ is finitary.

11 Conclusion

We anticipate a number of further investigations. For example, about the computability of the unification problem in Contact Logic. It is proved in Balbiani and Gencer (2015) that with respect to most classes of frames, in Contact Logic, when one restricts the discussion to con-free formulas, elementary unification is NP-complete whereas unification with constants is decidable, its exact complexity being still unknown. By means of Propositions 9 and 11, one can prove that with respect to balanced classes of frames, elementary unification and unification with constants are decidable in Contact Logic when one does not restrict the discussion to con-free formulas, but its exact complexity is unknown. In this respect, we believe that arguments developed in Baader (1998) could be used.

Acknowledgements Special acknowledgement is heartily granted to Tinko Tinchev (Sofia University, Bulgaria) for his valuable remarks. We also make a point of thanking the referees for their feedback: their helpful comments and their useful suggestions have been essential for improving the correctness and the readability of a preliminary version of this paper.

References

Baader, F. 1998. On the complexity of Boolean unification. *Information Processing Letters* 67: 215–220.

Baader, F., S. Borgwardt, and B. Morawska. 2012. Extending unification in $\mathcal{EL}$ towards general TBoxes. In *Principles of Knowledge Representation and Reasoning*, 568–572. AAAI Press.

Baader, F., and S. Ghilardi. 2011. Unification in modal and description logics. *Logic Journal of the IGPL* 19: 705–730.

Baader, F., and B. Morawska. 2009. Unification in the description logic $\mathcal{EL}$. In *Rewriting Techniques and Applications*, ed. R. Treinen, 350–364. Berlin: Springer

Balbiani, P., and Ç. Gencer. 2015. Admissibility and unifiability in contact logics. In *Logic, Language, and Computation*, 44–60. Berlin: Springer

Balbiani, P., and Ç. Gencer. 2017. KD is nullary. *Journal of Applied Non-Classical Logics* 27: 196–205.

Balbiani, P., T. Tinchev, and D. Vakarelov. 2007. Modal logics for region-based theories of space. *Fundamenta Informaticæ* 81: 29–82.

Cohn, A., and J. Renz. 2008. Qualitative spatial representation and reasoning. In *Handbook of Knowledge Representation*, 551–596. Elsevier.

Dimov, G., and D. Vakarelov. 2006a. Contact algebras and region-based theory of space: a proximity approach—I. *Fundamenta Informaticæ* 74: 209–249.

Dimov, G., and D. Vakarelov. 2006b. Contact algebras and region-based theory of space: proximity approach—II. *Fundamenta Informaticæ* 74: 251–282.

Düntsch, I., and M. Winter. 2005. A representation theorem for Boolean contact algebras. *Theoretical Computer Science* 347: 498–512.

Dzik, W.: *Unification Types in Logic*. Wydawnicto Uniwersytetu Slaskiego (2007).

Galton, A. 2000. *Qualitative Spatial Change*. Oxford University Press.

Gencer, Ç. 2002. Description of modal logics inheriting admissible rules for $K4$. *Logic Journal of the IGPL* 10: 401–411.

Gencer, Ç., and D. de Jongh. 2009. Unifiability in extensions of $K4$. *Logic Journal of the IGPL* 17: 159–172.

Ghilardi, S. 2000. Best solving modal equations. *Annals of Pure and Applied Logic* 102: 183–198.
Ghilardi, S., and L. Sacchetti. 2004. Filtering unification and most general unifiers in modal logic. *The Journal of Symbolic Logic* 69: 879–906.
Jeřábek, E. 2015. Blending margins: the modal logic K has nullary unification type. *Journal of Logic and Computation* 25: 1231–1240.
Kontchakov, R., Y. Nenov, I. Pratt-Hartmann, and M. Zakharyaschev. 2013. Topological logics with connectedness over Euclidean spaces. *ACM Transactions on Computational Logic* 14. https://doi.org/10.1145/2480759.2480765.
Kontchakov, R., I. Pratt-Hartmann, F. Wolter, and M. Zakharyaschev. 2010a. Spatial logics with connectedness predicates. *Logical Methods in Computer Science* 6: 1–43.
Kontchakov, R., I. Pratt-Hartmann, and M. Zakharyaschev. 2010b. Interpreting topological logics over Euclidean spaces. In *Proceedings of the Twelfth International Conference on the Principles of Knowledge Representation and Reasoning*, 534–544. AAAI Press.
Renz, J. 2002. *Qualitative Spatial Reasoning with Topological Information*. Berlin: Springer.
Rybakov, V. 1997. *Admissibility of Logical Inference Rules*. Elsevier.
Tinchev, T., and D. Vakarelov. 2010. Logics of space with connectedness predicates: complete axiomatizations. In *Advances in Modal Logic*, 434–453. College Publications.
Vakarelov, D. 2007. Region-based theory of space: algebras of regions, representation theory, and logics. In *Mathematical Problems from Applied Logic. Logics for the XXIst Century. II*, 267–348. Berlin: Springer.
Wolter, F., and M. Zakharyaschev. 2000. Spatio-temporal representation and reasoning based on RCC-8. In *Proceedings of the Seventh International Conference on Principles of Knowledge Representation and Reasoning*, 3–14. Morgan Kaufmann.

A Multi-agent Default Theory of Permission

Huimin Dong

Abstract This paper formulates a topology of permission by extending Horty's prioritized default theory, in order to study permissions with different defeasible manners in a reason-based reasoning. To do so, we reconstruct the prioritized default theory augmented with permissive norms and agents, so as to examine the possible relationships between permissions and defeasible norms in a agent-relative feature. This, we argue, brings about two essential elements the variety of permission and the multi-agent interaction in analyzing two noticeable pieces in Chinese and German legislations.

Keywords Permission · Default theory · Argumentation

This paper studies permissions in Horty's prioritized default theory (Horty 2012). There are two main motivations for doing this. The first is to extend prioritized default theory with a richer typology of permissions, in order to study the relationship between permissions and defeasible norms. Only one type of permission has been studied in prioritized default theory: the one that we so far have called "weak permission," that is the dual of obligation (Horty 2012). Recall that this is the same type of permission as in standard deontic logic (SDL) (Hilpinen and MaNamara 2013). Weak permission does not, however, exhaust all types of permission (Makinson and van der Torre 2003; Hansson 2013). We have already encountered strong permission, free choice permission and bilateral permission. With the notable exception of Makinson and van der Torre (2003), however, little work has been done on the relation between a richer typology of permissions and defeasible obligations. This gap should be filled, and Horty's prioritized default theory, being a prominent, recent theory of defeasible obligation, is a natural framework to do so. So by studying permissions in default theory, we are taking another step towards reaching the main goal of this thesis, that is understanding how permission goes together with non-monotonic normative reasoning.

H. Dong (✉)
Department of Philosophy & CSLC, Zhejiang University, Hangzhou, China
e-mail: Huimin.Dong@xixilogic.org

S. Ju et al. (eds.), *Nonclassical Logics and Their Applications*,
Logic in Asia: Studia Logica Library,
https://doi.org/10.1007/978-981-15-1342-8_5

The second motivation for studying permissions in prioritized default theory is to have application to legal reasoning. A number of salient pieces of legislation involve both permissions and defeasible norms. Horty's theory has already been applied to Common Law reasoning (Horty 2011, 2013). We show in this paper that, equipped with a richer typology of permissions, it can capture and analyze two running examples, one from Chinese Tax Law, and one from the German Driving Regulations.

Example 1 (*China Tax Incentive*) In the season for businesses filing their tax returns, any foreign company c established in China should file the corporate income tax (CIT) returns with the local tax authority a. If, however, c is a solar energy company, and then it can benefit from a 15% preferential tax rate from the government, if it submits valid documentation.

Example 2 (*German Right of Way*) Germany uses the "priority to the right" rule at intersections with no further priority signs. A driver on a priority road making a left turn is permitted to do so, unless another vehicle coming from the opposite direction on the same road drives straight on or turns right.

The main contribution of this paper is thus to extend prioritized default theory with various notions of permission (Sect. 2), and to apply it to concrete cases in Chinese and German Law (Sect. 2). For the first task we draw from existing accounts of explicit and tacit permissions (Makinson and van der Torre 2003; Parent and van der Torre 2013). The second requires developing a multi-agent version of Horty's theory. This paper is organized as follows. Section 1 introduces Horty's prioritized default theory and its agent-relative extension, and Sect. 2 presents the various permissions and their applications. Section 3 concludes.

1 Multi-agent Prioritized Default Theory

We start by introducing the basics of Horty's prioritized default theory, generalized such that it allows for agent-relative duties and permissions. Prioritized default theory is a syntactic theory about deriving defeasible consequences from background information and default rules (Horty 2012). Although similar in some aspects, this syntactic theory is different than Makinson's general syntactic theory for non-monotonic consequence (Makinson 2003) and from Hansson's semantic model for preference (Hansson 2001).

The language $\mathcal{L}$ of default theory is a set of propositions inductively defined by a set of atomic propositions $p \in Atom$ (containing $\top$ as the true constant) and $\wedge, \vee, \Rightarrow$, and $\neg$ as the operations of conjunction, disjunction, material implication, and negation. The turnstile $\vdash$ indicates standard consequence in propositional logic. So here this consequence relation is entirely classical. Given $\varphi, \psi, \chi \cdots$ as propositions, and X as a set of propositions, then $X \vdash \varphi$ means that φ is derivable from X according to the rules of propositional logic. So $Cn(X) = \{\varphi \mid X \vdash \varphi\}$ is the logical closure of X.

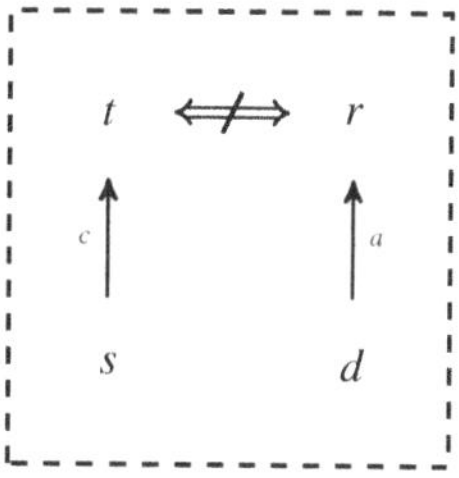

Fig. 1 A default theory of a solar energy business having to file the regular CIT return to the local tax authority

A default theory Δ is a triple $\langle \mathcal{W}, \mathcal{I}, \mathcal{D}, < \rangle$ where $\mathcal{W}$ is a set of propositions, $\mathcal{I}$ is a set of agents, $\mathcal{D}$ is a set of default rules, and $<$ is a strict partial order on $\mathcal{D}$. A default (or default rule) δ_i for agent i is a pair (φ, ψ) for φ, ψ formulas of our propositional language. We often denote such a default as[1]

$$\varphi \xrightarrow{\delta_i} \psi$$

We use the letters $\delta_i, \gamma_i, \ldots$ to denote defaults for agent i. Now we denote $\mathcal{D}_i$ as a subset of $\mathcal{D}$ such that $\mathcal{D}_i$ is a set of defaults δ_i for agent i. Given a default $\varphi \xrightarrow{\delta_i} \psi$, we say φ is the premise of δ_i, denoted as $Pre(\delta_i)$, and ψ is the conclusion of δ_i, denoted as $Con(\delta_i)$. Given a set $\mathcal{D}$ of defaults, the set of premises of $\mathcal{D}$ is defined as $Pre(\mathcal{D}) = \{Pre(\delta_i) \mid \delta_i \in \mathcal{D}\}$, and the set of conclusions of $\mathcal{D}$ is defined as $Con(\mathcal{D}) = \{Con(\delta_i) \mid \delta_i \in \mathcal{D}\}$. The order $<$ is used to represent the fact that certain defaults "override" others. So $\delta_i < \gamma_j$ indicates that the default γ_j for j has a higher priority than the default δ_i for i.

Horty adopts a reason-based reading of a given default rule $\varphi \xrightarrow{\delta_i} \psi$. In our agent-relative version of his theory this gives:

φ is a prima facie reason for i to ψ

Let us illustrate this by modeling part of the Chinese Tax Law Example. We come back to the German Driving Regulation later. Suppose that s means that this is the tax season in China, d means a company submits the valid documentation to prove that it is a solar energy business, t means that the company files the CIT, and r that the company gets the preferential tax rate. In this case we have two defaults: $s \xrightarrow{\delta_c} t$ and $d \xrightarrow{\delta_a} r$. The default theory $\Delta_{CIT} = \langle \mathcal{W}, \mathcal{I}, \mathcal{D}, < \rangle$ is then defined as follows: $\mathcal{W} = \{s, \neg t \Leftrightarrow r\}$, $\mathcal{I} = \{a, c\}$, $\mathcal{D} = \{s \xrightarrow{\delta_c} t, d \xrightarrow{\delta_a} r\}$, with the priority $\delta_c < \delta_a$, reflecting that in event that both defaults are triggered then the tax reduction gets priority. See Fig. 1.

A *scenario* based on a default theory $\Delta = \langle \mathcal{W}, \mathcal{I}, \mathcal{D}, < \rangle$ is a subset $\mathcal{S}$ of the set $\mathcal{D}$ of defaults in Δ. There are four scenarios in Δ_{CIT}: $\emptyset$, $\{\delta_c\}$, $\{\delta_a\}$, $\{\delta_c, \delta_a\}$. All things considered obligations for particular agents are derived from so-called *proper*

[1]Here we use $\Rightarrow$ for the material conditional, and $\rightarrow$ for default rules.

scenarios, which are computed in the same way as in Horty (2012). Proper scenarios should contain defaults that are be triggered, non-conflicted, and non-defeated at the same time.

Definition 1 *(Triggered)* Given a scenario $\mathcal{S}$ based on $\Delta = \langle \mathcal{W}, \mathcal{I}, \mathcal{D}, < \rangle$, a default $(\varphi, \psi) \in \mathcal{D}$ is triggered in this scenario is defined as

$$(\varphi, \psi) \in Triggered_{\mathcal{W},\mathcal{D}}(\mathcal{S}) \Leftrightarrow \varphi \in Cn(\mathcal{W} \cup Con(\mathcal{S}))$$

In Δ_{CIT} only δ_c triggered in all four scenarios. The default δ_a is never triggered, because its premise d is neither in the background information set $\mathcal{W}$ nor a conclusion of a default that could be included in a scenario.

Definition 2 *(Conflicted)* Given a scenario $\mathcal{S}$ based on $\Delta = \langle \mathcal{W}, \mathcal{I}, \mathcal{D}, < \rangle$, a default $(\varphi, \psi) \in \mathcal{D}$ is conflicted in this scenario is defined as

$$(\varphi, \psi) \in Conflicted_{\mathcal{W},\mathcal{D}}(\mathcal{S}) \Leftrightarrow \neg\psi \in Cn(\mathcal{W} \cup Con(\mathcal{S}))$$

In our example the default δ_a is conflicted in all scenarios that contain δ_c. This default is itself conflicted in all scenarios containing δ_a.

Definition 3 *(Defeated)* Given a scenario $\mathcal{S}$ based on $\Delta = \langle \mathcal{W}, \mathcal{I}, \mathcal{D}, < \rangle$, a default $(\varphi, \psi) \in \mathcal{D}$ is defeated in this scenario is defined as

$$\begin{aligned}(\varphi, \psi) \in Defeated_{\mathcal{W},\mathcal{D}}(\mathcal{S}) \Leftrightarrow\ & \exists(\varphi', \psi') \in Triggered_{\mathcal{W},\mathcal{D}}(\mathcal{S})\ s.t. \\ & (1)\ (\varphi, \psi) < (\varphi', \psi') \\ & (2)\ \neg\psi \in Cn(\mathcal{W} \cup \{\psi'\})\end{aligned}$$

In our example there is no defeated default, since the only potential defeater is δ_a but it is never triggered.

Definition 4 *(Proper Scenario)* Given a scenario $\mathcal{S}$ based on the theory $\Delta = \langle \mathcal{W}, \mathcal{I}, \mathcal{D}, < \rangle$. A default $(\varphi, \psi) \in \mathcal{D}$ is proper in this scenario is defined as

$$\begin{aligned}(\varphi, \psi) \in Proper_{\mathcal{W},\mathcal{D}}(\mathcal{S}) \Leftrightarrow\ & (\varphi, \psi) \in Triggered_{\mathcal{W},\mathcal{D}}(\mathcal{S}), \text{ and} \\ & (\varphi, \psi) \notin Conflicted_{\mathcal{W},\mathcal{D}}(\mathcal{S}), \text{ and} \\ & (\varphi, \psi) \notin Defeated_{\mathcal{W},\mathcal{D}}(\mathcal{S})\end{aligned}$$

We say a scenario $\mathcal{S}$ is a proper scenario iff $\mathcal{S} = Proper_{\mathcal{W},\mathcal{D}}(\mathcal{S})$, which involves all those defaults satisfying all three conditions for propriety. As a summary of all three conditions, $\{\delta_c\}$ is the unique proper scenario in the previous running example.

1.1 Agent-Relative Defeasible Obligations

Prioritized default theories allow to distinguish two types of defeasible obligation, the so-called conflict and disjunctive accounts. Intuitively, given a default theory Δ, φ is obligatory in the disjunctive sense if φ follows from all proper scenarios of Δ. On the other hand, φ is obligatory in the conflict sense if it follows from some proper scenarios of Δ. The labels "disjunctive" and "conflict" come from their logical behavior in case a given theory has multiple, incompatible scenarios, viz. some in which φ follows, and some in which ψ follows, with $\psi \xrightarrow{\delta_i} \neg\varphi$. On the conflict account both φ will be obligatory, and also $\neg\varphi$, although not their conjunction. On the disjunctive account only $\varphi \vee \psi$ will be obligatory, but none of the disjuncts will.

Definition 5 (*Agent-Neutral Obligations—Conflict and Disjunctive Accounts*) Let Δ be a default theory. Then:

- $\Delta \models O^c\varphi$, iff $\varphi \in Cn(\mathcal{W} \cup Con(\mathcal{S}))$ for some proper scenarios S of Δ
- $\Delta \models O^d\varphi$, iff $\varphi \in Cn(\mathcal{W} \cup Con(\mathcal{S}))$ for all proper scenarios S of Δ

While both types of duties are agent-neutral, in the legal applications that we have in mind duties and permissions are agent-relative. In Chinese Tax Law, the duty to submit the tax report bears on company c. There is no such duty for the tax authority. Similarly, in the Right of Way example one Driver's permission overrides the other's. More generally, the Hohfeldian typology (Hohfeld 1913) also rests on the notion of correlative duties, which are agent-relative.

We capture agent-relative duties by reducing them to a combination of agent-neutral duties and agent-relative reasons. Agent i's ought to make it the case that φ will then be interpreted as that it ought to be the case that i makes it the case that φ. This is similar to the reduction of "ought to do" to "ought to be" that is proposed in Horty (2000), and further discussed in Broome (2013). We define the latter notion using what we call *agent-relative reasons*. Informally, making it the case that φ whenever there is a reason for i in S to making it the case that φ, which comes down to saying that $\psi \xrightarrow{\delta_i} \varphi$ is in S, for some ψ. To illustrate such agent-relative reasons, we start from defining what are arguments and what are attacks between them. Agents will be added up after that. Intuitively speaking, an argument is a collection of agent-relative reasons.

Definition 6 (*Arguments*) An argument $\mathcal{D}' \subseteq \mathcal{D}$ is a set of default rules.

We have not distinguished *consistent* arguments from *inconsistent* ones. This can be done by proper scenarios. We will denote strong and weak arguments with additional piece of notation: given arguments $\mathcal{D}', \mathcal{D}'' \subseteq \mathcal{D}$, formula $(\mathcal{D}', \mathcal{D}'')_{\forall}\varphi$ is read as "argument $\mathcal{D}'$ is a strong attack to argument $\mathcal{D}''$ of concluding φ," and formula $(\mathcal{D}', \mathcal{D}'')_{\exists}\varphi$ "argument $\mathcal{D}'$ is a weak attack to argument $\mathcal{D}''$ of concluding φ."

Definition 7 (*Attacks*) Let Δ be a default theory and $\mathcal{D}', \mathcal{D}'' \subseteq \mathcal{D}$ be arguments. Then:

$$\Delta \models (\mathcal{D}', \mathcal{D}'')_{\forall}\varphi \text{ iff } (1)\ \varphi \in Cn(Con(\mathcal{D}')) \text{ and}$$
$$(2)\ \forall D \subseteq \mathcal{D}'' \text{ s.t. } \neg \bigwedge Con(D) \in Cn(\mathcal{W} \cup Con(\mathcal{D}'))$$
$$\Delta \models (\mathcal{D}', \mathcal{D}'')_{\exists}\varphi \text{ iff } (1)\ \varphi \in Cn(Con(\mathcal{D}')) \text{ and}$$
$$(2)\ \exists D \subseteq \mathcal{D}'' \text{ s.t. } \neg \bigwedge Con(D) \in Cn(\mathcal{W} \cup Con(\mathcal{D}'))$$

Notice that we can describe agent-relative attacks by encoding agents to each argument. For instance, formula $(\mathcal{D}'_i, \mathcal{D}''_j)_{\forall}\varphi$ can identify that agent i has a strong argument $\mathcal{D}'_i$ to attack agent j's argument $\mathcal{D}''_j$. And the above conditions have shown two important notions for attacking arguments. First condition (2) illustrates that there is a conflicts between $\mathcal{D}'$ and $\mathcal{D}''$, namely their union is not consistent. In addition, condition (1) means that $\mathcal{D}'$ is an argument to conclude result φ. If the reader is familiar with the attack models developed in Grossi (2013), they can observe that our weak attacks are the first-order counterparts of their modal formulas for attacking arguments. We leave the comparison between these two proposals in the future work. Further, we use $(\mathcal{D}', \mathcal{D}'')\varphi$ when there is no need to mention whether an attack is strong or weak. Back to the Chinese tax example, formula $(\mathcal{D}_c, \mathcal{D}_a)t$ delivers this idea: There is an argument $\mathcal{D}_c$ for the solar energy company c against the $\mathcal{D}_a$ for the local authority a of concluding the case that c files the regular CIT return.

Observe that our attacks of arguments are closed under conjunction and disjunction in the following sense:

Proposition 1 *Let Δ be a default theory and $\mathcal{D}_1, \mathcal{D}_2, \mathcal{D}', \mathcal{D}'' \subseteq \mathcal{D}$ be arguments. Then:*

1. $\Delta \models (\mathcal{D}_1, \mathcal{D}')_{\forall}\varphi \wedge (\mathcal{D}_2, \mathcal{D}')_{\forall}\psi \rightarrow (\mathcal{D}_1 \cup \mathcal{D}_2, \mathcal{D}')_{\forall}(\varphi \wedge \psi)$
2. $\Delta \models (\mathcal{D}_1, \mathcal{D}')_{\exists}\varphi \wedge (\mathcal{D}_2, \mathcal{D}')_{\exists}\psi \rightarrow (\mathcal{D}_1 \cup \mathcal{D}_2, \mathcal{D}')_{\exists}(\varphi \wedge \psi)$
3. $\Delta \models (\mathcal{D}_1, \mathcal{D}')_{\forall}\varphi \vee (\mathcal{D}_2, \mathcal{D}')_{\forall}\psi \rightarrow (\mathcal{D}_1 \cup \mathcal{D}_2, \mathcal{D}')_{\forall}(\varphi \vee \psi)$
4. $\Delta \models (\mathcal{D}_1, \mathcal{D}')_{\exists}\varphi \vee (\mathcal{D}_2, \mathcal{D}')_{\exists}\psi \rightarrow (\mathcal{D}_1 \cup \mathcal{D}_2, \mathcal{D}')_{\exists}(\varphi \vee \psi)$
5. $\Delta \models (\mathcal{D}_1, \mathcal{D}')_{\forall}\varphi \wedge (\mathcal{D}_2, \mathcal{D}'')_{\forall}\psi \rightarrow (\mathcal{D}_1 \cup \mathcal{D}_2, \mathcal{D}' \cup \mathcal{D}'')_{\forall}(\varphi \wedge \psi)$
6. $\Delta \models (\mathcal{D}_1, \mathcal{D}')_{\exists}\varphi \wedge (\mathcal{D}_2, \mathcal{D}'')_{\exists}\psi \rightarrow (\mathcal{D}_1 \cup \mathcal{D}_2, \mathcal{D}' \cup \mathcal{D}'')_{\exists}(\varphi \wedge \psi)$
7. $\Delta \models (\mathcal{D}_1, \mathcal{D}')_{\forall}\varphi \vee (\mathcal{D}_2, \mathcal{D}'')_{\forall}\psi \rightarrow (\mathcal{D}_1 \cup \mathcal{D}_2, \mathcal{D}' \cup \mathcal{D}'')_{\forall}(\varphi \vee \psi)$
8. $\Delta \models (\mathcal{D}_1, \mathcal{D}')_{\exists}\varphi \vee (\mathcal{D}_2, \mathcal{D}'')_{\exists}\psi \rightarrow (\mathcal{D}_1 \cup \mathcal{D}_2, \mathcal{D}' \cup \mathcal{D}'')_{\exists}(\varphi \vee \psi)$

Proof 1. Suppose $\Delta \models (\mathcal{D}_1, \mathcal{D}')_{\forall}\varphi$ and $\Delta \models (\mathcal{D}_2, \mathcal{D}')_{\forall}\psi$. This gives:

- $\varphi \in Cn(Con(\mathcal{D}_1))$ and $\forall D \subseteq \mathcal{D}' : \neg \bigwedge Con(D) \in Cn(\mathcal{W} \cup Con(\mathcal{D}_1))$, and
- $\psi \in Cn(Con(\mathcal{D}_2))$ and $\forall D \subseteq \mathcal{D}' : \neg \bigwedge Con(D) \in Cn(\mathcal{W} \cup Con(\mathcal{D}_2))$

They imply that $\varphi \vee \psi \in Cn(Con(\mathcal{D}_1 \cup \mathcal{D}_2))$ and $\forall D \subseteq \mathcal{D}' : \neg \bigwedge Con(D) \in Cn(\mathcal{W} \cup Con(\mathcal{D}_1 \cup \mathcal{D}_2))$. It then turns out to be $\Delta \models (\mathcal{D}_1 \cup \mathcal{D}_2, \mathcal{D}')_{\forall}(\varphi \wedge \psi)$.

2. This statement can be proven in the same way as the previous did.
3. Suppose either $\Delta \models (\mathcal{D}_1, \mathcal{D}')_{\forall}\varphi$ or $\Delta \models (\mathcal{D}_2, \mathcal{D}')_{\forall}\psi$. This gives:

- either $\varphi \in Cn(Con(\mathcal{D}_1))$ and $\forall D \subseteq \mathcal{D}' : \neg \bigwedge Con(D) \in Cn(\mathcal{W} \cup Con(\mathcal{D}_1))$
- or $\psi \in Cn(Con(\mathcal{D}_2))$ and $\forall D \subseteq \mathcal{D}' : \neg \bigwedge Con(D) \in Cn(\mathcal{W} \cup Con(\mathcal{D}_2))$

They imply that $\varphi \vee \psi \in Cn(Con(\mathcal{D}_1 \cup \mathcal{D}_2))$ and $\forall D \subseteq \mathcal{D}' : \neg \bigwedge Con(D) \in Cn(\mathcal{W} \cup Con(\mathcal{D}_1 \cup \mathcal{D}_2))$. It then turns out to be $\Delta \models (\mathcal{D}_1 \cup \mathcal{D}_2, \mathcal{D}')_\forall(\varphi \vee \psi)$.
4. Similarly, the statement here can be proven in the same way as the previous did.
5. Suppose $\Delta \models (\mathcal{D}_1, \mathcal{D}')_\forall\varphi$ and $\Delta \models (\mathcal{D}_2, \mathcal{D}'')_\forall\psi$. They imply that

   - $\varphi \in Cn(Con(\mathcal{D}_1))$ and $\forall D_1 \subseteq \mathcal{D}' : \neg \bigwedge Con(D_1) \in Cn(\mathcal{W} \cup Con(\mathcal{D}_1))$
   - $\psi \in Cn(Con(\mathcal{D}_2))$ and $\forall D_2 \subseteq \mathcal{D}'' : \neg \bigwedge Con(D_2) \in Cn(\mathcal{W} \cup Con(\mathcal{D}_2))$

   It then gives $\varphi \wedge \psi \in Cn(Con(\mathcal{D}_1 \cup \mathcal{D}_2))$. Further, suppose there is $D \subseteq \mathcal{D}' \cup \mathcal{D}''$ such that $\bigwedge Con(D) \in Cn(\mathcal{W} \cup Con(\mathcal{D}_1 \cup \mathcal{D}_2))$ (@). Let $D' \subseteq D \cap \mathcal{D}'$. By the first item concluded in above, we have $\neg \bigwedge Con(D') \in Cn(\mathcal{W} \cup Con(\mathcal{D}_1))$. So $\neg \bigwedge Con(D') \in Cn(\mathcal{W} \cup Con(\mathcal{D}_1 \cup \mathcal{D}_2))$, which contradicts (@). It concludes that $\forall D \subseteq \mathcal{D}' \cup \mathcal{D}'' : \neg \bigwedge Con(D) \in Cn(\mathcal{W} \cup Con(\mathcal{D}_1 \cup \mathcal{D}_2))$.
6. This is proven in a very similar way as the proof of the previous statement.
7. The proof is similar to that for item 5.
8. Similarly, the proof is similar to that for item 5. □

In fact, we will generalize Horty's idea of prima facie reasons into the level of arguments, by keeping the core notion of proper scenario for evaluation. We will add up a type of formula $R^{\mathcal{D}',-\mathcal{D}''}_{\triangle}\varphi$, for the idea "Argument $\mathcal{D}'$ is a (strong/weak) reason against $\mathcal{D}''$ of concluding φ," in the mean while, and a type of formula $R^{\mathcal{D}'}_{\triangle}\varphi$ for "Argument $\mathcal{D}'$ is a (strong/weak) reason of concluding φ." Whether such a formula holds is evaluated by the proper scenarios in a given default theory.

Definition 8 (Prima Facie *Reasons*) Let Δ be a default theory, $\mathcal{S}$ a proper scenario of Δ. Then

- Argument $\mathcal{D}'$ is a (strong/weak) prima facie reason against argument $\mathcal{D}''$ of concluding φ at $\mathcal{S}$ in Δ, denoted as $\Delta, \mathcal{S} \models R^{\mathcal{D}',-\mathcal{D}''}_{\triangle}\varphi$, iff

$$\exists D = \mathcal{D}' \cap \mathcal{S} \neq \emptyset \text{ s.t. } \Delta \models (D, \mathcal{D}'')_\triangle\varphi$$

  where $\triangle \in \{\forall, \exists\}$.
- Argument $\mathcal{D}'$ is a (strong/weak) prima facie reason of concluding φ at $\mathcal{S}$ in Δ, denoted as $\Delta, \mathcal{S} \models R^{\mathcal{D}'}_{\triangle}\varphi$, iff

$$\exists \mathcal{D}'' \subseteq \mathcal{D} \text{ s.t. } \Delta, \mathcal{S} \models R^{\mathcal{D}',-\mathcal{D}''}_{\triangle}\varphi$$

  where $\triangle \in \{\forall, \exists\}$.

The denotations D, D', $\forall$, $\exists$ can be ignored when no confusion is arisen.

Notice that our prima facie reasons in the level of arguments still fit the intuition of Horty's reasons in default theory. The hard core of both types of reasons is that the arguments/rules are contained in certain proper scenarios.

Indeed our reasons can be naturally treated as agent-relative by encoding agent-relative arguments. Let us illustrate this in our example again. Although the local tax office a has an argument $\mathcal{D}_a$ against $\mathcal{D}_c$ of the company c of concluding the case that c does not file the CIT tax return, namely $(\mathcal{D}_a, \mathcal{D}_c)\neg t$, this argument $\mathcal{D}_a$ is not a reason. But its contra $(\mathcal{D}_c, \mathcal{D}_a)t$ is. In addition, the argument $\mathcal{D}_c$ of the company c against $\mathcal{D}_a$ of concluding the case that c does benefit the preferential tax rate $\neg r$, namely $(\mathcal{D}_c, \mathcal{D}_a)\neg r$, is also another reason. $\mathcal{S} = \mathcal{D}_c$ is the only proper scenario in Example 1, and $\mathcal{D}_a$ is conflicted with it. So we get that $\Delta, \mathcal{S} \models R_{\triangle}^{\mathcal{D}_c, -\mathcal{D}_a} t$ and $\Delta, \mathcal{S} \models R_{\triangle}^{\mathcal{D}_c, -\mathcal{D}_a} \neg r$.

Observe that our prima facie reasons w.r.t two arguments are still closed under conjunction and disjunction, which are shown as follow:

Proposition 2 *Let* Δ *be a default theory and* $\mathcal{S}$ *a proper scenario of* Δ. *Then*

1. $\Delta, \mathcal{S} \models R_{\triangle}^{\mathcal{D}', -\mathcal{D}''} \varphi \wedge R_{\triangle}^{\mathcal{D}', -\mathcal{D}''} \psi \Rightarrow \Delta, \mathcal{S} \models R_{\triangle}^{\mathcal{D}', -\mathcal{D}''} (\varphi \wedge \psi)$
2. $\Delta, \mathcal{S} \models R_{\triangle}^{\mathcal{D}', -\mathcal{D}''} \varphi \vee R_{\triangle}^{\mathcal{D}', -\mathcal{D}''} \psi \Rightarrow \Delta, \mathcal{S} \models R_{\triangle}^{\mathcal{D}', -\mathcal{D}''} (\varphi \vee \psi)$
3. $\Delta, \mathcal{S} \models R_{\triangle}^{\mathcal{D}', -\mathcal{D}''} \varphi \wedge R_{\triangle}^{\mathcal{D}', -\mathcal{D}''} \psi \Rightarrow \Delta, \mathcal{S} \models R_{\triangle}^{\mathcal{D}', -\mathcal{D}''} (\varphi \vee \psi)$

Proof Our proof is a simple application of Proposition 1. Please see it as follows:

1. The closure of conjunction is proven as follows:

$$\begin{aligned}
&\Delta, \mathcal{S} \models R_{\triangle}^{\mathcal{D}', -\mathcal{D}''} \varphi \wedge R_{\triangle}^{\mathcal{D}', -\mathcal{D}''} \psi \\
\Leftrightarrow\ & \exists \mathcal{D}_1 = \mathcal{D}' \cap \mathcal{S} \neq \emptyset \text{ s.t. } \Delta \models (\mathcal{D}_1, \mathcal{D}'')_{\triangle} \varphi \\
& \text{and } \exists \mathcal{D}_2 = \mathcal{D}' \cap \mathcal{S} \neq \emptyset \text{ s.t. } \Delta \models (\mathcal{D}_2, \mathcal{D}'')_{\triangle} \psi \\
\Rightarrow\ & \exists (\mathcal{D}_1 \cup \mathcal{D}_2) \cap \mathcal{S} \neq \emptyset \text{ and } \Delta \models (\mathcal{D}_1 \cup \mathcal{D}_2, \mathcal{D}'')_{\triangle} (\varphi \wedge \psi) \qquad \text{by Proposition 1} \\
\Leftrightarrow\ & \Delta, \mathcal{S} \models R_{\triangle}^{\mathcal{D}', -\mathcal{D}''} (\varphi \wedge \psi)
\end{aligned}$$

2. The case of the closure of disjunction:

$$\begin{aligned}
&\Delta, \mathcal{S} \models R_{\triangle}^{\mathcal{D}', -\mathcal{D}''} \varphi \vee R_{\triangle}^{\mathcal{D}', -\mathcal{D}''} \psi \\
\Leftrightarrow\ & \exists \mathcal{D}_1 = \mathcal{D}' \cap \mathcal{S} \neq \emptyset \text{ s.t. } \Delta \models (\mathcal{D}_1, \mathcal{D}'')_{\triangle} \varphi \text{ or} \\
& \exists \mathcal{D}_2 = \mathcal{D}' \cap \mathcal{S} \neq \emptyset \text{ s.t. } \Delta \models (\mathcal{D}_2, \mathcal{D}'')_{\triangle} \psi \\
\Rightarrow\ & \exists (\mathcal{D}_1 \cup \mathcal{D}_2) \cap \mathcal{S} \neq \emptyset \text{ and } \Delta \models (\mathcal{D}_1 \cup \mathcal{D}_2, \mathcal{D}'')_{\triangle} (\varphi \vee \psi) \qquad \text{by Proposition 1} \\
\Leftrightarrow\ & \Delta, \mathcal{S} \models R_{\triangle}^{\mathcal{D}', -\mathcal{D}''} (\varphi \vee \psi)
\end{aligned}$$

3. By using the strategy in the first proposition, this can be proven straightforward. □

Both strong and weak agent-relative reasons satisfy the so-called "agglomeration" principle (Horty 2012, Chap. 4) for groups of agents. That is to say, having both $\psi \xrightarrow{\delta_i} \varphi$ and $\psi' \xrightarrow{\delta_j} \varphi'$ entails that $\varphi \wedge \varphi'$ is still in the conclusion of this proper scenario. In addition, our agent-relative reasons follow a principle of distribution. Either $\psi \xrightarrow{\delta_i} \varphi$ or $\psi' \xrightarrow{\delta_j} \varphi'$ can entail that $\varphi \vee \varphi'$ as a conclusion of the same proper scenario.

Similarly the prima facie reasons w.r.t single argument are also closed under these principles of agglomeration and distribution.

Proposition 3 *Let Δ be a default theory and $\mathcal{S}$ a proper scenario of Δ. Then*

1. $\Delta, \mathcal{S} \models R^{\mathcal{D}'}_{\triangle}\varphi \wedge R^{\mathcal{D}'}_{\triangle}\psi \Rightarrow \Delta, \mathcal{S} \models R^{\mathcal{D}'}_{\triangle}(\varphi \wedge \psi)$
2. $\Delta, \mathcal{S} \models R^{\mathcal{D}'}_{\triangle}\varphi \vee R^{\mathcal{D}'}_{\triangle}\psi \Rightarrow \Delta, \mathcal{S} \models R^{\mathcal{D}'}_{\triangle}(\varphi \vee \psi)$
3. $\Delta, \mathcal{S} \models R^{\mathcal{D}'}_{\triangle}\varphi \wedge R^{\mathcal{D}'}_{\triangle}\psi \Rightarrow \Delta, \mathcal{S} \models R^{\mathcal{D}'}_{\triangle}(\varphi \vee \psi)$

Proof We prove the results by using Proposition 1 again.

1. In the case of the closure of conjunction, the left-to-right direction is proved as follows:

$$
\begin{aligned}
& \Delta, \mathcal{S} \models R^{\mathcal{D}'}_{\triangle}\varphi \wedge R^{\mathcal{D}'}_{\triangle}\psi \\
\Leftrightarrow\ & \exists\mathcal{D}'' \subseteq \mathcal{D} \text{ s.t. } \Delta, \mathcal{S} \models R^{\mathcal{D}', -\mathcal{D}''}_{\triangle}\varphi \text{ and} \\
& \exists\mathcal{D}''' \subseteq \mathcal{D} \text{ s.t. } \Delta, \mathcal{S} \models R^{\mathcal{D}', -\mathcal{D}'''}_{\triangle}\psi \\
\Leftrightarrow\ & \exists\mathcal{D}'' \subseteq \mathcal{D}\exists\mathcal{D}_1 = \mathcal{D}' \cap \mathcal{S} \neq \emptyset \text{ s.t. } \Delta \models (\mathcal{D}_1, \mathcal{D}'')_{\triangle}\varphi \text{ and} \\
& \exists\mathcal{D}''' \subseteq \mathcal{D}\exists\mathcal{D}_2 = \mathcal{D}' \cap \mathcal{S} \neq \emptyset \text{ s.t. } \Delta \models (\mathcal{D}_2, \mathcal{D}''')_{\triangle}\psi \\
\Rightarrow\ & \exists\mathcal{D}'' \cup \mathcal{D}''' \subseteq \mathcal{D}\exists(\mathcal{D}_1 \cup \mathcal{D}_2) = \mathcal{D}' \cap \mathcal{S} \neq \emptyset \text{ s.t.} \\
& \Delta \models (\mathcal{D}_1 \cup \mathcal{D}_2, \mathcal{D}'' \cup \mathcal{D}''')_{\triangle}(\varphi \wedge \psi) \qquad \text{by Proposition 1} \\
\Leftrightarrow\ & \exists\mathcal{D}'' \cup \mathcal{D}''' \subseteq \mathcal{D} \text{ s.t. } \Delta, \mathcal{S} \models R^{\mathcal{D}', -\mathcal{D}'' \cup \mathcal{D}'''}_{\triangle}(\varphi \wedge \psi) \\
\Leftrightarrow\ & \Delta, \mathcal{S} \models R^{\mathcal{D}'}_{\triangle}(\varphi \wedge \psi)
\end{aligned}
$$

2. The case of the closure of disjunction:

$$
\begin{aligned}
& \Delta, \mathcal{S} \models R^{\mathcal{D}'}_{\triangle}\varphi \vee R^{\mathcal{D}'}_{\triangle}\psi \\
\Leftrightarrow\ & \exists\mathcal{D}'' \subseteq \mathcal{D} \text{ s.t. } \Delta, \mathcal{S} \models R^{\mathcal{D}', -\mathcal{D}''}_{\triangle}\varphi \text{ or} \\
& \exists\mathcal{D}''' \subseteq \mathcal{D} \text{ s.t. } \Delta, \mathcal{S} \models R^{\mathcal{D}', -\mathcal{D}'''}_{\triangle}\psi \\
\Leftrightarrow\ & \exists\mathcal{D}'' \subseteq \mathcal{D}\exists\mathcal{D}_1 = \mathcal{D}' \cap \mathcal{S} \neq \emptyset \text{ s.t. } \Delta \models (\mathcal{D}_1, \mathcal{D}'')_{\triangle}\varphi \text{ or} \\
& \exists\mathcal{D}''' \subseteq \mathcal{D}\exists\mathcal{D}_2 = \mathcal{D}' \cap \mathcal{S} \neq \emptyset \text{ s.t. } \Delta \models (\mathcal{D}_2, \mathcal{D}''')_{\triangle}\psi \\
\Rightarrow\ & \exists\mathcal{D}'' \cup \mathcal{D}''' \subseteq \mathcal{D}\exists(\mathcal{D}_1 \cup \mathcal{D}_2) = \mathcal{D}' \cap \mathcal{S} \neq \emptyset \text{ s.t.} \\
& \Delta \models (\mathcal{D}_1 \cup \mathcal{D}_2, \mathcal{D}'' \cup \mathcal{D}''')_{\triangle}(\varphi \vee \psi) \qquad \text{by Proposition 1} \\
\Leftrightarrow\ & \exists\mathcal{D}'' \cup \mathcal{D}''' \subseteq \mathcal{D} \text{ s.t. } \Delta, \mathcal{S} \models R^{\mathcal{D}', -\mathcal{D}'' \cup \mathcal{D}'''}_{\triangle}(\varphi \wedge \psi) \\
\Leftrightarrow\ & \Delta, \mathcal{S} \models R^{\mathcal{D}'}_{\triangle}(\varphi \wedge \psi)
\end{aligned}
$$

3. Following the strategy of proving the 1st proposition by using Proposition 1, we can have this result. □

To define agent-relative obligations we simply use prima facie reasons to redefine the disjunctive and the conflict accounts:

Definition 9 *(Agent-Relative Obligations—Conflict and Disjunctive Accounts)*

- $\Delta \models O_i^c \varphi$, iff $\Delta, \mathcal{S} \models R^{\mathcal{D}_i} \varphi$ for some proper scenario S of Δ
- $\Delta \models O_i^d \varphi$, iff $\Delta, \mathcal{S} \models R^{\mathcal{D}_i} \varphi$ for all proper scenarios S of Δ

When in a given theory Δ the conflict and the disjunctive account agree on all φ, we just write $\Delta \models O_i \varphi$.

Consider the Chinese tax example again. $R_{\triangle}^{\mathcal{D}_c, -\mathcal{D}_a} t$ and $R_{\triangle}^{\mathcal{D}_c, -\mathcal{D}_a} \neg r$ hold as reasons at the only proper scenario $\mathcal{S}$ in Example 1. These translate directly into obligations: $\Delta \models O_c t$ and $\Delta \models O_c \neg r$.

In accordance with the previous propositions of reasons, it can be observed that our agent-relative obligations satisfy the principles of agglomeration and distribution in the multi-agent setting. This phenomenon is displayed in the following propositions:

Proposition 4 *Let Δ be a default theory and $\mathcal{S}$ a proper scenario of Δ. Then*

1. $\Delta \models O_i \varphi \wedge O_i \psi \Leftrightarrow \Delta \models O_i(\varphi \wedge \psi)$
2. $\Delta \models O_i \varphi \vee O_i \psi \Rightarrow \Delta \models O_i(\varphi \vee \psi)$
3. $\Delta \models O_i \varphi \wedge O_i \psi \Rightarrow \Delta \models O_i(\varphi \vee \psi)$

Proof These results are naturally derivation by using Proposition 3. □

Accordingly, we can find out that some standard behaviors of neutral obligation also appear in the case of agent-relative obligation. This is because what agent-relative reasons concluding are closed under the propositional logic consequence.

Proposition 5 *Let Δ be a default theory and $\mathcal{S}$ a proper scenario of Δ. Then*

1. *$\Delta \models O_i \varphi$ does imply $\Delta \models O_i(\varphi \vee \psi)$*
2. *$\Delta \models \neg(O_i \varphi \wedge O_j \neg\varphi)$.*

Proof 1. This can be implied from Proposition 4.
2. In any given proper scenario, there is no way for any set of rules in it to conclude contrary conclusions. □

However, our agent-relative obligations, although following the properties of agglomeration and distribution, do not behave like obligations in SDL. Because every reasons for agent-relative obligations are constituted by proper scenarios, which gives the following property:

Proposition 6 *Let Δ be a default theory and $\mathcal{S}$ a proper scenario of Δ. Then*

- *$\Delta \models O_i \top$ is not always true.*

Proof Take the empty set of default rules as a proper scenario into account. As there is no such a rule for i is contained in $\mathcal{S}$, we can see that $\Delta \models O_i \top$ fails. We cannot find any rules in an empty proper scenario to conclude tautology as the conjunction of their conclusions. □

As further logical property, it should be clear that any empty intersection with proper scenarios do not entail prima facie reasons.

Proposition 7 *Let Δ be a default theory and $\mathcal{S}$ a proper scenario of Δ. Then*

1. $\Delta \models R^{\mathcal{D}_i}\varphi$ implies $\Delta \models O_i\varphi$

Proof This is proven according to the definition of agent-relative obligations.

This property is reminiscent of, but different from, a well-known fact in default theories, namely that "what is, ought to be": If Δ is a default theory and $\varphi \in \mathcal{W}$ then $\Delta \models O\varphi$. This property holds for agent-neutral obligations here, but not for agent-relative ones, because unless a background assumption $\varphi \in \mathcal{W}$ is also the conclusion of an argument $\mathcal{D}'$ in a proper scenario $\mathcal{S}$ it will not appear as a reason $R^{\mathcal{D}_i}\varphi$.

2 Three Types of Permissions

In this section we consider three types of permissions that can be defined in default theories: weak, explicit and tacit permissions. This classification, and the intuitions underlying each of them, come from Makinson and van der Torre (2003).

In doing so, we emphasize how important is the multi-agent aspect for the representation of the standard correlative relations between claim rights and duties. We furthermore compare the resulting permissions with weak permissions in SDL, strong permission in von Wright's sense, and free choice permission.

2.1 Weak Permissions

We start with weak permissions, the usual dual of obligations. Something is weakly permitted whenever it is not the case that it is forbidden, i.e. that its negation is mandatory. This type of permission is of course definable in Horty's prioritized default theory, and its logical behavior will depend on whether it is defined as the dual of the disjunctive or the conflict account.

Definition 10 (*Weak Permissions*) Let Δ be a default theory with $\mathcal{D}$ for conclusive norms. Then we can define the conflict and disjunctive accounts of agent-relative weak permission:

- $\Delta \models P_i^c\varphi$ iff $\Delta, \mathcal{S} \models \neg R^{\mathcal{D}_i}\neg\varphi$ for all proper scenarios $\mathcal{S}$ based on Δ
- $\Delta \models P_i^d\varphi$ iff $\Delta, \mathcal{S} \models \neg R^{\mathcal{D}_i}\neg\varphi$ for some proper scenarios $\mathcal{S}$ based on Δ

Similar to agent-relative obligations, we write $\Delta \models P_i\varphi$ when one unique proper scenario can be generated from Δ.

Let us illustrate these notions of permissions in our running example. Recall that $\mathcal{S} = \mathcal{D}_c$ is the only proper scenario above, and that we have $\Delta \models O_c t$ and $\Delta \models O_a \neg r$. From that we get that $\Delta \models P_c t$ and $\Delta \models P_c \neg r$, but also $\Delta \models P_c d$, as expected.

Looking a bit further we do find, however, counter-intuitive weak permissions, which speak in favor of looking at stronger notions. We do get, for instance, that $\Delta \models P_a d$, which might appear unwelcome since only the company but not the tax authority is permitted to submit the document. We rather see it as revealing how weak is the notion of weak permission, and showing the need for a stronger, explicit notion. Weak permissions do not distinguish between, on the one hand, an action not being forbidden for an agent because there are *some* reasons, possibly not conclusive, for doing it and, on the other hand, the same action not being forbidden simply because reasons are silent when it comes to that agent doing that action. We come back to this in the next section.

Now for the logical properties of weak permissions. As expected from the definition, P_i^d and P_i^c are the respective duals of O_i^d and O_i^c. But just as for agent-relative obligations, few properties of the "diamond" P in SDL are retained here.

Proposition 8 *Let Δ be a default theory and $\mathcal{S}$ a proper scenario of Δ. Then*[2]

1. $\Delta \models O_i\varphi \leftrightarrow \neg P_i \neg\varphi$.
2. $\Delta \models O_i\varphi \rightarrow P_i\varphi$

The results above state that the weak permission satisfies the standard properties in SDL: it is the dual of obligation, and also it can be implied from obligation.

Weak permission further satisfies the so-called properties of "agglomeration" and "distribution" in a way similar to the agent-relative obligation, which gives some standard behavior in SDL.

Proposition 9 *Let Δ be a default theory and $\mathcal{S}$ a proper scenario of Δ. Then*

1. $\Delta \models P_i(\varphi \wedge \psi) \Rightarrow \Delta \models P_i\varphi \wedge P_i\psi$
2. $\Delta \models P_i(\varphi \vee \psi) \Leftrightarrow \Delta \models P_i\varphi \vee P_i\psi$

Proof They can be verified according to Proposition 4.

2.2 *Explicit Permissions*

Explicit permissions, as their name suggest, are permissions that are explicitly stated in a moral or legal code. These are the strong permissions in the sense of von Wright (1968, p. 31). Here we follow (Makinson and van der Torre 2003) and stay at the intuitive level. Intuitively, the law-giver can forbid φ without further legal

[2]Here obligation and permission have to be in the same account: either both in the conflict account or both in the disjunctive account.

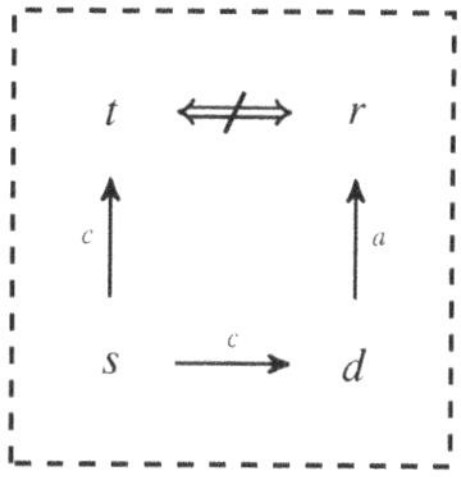

Fig. 2 A default theory of a solar energy business c's claim to the right to a tax incentive against the local tax authority

changes if φ is only weakly permitted, but not if it is explicitly permitted. This intuitive difference is also present in our running example. The permission for the foreign company c to submit the documents is explicit, while the same weak permission for the tax authority a is not. Such an example suggests the need to distinguish explicit from weak permissions.

In order to capture explicit permissions we extend Horty's account by splitting the set $\mathcal{D}$ into a set of permissive norms $\mathcal{P}$ and a set of mandatory norms O, with $<$ the same partial order on $\mathcal{D}$ as usual.

We illustrate this idea through our running example. The explicit permission here is the following: the solar energy business *may* submit the documentation if it is the tax season. This permissive norm takes also the form of a default $s \overset{\delta'_c}{\rightarrow} d$ for the company c. This default defeats the regular duty to pay full income tax. Formally, we get the default theory $\Delta'_{CIT} = \langle \mathcal{W}, \mathcal{I}, O \cup \mathcal{P}, < \rangle$ defined as follows: $\mathcal{W} = \{s, \neg t \Leftrightarrow r\}$, $\mathcal{I} = \{a, c\}, O = \{s \overset{\delta_c}{\rightarrow} t, d \overset{\delta_a}{\rightarrow} r\}, \mathcal{P} = \{s \overset{\delta'_c}{\rightarrow} d\}$, with the priority $\delta_c < \delta_a$, reflecting the priority of tax reduction. See Fig. 2. With the addition of this permissive norm, the unique proper scenario becomes $\{\delta'_c, \delta_a\}$, as desired.

Now we turn to the formal definition.

Definition 11 (*Explicit Permissions*) Given $\Delta = \langle \mathcal{W}, \mathcal{I}, O \cup \mathcal{P}, < \rangle$, we say that $P_e\varphi$ is an explicit permission on this default theory, denoted as $\Delta \models P^e_i\varphi$, iff

$$\Delta, \mathcal{S} \models R^{\mathcal{P}_i}\varphi \text{ for some proper scenarios } \mathcal{S} \text{ based on } \Delta$$

For example, with the permissive norm $s \overset{\delta'_c}{\rightarrow} d$ at hand, we get the explicit permission $P^e_c d$, indicating that the solar energy company may submit the valid documentation. Interestingly, the tax authority then has a correlative duty to grant the tax reduction. So this explicit permission $P^e_c d$ can be seen as a claim right, in Hohfeld's sense.

Explicit permissions give more intuitive results in this case than weak permissions. Recall that, as Fig. 2 shows, since there is no conclusive reason for the company that forbids granting the tax reduction, we obtained the rather counterintuitive result: it is a weak permission for the company to reduce the tax rate. But there is no such explicit permission here, for the simple reason that this action may only be implemented by the tax authority. So weak permissions are too weak to model this important piece of Chinese Law.

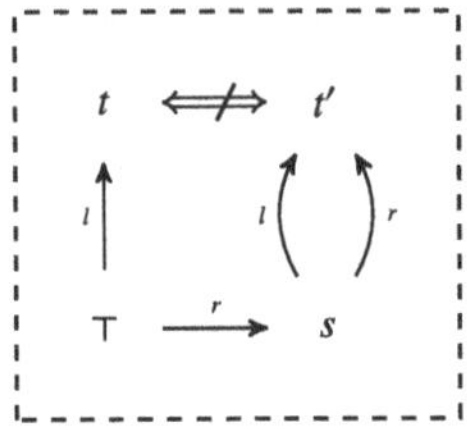

Fig. 3 A default theory of the driver r's signal for turning right being a claim right overriding driver l's permission to turn left

The Chinese Law example shows that there can be a weak permission without an explicit one. Now we should see that the converse is also possible, by looking back at the Right of Way example. In this example all the defaults are permissive. Suppose that l is the driver who wants to turn left at an intersection on the priority street, r is the driver who signals to turn right at the same intersection of this street, t means that driver l's turns left, s means that driver r signals to turn right, and t' means that driver r turns right. We have three permissive norms in this case: $\top \overset{\delta_l}{\rightarrow} t$, $\top \overset{\delta_r}{\rightarrow} s$, and $s \overset{\delta'_r}{\rightarrow} t$'. This situation can then be captured in the following default theory: $\Delta = \langle \mathcal{W}, \mathcal{I}, \mathcal{O} \cup \mathcal{P}, < \rangle$, where $\mathcal{W} = \{\top, t' \Leftrightarrow \neg t\}$, $\mathcal{O} = \{s \overset{\delta''_l}{\rightarrow} t\}$, and $\mathcal{P} = \{\top \overset{\delta_l}{\rightarrow} t, \top \overset{\delta_r}{\rightarrow} s, s \overset{\delta'_r}{\rightarrow} t'\}$ with $\delta_l < \delta'_r$, representing that the driver turning right has priority. See Fig. 3. In this default theory, the unique proper scenario is $\{\delta_r, \delta'_r, \delta''_l\}$, including permissive norms, which expresses that the driver r has the priority of turning right, as well as that the driver r has an explicit permission to turn right $P^e_r t'$, and the unique mandatory norm, which states that the driver l has the duty to let the other to turn right. However, he does not own a weak permission $P^c_r s$ to do so, for the simple reason that the set $\mathcal{O}_r$ for his conclusive reasons is empty.

Similarly to weak permission, explicit permission also satisfies the "agglomeration" and "distribution" properties. Though explicit permission and weak permission both have similar logical behavior, they are two independent categories of permission, as the previous arguments have shown, which do bring about different normative consequences in legal reasoning.

Proposition 10 *Let Δ be a default theory and $\mathcal{S}$ a proper scenario of Δ. Then*

1. $\Delta \models P^e_i \varphi \wedge P^e_i \psi \Leftrightarrow \Delta \models P^e_i(\varphi \wedge \psi)$
2. $\Delta \models P^e_i \varphi \vee P^e_i \psi \Rightarrow \Delta \models P^e_i(\varphi \vee \psi)$
3. $\Delta \models P^e_i \varphi \wedge P^e_i \psi \Rightarrow \Delta \models P^e_i(\varphi \vee \psi)$
4. $\Delta \models \neg(O_i \varphi \wedge P^e_j \neg\varphi)$.
5. $\Delta \models P^e_i \varphi \rightarrow P_i \varphi$.

Proof The first three statements can be proved by the similar strategy in Proposition 4. For the fourth it is not possible to have two reasons to conclude results contradicts each other in the same proper scenario. So this proposition holds. The last one is an implication of the fourth. □

2.3 Tacit Permissions

What we call "tacit permission" follows the basic idea of a "silent" type of permission in Makinson and van der Torre (2003, p. 398). The idea is that if prohibiting φ does cause incoherence, then φ must be permissible in advance. Translated into the present framework, this notion of incoherence bears on the non-defeated permissive norms. In other words, certain permissive norms silently let this permission undefeated by conclusive norms.

Definition 12 (*Tacit Permissions*) Given $\Delta = \langle \mathcal{W}, \mathcal{I}, \mathcal{D} = \mathcal{O} \cup \mathcal{P}, < \rangle$, we say that $P_i^t \varphi$ is a tacit permission based on Δ, denoted as $\Delta \models P_i^t \varphi$, iff for some proper scenarios $\mathcal{S}$ based on Δ s.t.

$$\Delta, \mathcal{S} \models \neg R^{\mathcal{O}_i} \neg \varphi \Rightarrow \Delta, \mathcal{S} \models \bigvee_{j \in \mathcal{I}} R^{\mathcal{P}_j} \varphi$$

Let us illustrate this idea in our two running examples. In the tax reduction case, we get that $P_c^t r$, which means that the permission for the company of granting the tax reduction by the authority is tacit. For the Right of Way example, driver r's permission to turn right is also tacit. This should show that tacit permission is neither weak nor explicit permission.

Observe that tacit permission has a complicate relation with weak and explicit permission. Also, an explicit permission should be a tacit one.

Proposition 11 *Let Δ be a default theory and $\mathcal{S}$ a proper scenario of Δ. Then*

1. $\Delta \models P_i^t \varphi \leftrightarrow (P_i \varphi \rightarrow \bigvee_{j \in \mathcal{I}} P_j^e \varphi)$.
2. $\Delta \models P_i^e \varphi \rightarrow P_i^t \varphi$.

Proof These results can be proven by the definition of tacit permission.

2.4 A Fourth Type: Protected Permissions

Having looked at weak, explicit and tacit permission, we now turn to notions of permissions arising from legal theory. This combines in a natural way the permissive norms introduced earlier and our multi-agent version of default theory. The crucial notion here is that of correlative permissions and duties. This can be illustrated in our Right of Way example. There, one's permissive right to turn at the crossroad correlates to other(s)' duty not to interfere. In short, this permissive right is protected. This reverses the standard relation between obligation and permission. As already mentioned in the introduction, in SDL obligation implies (weak) permission. This was the consistency requirement. With correlative duties this is the other way around: permission implies obligation. The so-called "protected permission" (Wenar et al. 2015; Raz 1994) generates obligation.

Definition 13 (*Protections: Implicit Duties*) Let a default theory Δ be a tuple of $\langle \mathcal{W}, \mathcal{I}, \mathcal{D} = \mathcal{O} \cup \mathcal{P}, < \rangle$. We say that $P_i^r \varphi$ is a protection or an implicit duty, denoted as $\Delta \models P_i^r \varphi$, iff

$$\Delta, \mathcal{S} \models R^{\mathcal{P}_i} \varphi \rightarrow \bigwedge_{j \in \mathcal{I}^{-i}} R^{\mathcal{O}_j} \varphi \text{ for some proper scenarios } \mathcal{S}' \text{ based on } \Delta$$

where $\mathcal{I}^{-i} = \mathcal{I} - \{i\}$.

Protected permission can be seen as an implicit type of duty, because it aims at ensuring the other's explicit permissions so as to be defended by all the others. That is to say, a protected permission indicates that such a permissive norm implies the correlative conclusive ones.

Proposition 12 *Given* $\Delta = \langle \mathcal{W}, \mathcal{I}, \mathcal{D} = \mathcal{O} \cup \mathcal{P}, < \rangle$, *we have*

$$\Delta \models P_i^r \varphi \leftrightarrow (P_i^e \varphi \rightarrow \bigwedge_{j \in \mathcal{I}^{-i}} O_j \varphi)$$

This type of protected permission is particularly salient in the Right of Way example. We have seen that $P_r^e t'$ is an explicit permission for the driver r to turn right. In terms of protected permission, we can then conclude that the other driver should not interfere with driver r's turning right.

3 Conclusion

This paper develops a formal theory for a topology of permissions by extending Horty's default theory. As a result, it is demonstrated that this formal theory is an appropriated tool for the analysis of two important pieces of Chinese and German Laws. Comparing to the pioneered Makinson and van der Torre work of various permissions, the permissions in our work are mutually exclusive, and each has its own significant logical property in Deontic Logic.

Aside from this, our formal theory has highlighted the multi-agent aspect in legal reasoning by the so-called attacking arguments (Definition 7). This turns to be an advantage for characterizing Hohfeldian correlative relations of claims and rights. We prove this viewpoint by displaying two case studies in explicit permission. However, a further argument for supporting this characterization should be given, in order to build up a firm foundation to link our normative concepts with Hohfeldian legal rights. We leave this to our future research.

Finally, our formal theory relies on the well known notion *defeasibility* in default theory and in argumentation theory. But two issues are left in our argument for the reason of lack of space. The first requires us a comparison of the representation of our *rebutting* defeater with the so-called *undercutting* defeater (Horty 2012) in analyzing legal cases. See Dong (2017) for a further reply to this issue, as well as a response to

Raz (1999) and Alchourrón and Bulygin (1981). The second issue concerns on the relation between our formal default theory and Grossi's attack semantics in modal logic (Grossi 2013). Right after Definition 7 we have mentioned a close relation between these two theories. For a precise comparison, we leave it for the future work.

Acknowledgements Huimin Dong is supported by the China Postdoctoral Science Foundation funded project [No. 2018M632494], the National Science Centre of Poland [No. UMO-2017/26/M/HS1/01092], the National Social Science Fund of China [No. 18ZDA290, No. 17ZDA026, No. 16AZX017], the MOE Project of Key Research Institute of Humanities and Social Sciences in Universities [No. 17JJD720008], and the Fundamental Research Funds for the Central Universities of China.

References

Alchourrón, C.E., and E. Bulygin. 1981. The expressive conception of norms. In *New Studies in Deontic Logic*, 95–124. Springer.
Broome, J. 2013. *Rationality Through Reasoning*. Wiley.
Dong, H. 2017. *Permission in Non-Monotonic Normative Reasoning*. PhD thesis, University of Bayreuth.
Grossi, D. 2013. Abstract argument games via modal logic. *Synthese* 190 (1): 5–29.
Hansson, S.O. 2001. *The Structure of Values and Norms*. Cambridge University Press.
Hansson, S.O. 2013. The varieties of permissions. In *Handbook of Deontic Logic and Normative Systems*, vol. 1, ed. D. Gabbay, J. Horty, X. Parent, R. van der Meyden, and L. van der Torre. College Publication.
Hilpinen, R., and P. MaNamara. 2013. Deontic logic: A historical survey and introduction. In *Handbook of Deontic Logic and Normative Systems*, vol. 1, ed. D. Gabbay, J. Horty, X. Parent, R. van der Meyden, and L. van der Torre. College Publication.
Hohfeld, W.N. 1913. Some fundamental legal conceptions as applied in judicial reasoning. *The Yale Law Journal* 23 (1): 16–59.
Horty, J.F. 2000. *Agency and Deontic Logic*. Oxford University Press.
Horty, J.F. 2011. Rules and reasons in the theory of precedent.
Horty, J.F. 2012. *Reasons as Defaults*. Oxford University Press.
Horty, J.F. 2013. Common law reasoning.
Makinson, D. 2003. Bridges between classical and nonmonotonic logic. *Logic Journal of IGPL* 11 (1): 69–96.
Makinson, D., and L. van der Torre. 2003. Permission from an input/output perspective. *Journal of Philosophical Logic* 32 (4): 391–416.
Parent, X., and L. van der Torre. 2013. Input/output logic. In *Handbook of Deontic Logic and Normative Systems*, vol. 1, ed. D. Gabbay, J. Horty, X. Parent, R. van der Meyden, and L. van der Torre. College Publication.
Raz, J. 1994. *Ethics in the Public Domain*. Oxford: Clarendon Press.
Raz, J. 1999. *Practical Reason and Norms*. OUP Oxford.
von Wright, G.H. 1968. *An Essay in Deontic Logic and the General Theory of Action*. North-Holland Publishing Company.
Wenar, L. 2015. Rights. In *The Stanford Encyclopedia of Philosophy*, ed. E.N. Zalta. Metaphysics Research Lab, Stanford University, fall 2015 edition.

Algebraic Semantics for Hybrid Logics

Willem Conradie and Claudette Robinson

Abstract We introduce hybrid algebras as algebraic semantics for hybrid languages with nominals and, possibly, the satisfaction operator. We establish a duality between hybrid algebras and the descriptive two-sorted general frames of ten Cate. We show that all axiomatic extensions of the basic hybrid logics, with or without the satisfaction operator, are complete with respect to their classes of hybrid algebras. Moreover, we show that by adding the usual non-orthodox rules to these logics, they become complete with respect to their classes of permeated hybrid algebras, corresponding to strongly descriptive two-sorted general frames.

Keywords Hybrid logic · Hybrid algebras · Algebraic semantics · Descriptive two-sorted general frames · Completeness

1 Introduction

Hybrid logics (Gargov and Goranko 1993; Blackburn 2000b; Areces and ten Cate 2006) extend modal logic with a second sort of atomic formulas, known as *nominals*, which are constrained to range over singleton subsets of Kripke frames and thus act as names for states in models. The expressive power of hybrid languages is further

W. Conradie (✉)
School of Mathematics, University of the Witwatersrand,
Johannesburg, South Africa
e-mail: willem.conradie@wits.ac.za

C. Robinson
Department of Pure and Applied Mathematics, University of Johannesburg,
Johannesburg, South Africa
e-mail: claudetter@uj.ac.za

S. Ju et al. (eds.), *Nonclassical Logics and Their Applications*,
Logic in Asia: Studia Logica Library,
https://doi.org/10.1007/978-981-15-1342-8_6

enhanced by the addition of various other connectives which capitalize on this naming power of the nominals.

Historically, hybrid logics can be traced back to the work of Arthur Prior (1957a, b). They were given their present form by Gargov and Goranko (1993), and have seen rapid development since the late 1990s.

In recent years modal logic has benefited greatly from the use of algebraic methods and the utilization of the duality between its relational and algebraic semantics (see e.g. Goldblatt 1989; Venema 2006). Strangely enough, given that modal and hybrid logic are such close cousins, there has been very little work on algebraic semantics for the latter. The only work in this direction of which we are aware is that of Litak (2006), who provides algebraic semantics for a very expressive hybrid language, which contains the 'down arrow' binder. The algebras introduced by Litak are akin the cylindric algebras used as semantics for first-order logic.

In this paper we consider hybrid languages without binders and, as a result, our algebras are much simpler. In fact, the algebraic semantics we introduce will be based on structures we call *hybrid algebras*, which are adapted versions of the Boolean algebras with operators (BAOs) which form the usual algebraic semantics of modal logic.

We consider two hybrid languages, namely, the language obtained by adding nominals to the basic modal language and the language obtained by additionally adding the satisfaction operator @. Our main results are general completeness results for axiomatic extensions of the basic logics associated with these languages, with respect to the corresponding classes of hybrid algebras.

Stepping through the looking glass of duality, the relational duals of hybrid algebras are already to be found in the Ph.D. thesis of Balder ten Cate (2005), in the form of *descriptive two sorted general frames*. Our inspiration for the definitions of hybrid algebras comes from these two sorted general frames. In fact, the completeness results presented in this paper could alternatively be obtained by properly establishing the duality between hybrid algebras and descriptive two sorted general frames, and then appealing to ten Cate's completeness results in ten Cate (2005). However, in our opinion, there is much value in presenting these results purely algebraically: firstly, it shows that a purely algebraic approach to the semantics of hybrid logic is feasible and fruitful; secondly, it provides the opportunity to develop and showcase a number of techniques and constructions on hybrid algebras that will prove very useful when this semantics is used to derive other results for hybrid logics like Sahlqvist-type theorems Conradie and Robinson (2017) and finite model properties Conradie and Robinson (2019).

The paper is organized as follows: Sect. 2 collects the necessary preliminaries on the syntax, relational semantics and axiomatics of the logics under consideration. In Sect. 3 we introduce hybrid algebras and permeated hybrid algebras as semantics for hybrid languages and prove some basic preliminary propositions. In Sect. 4 we prove the completeness of all axiomatic extensions of the basic logics in the hybrid language containing only nominals with respect to the corresponding classes of (permeated) hybrid algebras. In Sect. 5 we do the same for logics with the satisfaction operator and classes of (permeated) hybrid @-algebras. Section 6 considers completeness with respect to the global consequence relation. We conclude in Sect. 7.

2 Preliminaries

In this section we collect some essential preliminaries.

Syntax. Fix countably infinite disjoint sets PROP and NOM of *propositional variables* (denoted $p, q, r, \ldots$) and *nominals* (denoted $\mathbf{i}, \mathbf{j}, \mathbf{k}, \ldots$), respectively. Then the syntax of the languages $\mathcal{H}$ and $\mathcal{H}(@)$ is defined as follows:

$$\varphi ::= \bot \mid p \mid \mathbf{j} \mid \neg\varphi \mid \varphi \wedge \psi \mid \Diamond\varphi$$
$$\varphi ::= \bot \mid p \mid \mathbf{j} \mid \neg\varphi \mid \varphi \wedge \psi \mid \Diamond\varphi \mid @_{\mathbf{j}}\varphi$$

Here $p \in \mathsf{PROP}$ and $\mathbf{j} \in \mathsf{NOM}$. The Boolean connectives $\top, \vee, \rightarrow$ and $\leftrightarrow$ are defined as usual, and as usual, $\Box\varphi := \neg\Diamond\neg\varphi$.

Relational semantics. Like modal languages, the languages $\mathcal{H}$ and $\mathcal{H}(@)$ can be interpreted in various related structures. We will be concerned with Kripke frames, two sorted general frames, models, and hybrid algebras (to be defined later). A *(Kripke) frame* is a pair $\mathfrak{F} = (W, R)$ such that W is a non-empty set (called the *domain*) of objects called states, and $R \subseteq W^2$ is a binary *accessibility relation* on W. A *(Kripke) model* based on a frame $\mathfrak{F} = (W, R)$ is a pair $\mathfrak{M} = (\mathfrak{F}, V)$, where V is a *valuation* $V : \mathsf{PROP} \cup \mathsf{NOM} \rightarrow \mathcal{P}(W)$ such that $V(\mathbf{i})$ is a singleton subset of W.

The *truth* of a formula φ at a state w in a Kripke model $\mathfrak{M} = (W, R, V)$, denoted $\mathfrak{M}, w \Vdash \varphi$, is defined as usual. In particular, $\mathfrak{M}, w \Vdash \mathbf{i}$ iff $V(\mathbf{i}) = \{w\}$; $\mathfrak{M}, w \Vdash \Diamond\varphi$ iff there exists v such that wRv and $\mathfrak{M}, v \Vdash \varphi$; $\mathfrak{M}, w \Vdash @_{\mathbf{j}}\varphi$ iff $\mathfrak{M}, v \Vdash \varphi$ and $V(\mathbf{j}) = \{v\}$; Global truth in a model and local and global validity in frames are defined as usual.

Recall that a general frame for the basic modal language (see e.g. Blackburn et al. 2001) is a triple $\mathfrak{g} = (W, R, A)$ such that (W, R) is a Kripke frame, A is a non-empty collection of subsets of W (called *admissible subsets*) which is closed under finite intersection, relative complement, and under the operation $\langle R\rangle$ defined by $\langle R\rangle X := \{w \in W \mid \exists v \in X \text{ such that } wRv\}$. An *admissible valuation* on $\mathfrak{g}$ is a map $V : \mathsf{PROP} \rightarrow A$. We can think of Kripke frames as the special case of general frames for which $A = \mathcal{P}(W)$.

In order to interpret hybrid languages on general frames, suitable provision must be made for the nominals. The *two-sorted general frames* introduced by ten Cate (2005) do this in a natural way. Specifically, a *two-sorted general frame* is a structure $\mathfrak{g} = (W, R, A, B)$, where (W, R, A) is a general frame, and $B \subseteq W$ is a non-empty set such that $\{w\} \in A$ for all $w \in B$, called the *admissible points of* $\mathfrak{g}$. An *admissible valuation* on $\mathfrak{g}$ is then a map $V : \mathsf{PROP} \cup \mathsf{NOM} \rightarrow A$ such that $V(\mathbf{i}) \in \{\{w\} \mid w \in$

$B\}$ for each nominal $\mathbf{i} \in \mathsf{NOM}$. A *model based on* $\mathfrak{g} = (W, R, A, B)$ is a Kripke model (W, R, V), where V is an admissible valuation on $\mathfrak{g}$. A hybrid formula φ is valid at a point w in $\mathfrak{g}$, written $\mathfrak{g}, w \Vdash \varphi$, if $(W, R, V), w \Vdash \varphi$ for every admissible valuation V on $\mathfrak{g}$.

Recall (from e.g. Blackburn et al. 2001) that a general frame (W, R, A) is said to be *differentiated*, if for all $w, v \in W$ with $w \neq v$, there exists an $a \in A$ such that $w \in a$ and $v \notin a$. A general frame (W, R, A) is *tight*, if for all $u, v \in W$, it is the case that uRv iff $\forall a \in A(v \in a \Longrightarrow u \in \langle R \rangle a)$. It is *compact*, if $\bigcap A_0 \neq \varnothing$ for every subset A_0 of A which has the finite intersection property. We say that (W, R, A) is *descriptive* if it is differentiated, tight and compact.

Following ten Cate (2005), we say that a two-sorted general frame (W, R, A, B) is *descriptive* if the associated general frame (W, R, A) is descriptive. Finally, (W, R, A, B) is *strongly descriptive* if it is descriptive, and satisfies the following two conditions: (i) for all $a \in A$, if $a \neq \varnothing$, then there is a $w \in B$ such that $w \in a$; (ii) for all $a \in A$ and $u \in B$, if $\{v \in a \mid uRv\} \neq \varnothing$, then there is a $w \in B$ such that $w \in a$ and uRw.

Logics. We now recall axiomatizations of the minimal hybrid logics in the languages $\mathcal{H}$ and $\mathcal{H}(@)$. The systems given here are based on those in Blackburn et al. (2001) and ten Cate (2005). We will use the notation $\Diamond^n$ with $n \in \mathbb{N}$ to denote a string of n consecutive $\Diamond$'s. The notation $\Box^n$ is defined similarly.

The *minimal normal hybrid logic* $\mathbf{H}$ is the smallest set of $\mathcal{H}$-formulas containing all propositional tautologies and the axioms in Table 1, and which is closed under the inference rules in Table 1, except for (*Name*) and (*Paste*). $\mathbf{H}^+$ is defined similarly, closing in addition under (*Name*) and (*Paste*). If Σ is a set of $\mathcal{H}$-formulas, then $\mathbf{H} \oplus \Sigma$ and $\mathbf{H}^+ \oplus \Sigma$ are the normal hybrid logics *generated* by Σ. By *sorted substitution* we mean uniform substitution of formulas for propositional variables and nominals for nominals.

The rules (*Name*) and (*Paste*), as well as their @ versions, are known as 'non-orthodox' rules because of their syntactic side-conditions. These rules are admissible in the minimal hybrid logics obtained by omitting them. However, they are needed in order to obtain general frame-completeness result for axiomatic extensions with pure axioms (Blackburn and Tzakova 1999; Blackburn et al. 2001; ten Cate 2005).

Let Γ be a set of $\mathcal{H}$-formulas. A *derivation in* $\mathbf{H} \oplus \Sigma$ *from* Γ (respectively, in $\mathbf{H}^+ \oplus \Sigma$ from Γ) is a finite sequence $\varphi_1, \ldots, \varphi_n$ of formulas such that for every $i \in \{1, \ldots, n\}$, the formula φ_i is either an axiom or an assumption in Γ, or obtained from some of the preceding formulas by one of the inference rules of $\mathbf{H}$ (respectively, of $\mathbf{H}^+$). In a derivation, a formula φ_i depends on a formula φ_j if $i = j$ or φ_i is the conclusion of an application of a rule of inference, in which at least one of the premises depends on φ_j.

The formula φ is *locally derivable in* $\mathbf{H} \oplus \Sigma$ *from* Γ (respectively, locally derivable in $\mathbf{H}^+ \oplus \Sigma$ from Γ), denoted $\Gamma \vdash_{\mathbf{H} \oplus \Sigma} \varphi$ (respectively, $\Gamma \vdash_{\mathbf{H}^+ \oplus \Sigma} \varphi$), if there is a derivation in $\mathbf{H} \oplus \Sigma$ from Γ (respectively, in $\mathbf{H}^+ \oplus \Sigma$ from Γ) ending in φ and in which sorted substitution is applied only to axioms, and all rules other than modus ponens are applied only to formulas which do no depend on members of Γ. If we

Table 1 Axioms and inference rules of $\mathbf{H}$ and $\mathbf{H}^+$

Axioms	
(*Taut*)	All classical propositional tautologies φ
(*Dual*)	$\Diamond p \leftrightarrow \neg\Box\neg p$
(*K*)	$\Box(p \to q) \to (\Box p \to \Box q)$
(*Nom*)	$\Diamond^n(\mathbf{i} \wedge p) \to \Box^m(\mathbf{i} \to p)$ for all $n, m \in \mathbb{N}$
Rules of inference	
(*Modus ponens*)	From $\varphi \to \psi$ and φ, derive ψ
(*Sorted substitution*)	From φ derive φ', where φ' is obtained from φ by sorted substitution
(*Nec*)	From φ, derive $\Box\varphi$
(*NameLite*)	From $\neg\mathbf{i}$, derive $\vdash \bot$
(*Name*)	From $\mathbf{i} \to \varphi$, derive φ for $\mathbf{i}$ not occurring in φ
(*Paste*)	From $\Diamond^n(\mathbf{i} \wedge \Diamond(\mathbf{j} \wedge \varphi)) \to \psi$, derive $\Diamond^n(\mathbf{i} \wedge \Diamond\varphi) \to \psi$ for $n \in \mathbb{N}$, $\mathbf{i} \neq \mathbf{j}$, and $\mathbf{j}$ not occurring in φ or ψ

also allow the application in derivations of (*Nec*) to formulas which do depend on members of Γ, we obtain the notions of *global derivability* in $\mathbf{H}$ and $\mathbf{H}^+$, denoted $\Gamma \vdash^g_{\mathbf{H}\oplus\Sigma} \varphi$ and $\Gamma \vdash^g_{\mathbf{H}^+\oplus\Sigma} \varphi$, respectively. As in modal logic, a deduction theorem holds for local derivability, and so $\Gamma \vdash_{\mathbf{H}\oplus\Sigma} \varphi$ (respectively, $\Gamma \vdash_{\mathbf{H}^+\oplus\Sigma} \varphi$) iff there is a finite subset $\{\gamma_1, \ldots, \gamma_m\} \subseteq \Gamma$ such that $\vdash_{\mathbf{H}\oplus\Sigma} \gamma_1 \wedge \cdots \wedge \gamma_m \to \varphi$ (respectively, $\vdash_{\mathbf{H}^+\oplus\Sigma} \gamma_1 \wedge \cdots \wedge \gamma_m \to \varphi$).

The *minimal normal hybrid logic* $\mathbf{H}(@)$ is the smallest set of $\mathcal{H}(@)$-formulas containing all propositional tautologies and the axioms in Table 2, and which is closed under the inference rules in Table 2, except for (*Name*$_@$) and (*BG*$_@$). $\mathbf{H}^+(@)$ is defined similarly, closing in addition under (*Name*$_@$) and (*BG*$_@$). If Σ is a set of $\mathcal{H}(@)$-formulas, then $\mathbf{H} \oplus \Sigma$ and $\mathbf{H}^+ \oplus \Sigma$ are the normal hybrid logics *generated* by Σ.

The definitions of local and global derivation in $\mathbf{H}(@) \oplus \Sigma$ and $\mathbf{H}^+(@) \oplus \Sigma$ are the obvious analogues of those in $\mathbf{H} \oplus \Sigma$ and $\mathbf{H}^+ \oplus \Sigma$. As with the latter logics, in local $\mathbf{H}(@) \oplus \Sigma$ and $\mathbf{H}^+(@) \oplus \Sigma$-derivations modus ponens is the only rule that may be applied to formulas which depend on members of Γ, while global derivations also allow the application of (Nec) and (Nec$_@$) to such formulas.

As usual, a set Γ of formulas is *consistent with respect to a hybrid logic* $\mathbf{L}$ if $\Gamma \nvdash_{\mathbf{L}} \bot$ (or $\mathbf{L}$-*consistent*). Global consistency is defined similarly replacing $\vdash_{\mathbf{L}}$ with $\vdash^g_{\mathbf{L}}$.

Remark 2.1 One naturally wonders how the logics with and without the additional inference rules compare. The minimal hybrid logics $\mathbf{H}$ and $\mathbf{H}^+$ have the same theorems, since both are sound and strongly complete with respect to the class of all frames, see e.g. Gargov and Goranko (1993), Blackburn and Tzakova (1999) and Blackburn et al. (2001, Sect. 7.3). This picture changes when we extend these logics

Table 2 Axioms and inference rules of $\mathbf{H}(@)$ and $\mathbf{H}^+(@)$

Axioms	
(*Taut*)	All classical propositional tautologies φ
(*K*)	$\Box(p \to q) \to (\Box p \to \Box q)$
(*Dual*)	$\Diamond p \leftrightarrow \neg\Box\neg p$
($K_@$)	$@_\mathbf{j}(p \to q) \to (@_\mathbf{j} p \to @_\mathbf{j} q)$ for all $\mathbf{j} \in \mathsf{NOM}$
(*Selfdual*)	$\neg @_\mathbf{j} p \leftrightarrow @_\mathbf{j}\neg p$ for all $\mathbf{j} \in \mathsf{NOM}$
(*Intro*)	$\mathbf{j} \wedge p \to @_\mathbf{j} p$ for all $\mathbf{j} \in \mathsf{NOM}$
(*Ref*)	$@_\mathbf{j}\mathbf{j}$ for all $\mathbf{j} \in \mathsf{NOM}$
(*Agree*)	$@_\mathbf{i} @_\mathbf{j} p \to @_\mathbf{j} p$ for all $\mathbf{i}, \mathbf{j} \in \mathsf{NOM}$
(*Back*)	$\Diamond @_\mathbf{j} p \to @_\mathbf{j} p$ for all $\mathbf{j} \in \mathsf{NOM}$
Rules of inference	
(*Modus ponens*)	From $\varphi \to \psi$ and $\vdash \varphi$, derive $\vdash \psi$
(*Sorted substitution*)	From φ derive φ', where φ' is obtained from φ by sorted substitution
(*Nec*)	From φ, derive $\Box\varphi$
($Nec_@$)	From φ, derive $\vdash @_\mathbf{j}\varphi$
($Name_@$)	From $@_\mathbf{j}\varphi$, derive φ for $\mathbf{j}$ not occurring in φ
($BG_@$)	From $@_\mathbf{i}\Diamond\mathbf{j} \wedge @_\mathbf{j}\varphi \to \psi$, derive $\vdash @_\mathbf{i}\Diamond\varphi \to \psi$ for $\mathbf{j} \neq \mathbf{i}$ and $\mathbf{j}$ not occurring in φ and ψ

with additional axioms. Consider, for example, the set $\Sigma = \{\mathbf{j} \to \Box\bot\}$ and the formula $\varphi = \Diamond\top$. Let $\mathfrak{g} = (W, R, A, B)$ be the descriptive two-sorted general frame with $W = \{u, v\}$, $R = \{(u, u)\}$, $A = \mathcal{P}(W)$ and $B = \{v\}$. Then $\mathfrak{g}$ validates the members of Σ, and furthermore, $\Diamond\top$ is satisfied at u. By the soundness of $\mathbf{H} \oplus \Sigma$ with respect to its class of descriptive two-sorted general frames (see above), we have $\nvdash_{\mathbf{H}\oplus\Sigma} \Box\bot$. On the other hand, by applying the (*Name*) rule to $\mathbf{j} \to \Box\bot$, we see that $\vdash_{\mathbf{H}^+\oplus\Sigma} \Box\bot$. Incidentally, note that this also shows that $\mathbf{H} \oplus \Sigma$ is frame-incomplete.

Analogous arguments, using the same two-sorted descriptive general frame $\mathfrak{g}$, show that $\nvdash_{\mathbf{H}(@)\oplus\{@_\mathbf{j}\Box\bot\}} \Box\bot$ while $\vdash_{\mathbf{H}^+(@)\oplus\{@_\mathbf{j}\Box\bot\}} \Box\bot$.

General completeness with respect to two sorted general frames. Ten Cate has shown that, for every set Σ of $\mathcal{H}$-formulas (respectively, $\mathcal{H}(@)$-formulas), the logic $\mathbf{H} \oplus \Sigma$ (respectively, $\mathbf{H}(@) \oplus \Sigma$) is strongly complete with respect to the class of descriptive two-sorted general frames validating its axioms (ten Cate 2005, Theorem 5.3.1) (respectively, ten Cate 2005, Theorem 5.3.5). Moreover, $\mathbf{H}^+ \oplus \Sigma$ (respectively, $\mathbf{H}^+(@) \oplus \Sigma$) is strongly complete with respect to the class of *strongly* descriptive two-sorted general frames validating its axioms (ten Cate 2005, Theorem 5.3.10) (respectively, ten Cate 2005, Theorem 5.3.13).

Boolean algebras with operators. Boolean algebras with additional operators (BAOs) offer a natural and standard interpretation of modal languages. A *Boolean* *algebra with operators* is an algebra $\mathbf{A} = (A, \wedge, \vee, \neg, \bot, \top, \Diamond_i)_{i \in I}$ such that $(A, \wedge, \vee, \neg, \bot, \top)$ is a Boolean algebra and the $\Diamond_i$'s are *operators*, i.e., finitely join preserving and normal. In this paper we will mostly be concerned with unary operators.

An operation on a Boolean algebra is *monotonic*, if $a \leq b$ implies $\Diamond_i(a) \leq \Diamond_i(b)$. All operators are monotonic.

Let $\mathbf{A} = (A, \wedge, \vee, \neg, \bot, \top, \Diamond_i)_{i \in I}$ be a BAO. An *assignment* on $\mathbf{A}$ is a function v: $\mathsf{PROP} \to A$ associating an element of A with each propositional variable in PROP. Given such an assignment v, we calculate the *meaning* $\tilde{v}(\psi)$ of a formula ψ as follows: $\tilde{v}(\bot) = \bot$, $\tilde{v}(p) = v(p)$, $\tilde{v}(\neg\psi) = \neg\tilde{v}(\psi)$, $\tilde{v}(\psi_1 \wedge \psi_2) = \tilde{v}(\psi_1) \wedge \tilde{v}(\psi_2)$ and, for each $i \in I$, $\tilde{v}(\Diamond_i\psi) = \Diamond_i(\tilde{v}(\psi))$.

We say that an equation $\varphi \approx \psi$ is *valid* in a BAO $\mathbf{A}$ (denoted $\mathbf{A} \models \varphi \approx \psi$), if for all assignments θ, $\tilde{\theta}(\varphi) = \tilde{\theta}(\psi)$.

Let Σ be a set of formulas. We define $\Sigma^{\approx}$ to be the set $\{\sigma \approx \top \mid \sigma \in \Sigma\}$.

Every normal modal logic is sound and complete with respect to the class of all BAOs which validate its axioms (see e.g. Blackburn et al. 2001). This is no surprise as every normal logic is also sound and strongly complete with respect to its class of descriptive general frames Goldblatt (1976), and BAOs and descriptive general frames are duals of each other (see e.g. Blackburn et al. 2001).

Canonical extensions. Recall that the *canonical extension* of a BAO $\mathbf{A}$ (see e.g., Venema 2006) is complete and atomic and is the unique superalgebra $\mathbf{A}^\delta$ of $\mathbf{A}$ satisfying:

(density) every element of $\mathbf{A}^\delta$ can be expressed as both a join of meets and as a meet of joins of elements of $\mathbf{A}$, and

(compactness) for all subsets $S \subseteq A$ and $T \subseteq A$, if $\bigwedge S \leq \bigvee T$ in $\mathbf{A}^\delta$, then there exist finite sets $S_0 \subseteq S$ and $T_0 \subseteq T$ such that $\bigwedge S_0 \leq \bigvee T_0$.

Arbitrary meets (joins) of elements of $\mathbf{A}$ form the *closed* (*open*) *elements* of $\mathbf{A}^\delta$. The open and closed elements of $\mathbf{A}^\delta$ will be denoted by $\mathbb{O}(\mathbf{A}^\delta)$ and $\mathbb{K}(\mathbf{A}^\delta)$, respectively.

The following lemma is folklore. To keep the paper self-contained, we include a short proof.

Lemma 2.2 *Let* $\mathbf{A}$ *be a Boolean algebra or BAO. Every atom of* $\mathbf{A}$ *is also an atom of the canonical extension* $\mathbf{A}^\delta$.

Proof Let $a \in At\mathbf{A}$. Since $\mathbf{A}^\delta$ is atomic, there is an atom $b \in At\mathbf{A}^\delta$ such that $b \leq a$. Now, since $At\mathbf{A}^\delta \subseteq \mathbb{K}(\mathbf{A}^\delta)$ (see e.g. Venema 2006), $b = \bigwedge S$ where $S = \{b' \in \mathbf{A} \mid b \leq b'\}$. Then $a \leq b'$, for each $b' \in S$. Indeed, if it were the case that $a \not\leq b'$, then we would have $a \wedge b' = \bot$, which would mean that $a \wedge b = \bot$, contradicting the assumption that $b \leq a$. Therefore $a = b \in \mathbf{A}^\delta$. □

Adjoint pairs. In our general completeness proofs later we will give algebraic constructions that make crucial use of the adjoint of $\Box$ and $@_\mathrm{j}$, so let us recall some relevant preliminaries on adjoints. In what follows, $\mathbf{A}$ and $\mathbf{B}$ are two complete BAOs. The monotone maps $f\colon \mathbf{A} \to \mathbf{B}$ and $g\colon \mathbf{B} \to \mathbf{A}$ form an *adjoint pair* (denoted $f \dashv g$), if for all a in $\mathbf{A}$ and b in $\mathbf{B}$, $f(a) \leq b$ iff $a \leq g(b)$. If $f \dashv g$, f is called the *left adjoint* of g, while g is called the *right adjoint* of f.

An important property of adjoint pairs is that if a map is completely join-preserving (meet-preserving), then we can compute its right (left) adjoint pointwise from the map itself and the order relation on the BAO: for monotone maps $f: A \to B$ and $g: B \to A$ such that $f \dashv g$, $f(a) = \bigwedge\{b \in B \mid a \leq g(b)\}$ and $g(b) = \bigvee\{a \in A \mid f(a) \leq b\}$. Moreover, for any map $f: A \to B$, f is completely join-preserving (completely meet-preserving) iff it has a right adjoint (left adjoint). For the proofs of this see Davey and Priestley (2002).

We know that $\mathbf{A}^\delta$ is perfect (see e.g. Venema 2006), so the operations $\Diamond$ and $\Box$ are completely join- and meet-preserving, respectively, and therefore have right and left adjoints, respectively. We will denote the right adjoint of $\Diamond$ by $\Box^{-1}$, and the left adjoint of $\Box$ by $\Diamond^{-1}$. Alternatively, $\mathbf{A}^\delta$ is isomorphic to the complex algebra of the ultrafilter frame of $\mathbf{A}$, where $\Box^{-1}$ and $\Diamond^{-1}$ are the 'tense' modalities interpreted with the inverse of the accessibility relation in that frame.

Similarly, in complex algebras of frames, if we fix the the nominal coordinate, @ is both completely join- and meet-preserving, and therefore has both a left and a right adjoint. We will denote the left adjoint of $@_\mathbf{j}$ by $@_\mathbf{j}^{-1}$.

3 Algebraic Semantics

In this section we introduce two possible algebraic semantics for each of the languages $\mathcal{H}$ and $\mathcal{H}(@)$. We introduce *hybrid algebras* and also introduce the so-called *grounded* and *degenerate* versions of these structures and study certain truth preserving operations on them. We define *orthodox interpretations* which treat nominals as constants, thus allowing us to fall back on much of the existing theory for modal logic as interpreted in BAOs.

3.1 Algebraic Semantics for the Language $\mathcal{H}$

Before defining hybrid algebras, we define an intermediate semantics that will be used in some of the proofs of our main results. This type of algebraic semantics for $\mathcal{H}$ is called an *orthodox interpretation*. An orthodox interpretation of $\mathcal{H}$ is an algebra $\mathbf{A} = (A, \wedge, \vee, \neg, \bot, \top, \Diamond, \{s_\mathbf{i}\}_{\mathbf{i} \in \mathsf{NOM}})$, where $(A, \wedge, \vee, \neg, \bot, \top, \Diamond)$ is a BAO, and each $s_\mathbf{i}$ is the interpretation of the nominal $\mathbf{i}$ as a *constant*. Moreover, $\mathbf{A}$ is required to validate the inequality $\Diamond^n(s_\mathbf{i} \wedge a) \leq \Box^m(\neg s_\mathbf{i} \vee a)$ for all $\mathbf{i} \in \mathsf{NOM}$, $a \in A$ and $n, m \in \mathbb{N}$. We use the term "orthodox" since it is really the 'standard' algebraic semantics for modal logics with constants. However, for us it is 'non-standard' since it is not appropriately dual to the intended relational semantics of hybrid logic. Moreover, the rule (*Sorted substitution*) is not generally sound on orthodox interpretations, in the sense that for an orthodox interpretation $\mathbf{A}$ and $\mathcal{H}$-formula ψ, it may happen that $\mathbf{A} \models \psi \approx \top$ but that $\mathbf{A} \not\models \psi' \approx \top$ for some sorted substitution instance ψ' of ψ. For example, consider the orthodox interpretation $\mathbf{A} = (\mathbf{2}, \Diamond, \{s_\mathbf{j}\}_{\mathbf{j} \in \mathsf{NOM}})$, where $\mathbf{2}$ is the two-element Boolean algebra, $\Diamond 0 = 0$, $\Diamond 1 = 1$, $s_\mathbf{j} = 0$, and $s_\mathbf{i} = 1$ for $\mathbf{i} \neq \mathbf{j}$. Then $\mathbf{A} \models \Diamond \mathbf{i} \approx \top$ but $\mathbf{A} \not\models \Diamond \mathbf{j} \approx \top$. However, this will not be a concern to us, as in the ensuing we will always require that an orthodox interpretation validates (all theorems of) a logic $\mathbf{H} \oplus \Sigma$, which is by definition closed under sorted substitution already.

The second and main type of algebraic structures for $\mathcal{H}$ which we consider are *hybrid algebras*, defined below. Unlike orthodox interpretations they do not hold nominals constant and (*Sorted substitution*) is sound for hybrid algebras. They also enjoy a duality with two-sorted general frames.

Definition 3.1 A *hybrid algebra* is a pair $\mathfrak{A} = (\mathbf{A}, X_A)$, where $\mathbf{A} = (A, \wedge, \vee, \neg, \bot, \top, \Diamond)$ such that $(A, \wedge, \vee, \neg, \bot, \top, \Diamond)$ is a BAO containing at least one atom and X_A is a non-empty subset of the set $At\mathbf{A}$ of atoms of $\mathbf{A}$. We will often refer to X_A as a *set of designated atoms* of the algebra. We also make the following convention: $\Box a := \neg\Diamond\neg a$.

We say that a hybrid algebra $\mathfrak{A} = (\mathbf{A}, X_A)$ is *permeated*, if for each $\bot \neq a \in A$, there is an atom $x \in X_A$ such that $x \leq a$, and for all $x \in X_A$ and $a \in A$, if $x \leq \Diamond a$, then there exists a $y \in X_A$ such that $y \leq a$ and $x \leq \Diamond y$.

An *assignment* on a hybrid algebra $\mathfrak{A} = (\mathbf{A}, X_A)$ is a map $v\colon \mathsf{PROP} \cup \mathsf{NOM} \to A$ associating an element of A with each propositional variable in PROP and an atom in X_A with each nominal in NOM. Given such an assignment v, we calculate the *meaning* $\tilde{v}(\psi)$ of a formula ψ as follows: $\tilde{v}(\bot) = \bot$, $\tilde{v}(p) = v(p)$, $\tilde{v}(\mathbf{j}) = v(\mathbf{j})$, $\tilde{v}(\neg\psi) = \neg\tilde{v}(\psi)$, $\tilde{v}(\psi_1 \wedge \psi_2) = \tilde{v}(\psi_1) \wedge \tilde{v}(\psi_2)$, and $\tilde{v}(\Diamond\psi) = \Diamond\tilde{v}(\psi)$.

An equation $\varphi \approx \psi$ is *valid* in a hybrid algebra $\mathfrak{A}$ (denoted $\mathfrak{A} \models \varphi \approx \psi$), if for all assignments v, $\tilde{v}(\varphi) = \tilde{v}(\psi)$. A set E of equations is valid in a hybrid algebra $\mathfrak{A}$ (denoted $\mathfrak{A} \models E$), if each equation in E is valid in $\mathfrak{A}$.

We now turn our attention to products of hybrid algebras. Let $\mathfrak{A} = (\mathbf{A}, X_A)$ and $\mathfrak{B} = (\mathbf{B}, X_B)$ be two hybrid algebras. The *pseudo-product*[1] $\mathfrak{A} \times \mathfrak{B}$ of $\mathfrak{A}$ and $\mathfrak{B}$ is given by $(\mathbf{A} \times \mathbf{B}, X_{A\times B})$, where $\mathbf{A} \times \mathbf{B}$ is defined in the usual way and $X_{A\times B} = \{(x, \bot^{\mathfrak{B}}) \mid x \in X_A\} \cup \{(\bot^{\mathfrak{A}}, y) \mid y \in X_B\}$.

Validity is *not* generally preserved under pseudo-products of hybrid algebras. This mirrors the fact that validity of hybrid formulas is not preserved under taking disjoint unions of frames. Consider the hybrid algebra $\mathfrak{A} = (\mathbf{2}, \Diamond, \{1\})$, where $\mathbf{2}$ is the two-element Boolean algebra, $\Diamond 0 = 0$, and $\Diamond 1 = 1$. Then $\mathfrak{A} \models \Diamond\mathbf{i} \approx \top$ but $\mathfrak{A}^2 \not\models \Diamond\mathbf{i} \approx \top$.

We can easily fix this by simply adding $\bot$ to the sets of designated atoms. A *grounded hybrid algebra* is just like a hybrid algebra $\mathfrak{A} = (\mathbf{A}, X_A)$, except that the bottom element of the algebra is also included in the set X_A. $\mathcal{H}$-equations are interpreted in grounded hybrid algebras as they are in hybrid algebras with nominals ranging over the elements in the designated set of atoms and $\bot$. Given a

[1] We call this construction a pseudo-product rather than a product, as its interaction with projections does not satisfy the category-theoretic notion of a product.

hybrid algebra $\mathfrak{B} = (\mathbf{B}, X_B)$, the *associated grounded hybrid algebra* is the structure $\mathfrak{B}_0 = (\mathbf{B}, X_B \cup \{\perp\})$.

Although validity is not preserved under taking pseudo-products of hybrid algebras, we can show that if each of the associated grounded hybrid algebras of two hybrid algebras validates an $\mathcal{H}$-formula, then the pseudo-product of the original hybrid algebras does too. But first, recall that the *projection map* on the ith coordinate of $A_1 \times A_2$ is the map $\pi_i \colon A_1 \times A_2 \to A_i$ defined by $\pi_i(a_1, a_2) = a_i$.

Proposition 3.2 *If $\mathfrak{A}$ and $\mathfrak{B}$ are hybrid algebras such that $\mathfrak{A}_0 \models \varphi \approx \psi$ and $\mathfrak{B}_0 \models \varphi \approx \psi$, then $\mathfrak{A} \times \mathfrak{B} \models \varphi \approx \psi$.*

Proof We prove the contrapositive, so assume $\mathfrak{A} \times \mathfrak{B} \not\models \varphi \approx \psi$. Then there is an assignment ν: PROP $\cup$ NOM $\to A \times B$ such that $\nu(\varphi) \neq \nu(\psi)$. But then $\pi_1(\nu(\varphi)) \neq \pi_1(\nu(\psi))$ or $\pi_2(\nu(\varphi)) \neq \pi_2(\nu(\psi))$. If $\pi_1(\nu(\varphi)) \neq \pi_1(\nu(\psi))$, consider the assignment ι: PROP $\cup$ NOM $\to A$ defined by $\iota(p) = \pi_1(\nu(p))$ and $\iota(\mathbf{j}) = \pi_1(\nu(\mathbf{j}))$ for each $p \in$ PROP and $\mathbf{j} \in$ NOM. Note that $\iota(\mathbf{j})$ might be $\perp$ since some of the atoms in $X_{A\times B}$ are of the form $(\perp, y)$. Using structural induction, we can show that $\iota(\gamma) = \pi_1(\nu(\gamma))$ for all $\mathcal{H}$-formulas γ. Hence, $\iota(\varphi) = \pi_1(\nu(\pi)) \neq \pi_1(\nu(\psi)) = \iota(\psi)$, and so $\mathfrak{A}_0 \not\models \varphi \approx \psi$. The case where $\pi_2(\nu(\varphi)) \neq \pi_2(\nu(\psi))$ is similar. □

In the next section, we would need to take the pseudo-product of two algebras $\mathfrak{A} = (\mathbf{A}, X_A)$ and $\mathfrak{B} = (\mathbf{B}, X_B)$, where either X_A or X_B is empty. But since this goes beyond the definition of a hybrid algebra, we will refer to these structures as *degenerate hybrid algebras*. A *degenerate hybrid algebra* is a pair $\mathfrak{A} = (\mathbf{A}, X_A)$, where $\mathbf{A}$ is a BAO and $X_A = \varnothing$. Given a degenerate hybrid algebra $\mathfrak{A} = (\mathbf{A}, X_A)$, the *associated grounded degenerate hybrid algebra* is the structure $\mathfrak{A}_0 = (\mathbf{A}, \{\perp\})$. We then have the following useful preservation result:

Proposition 3.3 *Let $\mathfrak{A}$ be a degenerate hybrid algebra and $\mathfrak{B}$ a hybrid algebra. If $\mathfrak{A}_0 \models \varphi \approx \psi$ and $\mathfrak{B} \models \varphi \approx \psi$, then $\mathfrak{A} \times \mathfrak{B} \models \varphi \approx \psi$.*

Proof The proof is similar to that of Proposition 3.2. □

Later we will need the fact that the pseudo-product of two permeated hybrid algebras is also permeated. We now show that this is indeed the case.

Proposition 3.4 *Let $\mathfrak{A} = (\mathbf{A}, X_A)$ and $\mathfrak{B} = (\mathbf{B}, X_B)$ be two permeated hybrid algebras. Then $\mathfrak{A} \times \mathfrak{B}$ is also permeated.*

Proof For the first condition, let $(a, b) \in A \times B$ such that $(a, b) \neq (\perp, \perp)$. Then $a \neq \perp$ or $b \neq \perp$. Assume $a \neq \perp$. Now, since $\mathfrak{A}$ is permeated, there is some $x \in X_A$ such that $x \leq a$. So we know that $(x, \perp) \leq (a, b)$. Furthermore, $x \in X_A$, hence $(x, \perp) \in X_{A\times B}$ by definition. Similarly for $b \neq \perp$. For the second condition, we have two consider two cases:

Case 1: $(\perp, y) \in X_{A\times B}$ such that $(\perp, y) \leq \Diamond(a, b)$. Then $(\perp, y) \leq (\Diamond a, \Diamond b)$, and so, $y \leq \Diamond b$. But we know that $\mathfrak{B}$ is permeated, so there is a $y' \in X_B$ such that $y' \leq b$ and $y \leq \Diamond y'$. Hence, $(\perp, y') \leq (a, b)$ and $(\perp, y) \leq (\perp, \Diamond y') = (\Diamond\perp, \Diamond y') = \Diamond(\perp, y')$. Furthermore, since $y' \in X_B$, $(\perp, y') \in X_{A\times B}$.

Case 2: $(x, \perp) \in X_{A\times B}$ such that $(x, \perp) \leq \Diamond(a, b)$. Similar to Case 1. □

3.2 Algebraic Semantics for the Language $\mathcal{H}(@)$

As for the language $\mathcal{H}$, we present two possible algebraic semantics for $\mathcal{H}(@)$. Again the first involves interpreting nominals as constants, and as for $\mathcal{H}$, we will refer to these algebras as orthodox interpretations. An *orthodox interpretation* of $\mathcal{H}(@)$ is an algebra $\mathbf{A} = (A, \wedge, \vee, \neg, \bot, \top, \Diamond, @, \{s_\mathbf{i}\}_{\mathbf{i} \in \mathsf{NOM}})$, where $(A, \wedge, \vee, \neg, \bot, \top, \Diamond)$ is a BAO, $@$ is a binary operator: $@$ is antitone in its first coordinate, each $s_\mathbf{i}$ is the interpretation of the nominal $\mathbf{i}$ as a *constant*, and $\mathbf{A}$ is required to validate the following for all $\mathbf{i}, \mathbf{j} \in \mathsf{NOM}$ and all $a, b \in A$:

$$@_{s_\mathbf{i}}(\neg a \vee b) \leq \neg @_{s_\mathbf{i}} a \vee @_{s_\mathbf{i}} b \qquad \neg @_{s_\mathbf{i}} a = @_{s_\mathbf{i}} \neg a \qquad @_{s_\mathbf{i}} @_{s_\mathbf{j}} a \leq @_{s_\mathbf{j}} a$$

$$@_{s_\mathbf{i}} s_\mathbf{i} = \top \qquad s_\mathbf{i} \wedge a \leq @_{s_\mathbf{i}} a \qquad \Diamond @_{s_\mathbf{i}} a \leq @_{s_\mathbf{i}} a$$

The behaviour of the $@$-operator in an orthodox interpretation of $\mathcal{H}(@)$ is characterized by Proposition 3.5. To prove this proposition, we make use of the fact that, in orthodox interpretations, the satisfaction operator preserves finite (including empty) meets and joins in its second coordinate, and is consequently also monotone in that coordinate. The proofs of both these facts are left for the reader.

Proposition 3.5 *Let* $\mathbf{A}$ *be an orthodox interpretation of* $\mathcal{H}(@)$*, and let* a *be an element of* A *and* $s_\mathbf{i}$ *the constant interpretation of* $\mathbf{i}$*. Then* $@_{s_\mathbf{i}} a = \top$ *iff* $s_\mathbf{i} \leq a$ *and* $@_{s_\mathbf{i}} a = \bot$ *iff* $s_\mathbf{i} \leq \neg a$.

Proof First, assume $@_{s_\mathbf{i}} a = \top$. Then $\neg @_{s_\mathbf{i}} a = \bot$, and so $@_{s_\mathbf{i}} \neg a = \bot$. But $s_\mathbf{i} \wedge \neg a \leq @_{s_\mathbf{i}} \neg a$, so $s_\mathbf{i} \wedge \neg a \leq \bot$, which implies that $s_\mathbf{i} \leq a$.

For the converse, assume $s_\mathbf{i} \leq a$. Then $s_\mathbf{i} \vee a = a$, and so $@_{s_\mathbf{i}}(s_\mathbf{i} \vee a) = @_{s_\mathbf{i}} a$. Hence, since $@$ preserves finite joins in its second coordinate, $@_{s_\mathbf{i}} s_\mathbf{i} \vee @_{s_\mathbf{i}} a = @_{s_\mathbf{i}} a$. But $@_{s_\mathbf{i}} s_\mathbf{i} = \top$, so $@_{s_\mathbf{i}} a = \top$.

Next, assume $@_{s_\mathbf{i}} a = \bot$. Then we have $s_\mathbf{i} \wedge a \leq @_{s_\mathbf{i}} a = \bot$, so $s_\mathbf{i} \leq \neg a$, as required.

Conversely, suppose $s_\mathbf{i} \leq \neg a$. By the monotonicity of $@$, $@_{s_\mathbf{i}} s_\mathbf{i} \leq @_{s_\mathbf{i}} \neg a$, and so, since $@_{s_\mathbf{i}} s_\mathbf{i} = \top$, $@_{s_\mathbf{i}} \neg a = \top$. Hence, $\neg @_{s_\mathbf{i}} a = \top$, and so $@_{s_\mathbf{i}} a = \bot$. □

Observe that in an orthodox interpretation of $\mathcal{H}(@)$, if $s_\mathbf{i} \neq \bot$, then $@_{s_\mathbf{i}} \bot = \bot$ and $@_{s_\mathbf{i}} \top = \top$ for all $\mathbf{i} \in \mathsf{NOM}$.

We now extend the hybrid algebras introduced in the previous subsection to accommodate the interpretation of the $@$-operator.

Definition 3.6 A hybrid $@$-algebra is a pair $\mathfrak{A} = (\mathbf{A}, X_A)$, where $\mathbf{A} = (A, \wedge, \vee, \neg, \bot, \top, \Diamond, @)$ with $(A, \wedge, \vee, \neg, \bot, \top, \Diamond)$ a BAO containing at least one atom, X_A is a non-empty subset of atoms of $\mathbf{A}$, $@$ is a binary operator whose first coordinate ranges over X_A and the second coordinate over all elements of the algebra, and for all $a, b \in A$ and all $x, y \in X_A$ the following hold:

$(K@) @_x(\neg a \vee b) \leq \neg @_x a \vee @_x b$, (*ref*) $@_x x = \top$,
(*self-dual*) $\neg @_x a = @_x \neg a$, (*introduction*) $x \wedge a \leq @_x a$, and
(*agree*) $@_x @_y a \leq @_y a$, (*back*) $\Diamond @_x a \leq @_x a$.

A hybrid @-algebra $\mathfrak{A} = (\mathbf{A}, X_A)$ is said to be *permeated* if its hybrid algebra reduct is permeated.

As before, an *assignment* on a hybrid @-algebra $\mathfrak{A}$ is a map v: PROP ∪ NOM → A associating an element of A with each propositional variable in PROP and an atom in X_A with each nominal in NOM. Given such an assignment v, we calculate the *meaning* $\tilde{v}(\psi)$ of a formula ψ as before. In particular, $\tilde{v}(@_{\mathrm{j}}\psi) = @_{\tilde{v}(\mathrm{j})}\tilde{v}(\psi)$. Truth is defined as for hybrid algebras.

Proposition 3.7 below characterizes the behaviour of the @-operator in hybrid @-algebras.

Proposition 3.7 *Let* $\mathfrak{A} = (\mathbf{A}, X_A)$, *where* $\mathbf{A} = (A, \wedge, \vee, \neg, \bot, \top, \Diamond, @)$ *with* $(A, \wedge, \vee, \neg, \bot, \top, \Diamond)$ *a BAO*, X_A *is a non-empty subset of atoms of* $\mathbf{A}$, *and* @ *is a binary operator with first coordinate ranging over* X_A *and second coordinate over all elements of the algebra. Then* $\mathfrak{A}$ *is a hybrid @-algebra iff for all* $x \in X_A$ *and* $a \in A$, $@_x a = \top$ *iff* $x \leq a$ *and* $@_x a = \bot$ *iff* $x \nleq a$.

Proof The proof of the left-to-right direction is similar to that of Proposition 3.5. For the converse direction, we have to show that @ satisfies $(K@)$, (*self-dual*), (*agree*), (*ref*), (*introduction*) and (*back*).

For $(K@)$, let $x \in X_A$ and $a, b \in A$, and assume $x \leq \neg a \vee b$. Then $@_x(\neg a \vee b) = \top$. But since x is an atom, $x \leq \neg a$ or $x \leq b$. Hence, $x \nleq a$ or $x \leq b$, and so $@_x a = \bot$ or $@_x b = \top$. We then have $\neg @_x a = \top$ or $@_x b = \top$, which means that $\neg @_x a \vee @_x b = \top$. Therefore, $@_x(\neg a \vee b) \leq \neg @_x a \vee @_x b$, as required. On the other hand, if $x \nleq \neg a \vee b$, $@_x(\neg a \vee b) = \bot$, which gives $@_x(\neg a \vee b) \leq \neg @_x a \vee @_x b$.

For (*self-dual*), let $x \in X_A$ and $b \in A$, and assume $x \leq b$. Then $@_x b = \top$, and so $\neg @_x b = \bot$. But $x \leq b$ implies $x \nleq \neg b$, which means that $@_x \neg b = \bot$. Hence, $\neg @_x b = @_x \neg b$. On the other hand, if $x \nleq b$, then $@_x b = \bot$, which gives $\neg @_x b = \top$. But we know from $x \nleq b$ that $x \leq \neg b$, so $@_x \neg b = \top$.

To show that @ satisfies (*agree*), let $x, y \in X_A$ and $b \in A$, and assume $y \leq b$. Then $@_y b = \top$, and so $@_x @_y b \leq @_y b$. On the other hand, if $y \nleq b$, $@_y b = \bot$, which means that $@_x @_y b = @_x \bot = \bot$. Hence, $@_x @_y b \leq @_y b$.

For (*ref*), note that we have $x \leq x$ for all $x \in X_A$, so $@_x x = \top$.

(*introduction*) Let $x \in X_A$ and $b \in A$, and assume $x \leq b$. Then $@_x b = \top$, and so $x \wedge b \leq @_x b$. If $x \nleq b$, $x \wedge b = \bot$ and $@_x b = \bot$, which gives $x \wedge b \leq @_x b$.

Finally, to show that @ satisfies (*back*), let $x \in X_A$ and $b \in A$. If $x \leq b$, then $@_x b = \top$, which means that $\Diamond @_x b \leq @_x b$. On the other hand, assume $x \nleq b$. Then we have $@_x b = \bot$, and so we get $\Diamond @_x b = \Diamond \bot = \bot$. Therefore, $\Diamond @_x b \leq @_x b$. □

Remark 3.8 In an orthodox interpretation for $\mathcal{H}(@)$ it is not necessarily the case that $@_{s_{\mathrm{i}}} a = \bot$ iff $s_{\mathrm{i}} \nleq a$. This reflects the fact that the constant interpretations of nominals need not be atoms. Indeed, consider the orthodox interpretation $\mathbf{A} = (\mathbf{2}, \Diamond, @, \{s_{\mathbf{j}}\}_{\mathbf{j} \in \mathsf{NOM}})$, where $\mathbf{2}$ is the two-element Boolean algebra, $\Diamond 0 = 0$, $\Diamond 1 = 1$, $s_{\mathbf{j}} = 0$, and $s_{\mathbf{i}} = 0$ for $\mathbf{i} \neq \mathbf{j}$. Then $@_{s_{\mathbf{j}}} \neg s_{\mathbf{j}} = \neg @_{s_{\mathbf{j}}} s_{\mathbf{j}} = \neg 1 = 0$ but $s_{\mathbf{j}} = 0 \leq 1 = \neg s_{\mathbf{j}}$. This further motivates our choice to work with hybrid algebras instead of orthodox interpretations.

3.3 *Duality with Two Sorted-General Frames*

As already indicated, the duality between Boolean algebras with operators and descriptive general frames extends naturally to a duality between hybrid algebras and descriptive two-sorted general frames. In this subsection we will formulate these results more precisely. We will omit proofs, as these are relatively straightforward extensions of the corresponding proofs in the modal case which can be found e.g. in Blackburn et al. (2001, Sect. 5.4). For completely worked out proofs and also the duality of morphisms, we refer the interested reader to Robinson (2015, Chap. 2).

We turn two-sorted general frames into hybrid algebras, and vice versa, in a natural way. The underlying hybrid algebra of a two-sorted general frame consists of its algebra of admissible sets together with the singletons containing its admissible points:

Definition 3.9 Let $\mathfrak{g} = (W, R, A, B)$ be a two-sorted general frame. The *underlying hybrid algebra of* $\mathfrak{g}$ is the structure

$$\mathfrak{g}^* = (A, \cap, \cup, -, \varnothing, W, \langle R \rangle, X_B),$$

where $X_B = \{\{w\} \mid w \in B\}$. We can easily extend this to accommodate the interpretation of the @-operator by setting

$$@_{\{w\}} a = \begin{cases} W \text{ if } \{w\} \subseteq a \\ \varnothing \text{ otherwise.} \end{cases}$$

The two-sorted general ultrafilter frame of a hybrid algebra (hybrid @-algebra) is the (ordinary) ultrafilter frame of its BAO part, augmented with the set of principle ultrafilters generated by the designated atoms:

Definition 3.10 Let $\mathfrak{A} = (\mathbf{A}, X_A)$ be a hybrid algebra (hybrid @-algebra). Then the *two-sorted general ultrafilter frame of* $\mathfrak{A}$ is defined as

$$\mathfrak{A}_* = (Uf\mathfrak{A}, Q_\Diamond, \widehat{A}, X_A\!\uparrow),$$

where

- $Uf\mathfrak{A}$ is the set of all ultrafilters of $\mathbf{A}$,
- $Q_\Diamond \subseteq Uf\mathfrak{A} \times Uf\mathfrak{A}$ such that $Q_\Diamond uv$ iff $\Diamond a \in u$ for all $a \in v$,

- $\widehat{A} := \{\widehat{a} \mid a \in A\}$ where $\widehat{a} = \{u \in Uf\mathfrak{A} \mid a \in u\}$, and
- $X_A\!\uparrow = \{x\!\uparrow \mid x \in X_A\}$ is the set of principle ultrafilters generated by the elements of X_A.

These two constructions are systematically connected by the following theorems:

Theorem 3.11 *Let* $\mathfrak{A} = (\mathbf{A}, X_A)$ *be a hybrid algebra (hybrid @-algebra), and* $\mathfrak{g} = (W, R, A, B)$ *a two-sorted general frame. Then*

$$\mathfrak{g} \Vdash \psi \text{ iff } \mathfrak{g}^* \models \psi \approx \top;$$
$$\mathfrak{A} \models \psi \approx \top \text{ iff } \mathfrak{A}_* \Vdash \psi.$$

Theorem 3.12 *Let* $\mathfrak{A} = (\mathbf{A}, X_A)$ *be a hybrid algebra (hybrid @-algebra), and* $\mathfrak{g} = (W, R, A, B)$ *a two-sorted general frame. Then*

1. $\mathfrak{A}_*$ *is a descriptive two-sorted general frame,*
2. $\mathfrak{g}^*$ *is a hybrid algebra (hybrid @-algebra),*
3. $(\mathfrak{A}_*)^* \cong \mathfrak{A}$,
4. $(\mathfrak{g}^*)_* \cong \mathfrak{g}$ *iff* $\mathfrak{g}$ *is descriptive,*
5. *if* $\mathfrak{A}$ *is permeated, then* $\mathfrak{A}_*$ *is strongly descriptive, and*
6. *if* $\mathfrak{g}$ *is strongly descriptive, then* $\mathfrak{g}^*$ *is permeated.*

This duality extends to morphism (generated subframes and bounded morphic images on the side of frames with homomorphic images and embeddings on the side of hybrid algebras) in a straightforward way. For details the reader is refereed to Robinson (2015).

4 Algebraic Completeness: Axiomatic Extensions of H and H$^+$

In this section we prove completeness of all axiomatic extensions of the logics **H** and **H**$^+$ with respect to the hybrid algebras introduced in Sect. 3. The general pattern is as follows: the axiomatizations without the additional 'non-orthodox' rules are complete with respect to the class of hybrid algebras, whereas the axiomatizations with the additional 'non-orthodox' rules are complete with respect to the class of permeated hybrid algebras.

Definition 4.1 A hybrid logic Λ is *strongly complete* with respect to a class C of hybrid algebras, if for every Λ-consistent set of formulas Γ, there is a hybrid algebra $\mathfrak{A} \in \mathsf{C}$, an assignment ν on $\mathfrak{A}$ and an element $a \neq \bot$ in A such that $\widetilde{\nu}(\gamma) \geq a$ for all $\gamma \in \Gamma$. Weak completeness is the case when Γ is a singleton.

4.1 Completeness of $\mathbf{H} \oplus \Sigma$

The standard proof of the completeness of modal logics with respect to classes of BAOs proceeds by the well-known Lindenbaum-Tarski construction, see e.g. Blackburn et al. (2001). In proving the completeness of logics $\mathbf{H} \oplus \Sigma$ with respect to classes of hybrid algebras we will also make use of this construction with, however, some significant complications. The main hurdle is the fact that the equivalence classes $[\mathbf{i}]$ of nominals need not be atoms of the Lindenbaum-Tarski algebra. In order to be able to fall back on the established theory of modal logic, we temporarily interpret the nominals as modal constants and work with the orthodox Lindenbaum-Tarski-algebra of $\mathbf{H} \oplus \Sigma$ over PROP.[2] This algebra then requires a certain amount of sculpting to change it into a hybrid algebra of the right kind, as we will soon see. We first state and prove the main theorem, and consequently prove the lemmas needed in its proof.

Theorem 4.2 *For any set Σ of $\mathcal{H}$-formulas, the logic $\mathbf{H} \oplus \Sigma$ is sound and strongly complete with respect to the class of all hybrid algebras which validate Σ.*

Proof It is straightforward to check the soundness direction of the above. For the completeness direction, suppose Γ is a local $\mathbf{H} \oplus \Sigma$-consistent set of formulas. For the purpose of this proof, we will temporarily treat the nominals as modal constants and work with orthodox interpretations of $\mathcal{H}$. Using the Lindenbaum Lemma (see e.g. Blackburn et al. 2001, Lemma 4.20), we can extend this to a maximal local consistent set Γ^+.

Now, consider the usual Lindenbaum-Tarski algebra (see e.g. Blackburn et al. 2001) of the logic $\mathbf{H} \oplus \Sigma$ over PROP with the nominals treated like constants or 0-ary modalities. For the sake of brevity, we will denote this algebra simply by $\mathbf{A}$. The fact that $\mathbf{A}$ validates precisely the theorems of $\mathbf{H} \oplus \Sigma$ is proved in the usual way, which implies that $\mathbf{A}$ is an orthodox interpretation of $\mathcal{H}$.

By standard propositional reasoning, we can show that the set $[\Gamma^+] = \{[\gamma] \mid \gamma \in \Gamma^+\}$ is a filter of $\mathbf{A}$. We can thus use the finite meet property to conclude that for every finite subset $\Gamma' \subseteq \Gamma^+$, it holds that $\bigwedge[\Gamma'] > \bot$ in $\mathbf{A}$.

Next we consider the canonical extension $\mathbf{A}^\delta$ of $\mathbf{A}$ (seen as a BAO). First, note that since all axioms of $\mathbf{H}$ are Sahlqvist under the orthodox interpretation, it follows from the canonicity of Sahlqvist equations that $\mathbf{A}^\delta \models \mathbf{H}^\approx$. However, the validity of the equations in $\Sigma^\approx$ is not necessarily preserved in passing from $\mathbf{A}$ to $\mathbf{A}^\delta$. Second, note that in $\mathbf{A}^\delta$ we have $\bigwedge[\Gamma^+] > \bot$. To see this, suppose $\bigwedge[\Gamma^+] \leq \bot$. But by the compactness of the embedding of $\mathbf{A}$ into $\mathbf{A}^\delta$, there is a finite subset $\Gamma' \subseteq \Gamma^+$ such that $\bigwedge[\Gamma'] \leq \bot$ in $\mathbf{A}$, contradicting the claim above.

Since $\bigwedge[\Gamma^+] > \bot$ in $\mathbf{A}^\delta$ and $\mathbf{A}^\delta$ is atomic, there is some atom d in $\mathbf{A}^\delta$ such that $d \leq \bigwedge[\Gamma^+]$. Next define a sequence $d_0, d_1, d_2, \ldots$ of elements of $\mathbf{A}^\delta$, beginning with $d_0 = d$. Suppose d_n is already defined, then let $d_{n+1} = \Diamond^{-1} d_n$. Set

[2]This strategy of treating nominals as modal constants and exploiting the classical theory has been widely used in hybrid logic. It can already be found in the work of Gargov et al. (1987).

$$D = \bigvee_{n \in \mathbb{N}} d_n.$$

Define $\mathbf{A}_D = (A_D, \wedge^D, \vee^D, \neg^D, \bot^D, \top^D, \Diamond^D, \{s_{\mathbf{j}}^D\}_{\mathbf{j} \in \mathsf{NOM}})$, where $A_D = \{a \wedge D \mid a \in A\}$, $\wedge^D$ and $\vee^D$ are the restriction of $\wedge$ and $\vee$ to A_D, and

$$\neg^D a = \neg a \wedge D \qquad \Diamond^D a = \Diamond a \wedge D \qquad s_{\mathbf{j}}^D = s_{\mathbf{j}} \wedge D$$
$$\top^D = D \qquad \bot^D = \bot$$

Now, by Lemma 4.4, $\mathbf{A}_D$ is an algebra. In fact, $\mathbf{A}_D$ is the algebraic dual of the submodel of the atom structure (see e.g. Venema 2006, Definition 41) of $\mathbf{A}^\delta$ generated by d. We also know from Lemma 4.5 that $\mathbf{A}_D \models \mathbf{H} \oplus \Sigma^{\approx}$, which means $\mathbf{A}_D$ is an orthodox interpretation of $\mathcal{H}$. Consider the assignment ν_D: $\mathsf{PROP} \to A_D$ given by $\nu_D(p) = h(\nu(p))$, where h is the homomorphism from $\mathbf{A}$ onto $\mathbf{A}_D$ defined in Lemma 4.5, and ν is the natural map taking p to $[p]$. We can show by structural induction that $\widetilde{\nu}_D(\psi) = h(\widetilde{\nu}(\psi))$ for all $\mathcal{H}$-formulas ψ that use variables from PROP. Since $d \leq D$ and $d \leq \bigwedge[\Gamma^+]$, we have $\widetilde{\nu}_D(\gamma) = h(\widetilde{\nu}(\gamma)) = \widetilde{\nu}(\gamma) \wedge D = [\gamma] \wedge D \geq \bigwedge[\Gamma^+] \wedge D \geq d > \bot$, for all $\gamma \in \Gamma$.

Next we want to produce a *hybrid* algebra out of $\mathbf{A}_D$. The desired result could be obtained if we could show that the constant interpretations of the nominals in $\mathbf{A}_D$ are atoms. This would allow us to drop the constants from the signature of $\mathbf{A}_D$ and replace them with a designated set of atoms consisting of all $s_{\mathbf{i}}^D$, $\mathbf{i} \in \mathsf{NOM}$. However, this is not necessarily the case. By Lemma 4.7, $s_{\mathbf{i}}^D$ is either an atom or $\bot$. So let $\mathfrak{A}_D = (\mathbf{A}_D^-, X_{A_D})$, where $\mathbf{A}_D^-$ is the modal algebra reduct of $\mathbf{A}_D$ obtained by omitting the constant interpretations of nominals, and $X_{A_D} = \{s_{\mathbf{i}}^D \mid s_{\mathbf{i}}^D > \bot\}$. But this is still not necessarily a hybrid algebra since it is possible that X_{A_D} can be empty. In fact, we have three possibilities, corresponding to the following three cases:

Case 1: $s_{\mathbf{i}}^D > \bot$ for all $\mathbf{i} \in \mathsf{NOM}$. This is the simplest case. Since $X_{A_D} \neq \varnothing$, it follows from the foregoing that $\mathfrak{A}_D$ is a hybrid algebra. Furthermore, since $\mathbf{H} \oplus \Sigma$ is closed under (*Sorted substitution*), $\mathfrak{A}_D \models \Sigma^{\approx}$. Consider the assignment ν'_D which extends ν_D from PROP to $\mathsf{PROP} \cup \mathsf{NOM}$, obtained by simply setting $\nu'_D(\mathbf{i}) = s_{\mathbf{i}}^D$ for each $\mathbf{i} \in \mathsf{NOM}$. It is clear that $\widetilde{\nu}'_D(\psi) = \widetilde{\nu}_D(\psi)$ for all $\mathcal{H}$-formulas ψ, and hence $\widetilde{\nu}'_D(\gamma) \geq d > \bot$, for all $\gamma \in \Gamma$.

Case 2: $s_{\mathbf{i}}^D = \bot$ for some $\mathbf{i} \in \mathsf{NOM}$ but not all. From the foregoing, we know that $(\mathfrak{A}_D)_0 \models \mathbf{H} \oplus \Sigma^{\approx}$, and hence, by Proposition 3.2, $\mathfrak{A}_D \times \mathfrak{A}_D \models \mathbf{H} \oplus \Sigma^{\approx}$. Now, let $\mathbf{j} \in \mathsf{NOM}$ such that $s_{\mathbf{j}}^D \neq \bot$. Consider the assignment ν''_D obtained by setting $\nu''_D(p) = (\nu_D(p), \nu_D(p))$ for all propositional variables $p \in \mathsf{PROP}$, and

$$\nu''_D(\mathbf{i}) = \begin{cases} (s_{\mathbf{i}}^D, \bot) & \text{if } s_{\mathbf{i}}^D > \bot \\ (\bot, s_{\mathbf{j}}^D) & \text{if } s_{\mathbf{i}}^D = \bot \end{cases}$$

for all nominals $\mathbf{i} \in \mathsf{NOM}$. It is straightforward to show (using structural induction) that for any $\mathcal{H}$-formula ψ, we have $\widetilde{\nu}''_D(\psi) = (\widetilde{\nu}_D(\psi), a_\psi)$, where a_ψ is some element of $\mathfrak{A}_D$. But then $\widetilde{\nu}''_D(\gamma) = (\widetilde{\nu}_D(\gamma), a_\gamma) \geq (d, a_\gamma) > (\bot, \bot)$, for all $\gamma \in \Gamma$.

Case 3: $s_{\mathbf{i}}^D = \bot$ for all $\mathbf{i} \in \mathsf{NOM}$. In this case, $X_{A_D} = \varnothing$, so $\mathfrak{A}_D$ is not a hybrid algebra. So we need another strategy for constructing a hybrid algebra that will work here. First, we claim that $[\mathbf{i}] > \bot$ in $\mathbf{A}$ for all $\mathbf{i} \in \mathsf{NOM}$. To see this, suppose $[\mathbf{i}] = \bot$. Then $\vdash \mathbf{i} \leftrightarrow \bot$, and so $\vdash \neg\mathbf{i} \leftrightarrow \top$. Hence, $\vdash \neg\mathbf{i}$, which means that $\vdash \bot$ by the (*NameLite*) rule. However, this is a contradiction. We thus also have that $[\mathbf{i}] > \bot$ in $\mathbf{A}^\delta$ for all $\mathbf{i} \in \mathsf{NOM}$. So choose some nominal $\mathbf{j} \in \mathsf{NOM}$. Since $\mathbf{A}^\delta$ is atomic, there is an atom d' in $\mathbf{A}^\delta$ such that $d' \leq [\mathbf{j}]$. Now, using d' instead of d, we define D' and $\mathbf{A}_{D'}$ in the same way as D and $\mathbf{A}_D$, i.e. by setting $d'_0 = d'$, $d'_{n+1} = \Diamond^{-1} d'_n$ and $D' = \bigvee_{n\in\mathbb{N}} d'_n$. In the same way as for $\mathbf{A}_D$, we can prove that $\mathbf{A}_{D'} \models \mathbf{H} \oplus \Sigma^{\approx}$ and that $s_{\mathbf{i}}^{D'}$ is either $\bot$ or an atom of $\mathbf{A}_{D'}$. Define $\nu_{D'}$ in the same way as ν_D. Note that we do not know if $\widetilde{\nu}_{D'}(\gamma) > \bot$, for all $\gamma \in \Gamma$. But this is not a problem, as we will soon see. Now, let $\mathfrak{A}_{D'} = (\mathbf{A}^-_{D'}, X_{A_{D'}})$ where $\mathbf{A}^-_{D'}$ is the reduct of $\mathbf{A}_{D'}$ obtained by omitting the constant interpretations of nominals and $X_{A_{D'}} = \{s_{\mathbf{i}}^{D'} \mid s_{\mathbf{i}}^{D'} > \bot\}$. We know that $X_{A_{D'}} \neq \varnothing$ since at least $s_{\mathbf{j}}^{D'} \neq \bot$. Furthermore, $(\mathfrak{A}_{D'})_0 \models \mathbf{H} \oplus \Sigma^{\approx}$, and so, since $(\mathfrak{A}_D)_0 \models \mathbf{H} \oplus \Sigma^{\approx}$, $\mathfrak{A}_D \times \mathfrak{A}_{D'} \models \mathbf{H} \oplus \Sigma^{\approx}$ by Proposition 3.2. Finally, let ν''' be defined by $\nu'''_D(p) = (\nu_D(p), \nu_{D'}(p))$ for all propositional variables $p \in \mathsf{PROP}$, and

$$\nu'''_D(\mathbf{i}) = \begin{cases} (\bot, s_{\mathbf{i}}^{D'}) & \text{if } s_{\mathbf{i}}^{D'} > \bot \\ (\bot, s_{\mathbf{j}}^{D'}) & \text{if } s_{\mathbf{i}}^{D'} = \bot \end{cases}$$

for all nominals $\mathbf{i} \in \mathsf{NOM}$. Using structural induction, we can show that for all $\mathcal{H}$-formulas ψ, $\widetilde{\nu}'''_D(\psi) = (\widetilde{\nu}_D(\psi), a'_\psi)$, where a'_ψ is some element in $\mathfrak{A}_{D'}$. But then $\widetilde{\nu}'''_D(\gamma) = (\widetilde{\nu}_D(\gamma), a'_\gamma) \geq (d, a'_\gamma) > (\bot, \bot')$, for all $\gamma \in \Gamma$. □

We will now prove the lemmas used in the proof of the above theorem. Unless stated otherwise, in what follows $\mathbf{A}$, $\mathbf{A}^\delta$, $\mathbf{A}_D$, ν and ν_D will be as in the proof of Theorem 4.2. The first lemma we need is that $\mathbf{A}_D$ is an algebra. To prove this, we have to show that A_D is closed under the operations defined in the proof of Theorem 4.2. But first we need the following lemma:

Lemma 4.3 $D \leq \Box D$.

Proof By the definition of D, $\Box D = \Box \bigvee_{n\in\mathbb{N}} d_n$, and so, since $\Box$ is monotone, $\Box D \geq \bigvee_{n\in\mathbb{N}} \Box d_n$. But by the definition of d_n, $\bigvee_{n\in\mathbb{N}} \Box d_n = \left(\bigvee_{n\in\mathbb{N}-\{0\}} \Box\Diamond^{-1} d_{n-1}\right) \vee \Box d_0$, so $\Box D \geq \left(\bigvee_{n\in\mathbb{N}-\{0\}} \Box\Diamond^{-1} d_{n-1}\right) \vee \Box d_0$. Hence, $\Box D \geq \left(\bigvee_{n\in\mathbb{N}-\{0\}} \Box\Diamond^{-1} d_{n-1}\right)$. Since $\Box$ and $\Diamond^{-1}$ are adjoint, we have $\Box\Diamond^{-1} d_i \geq d_i$, and therefore $\Box D \geq \bigvee_{n\in\mathbb{N}-\{0\}} d_{n-1}$. We thus have that $\Box D \geq D$ by the definition of D. □

Lemma 4.4 *A_D is closed under the operations $\wedge^D$, $\vee^D$, $\neg^D$, and $\Diamond^D$.*

Proof The cases for $\wedge^D$ and $\vee^D$ are straightforward, so we consider the cases for $\neg^D$ and $\Diamond^D$. Let $a \in A_D$. Then $a = a' \wedge D$ for some $a' \in A$. Firstly, $\neg^D a = \neg(a' \wedge$

$D) \wedge D = (\neg a' \vee \neg D) \wedge D = (\neg a' \wedge D) \vee \bot = \neg a' \wedge D$. But A is closed under $\neg$, so $\neg a' \in A$, which means that $\neg^D a \in A_D$.

Next, $\Diamond^D a = \Diamond(a' \wedge D) \wedge D \leq \Diamond a' \wedge D$, and conversely, $\Diamond^D a = \Diamond(a' \wedge D) \wedge D \geq \Diamond a' \wedge \Box D \wedge D = \Diamond a' \wedge D$, where the last step follows from Lemma 4.3. Hence, $\Diamond^D a = \Diamond a' \wedge D$, and so, since A is closed under $\Diamond$, $\Diamond a' \in A$. We thus have that $\Diamond^D a \in A_D$. □

To show that the algebra $\mathbf{A}_D$ validates the equations in $\mathbf{H} \oplus \Sigma^{\approx}$, we prove that $\mathbf{A}_D$ is a homomorphic image of $\mathbf{A}$.

Lemma 4.5 *The map $h\colon A \to A_D$ defined by $h(a) = a \wedge D$ is a surjective homomorphism from $\mathbf{A}$ onto $\mathbf{A}_D$.*

Proof First, h is clearly surjective. In verifying that h is a homomorphism, all cases except those for $\neg$ and $\Diamond$ are straightforward. The case for $\neg$ is proved as follows: $h(\neg a) = \neg a \wedge D = (\neg a \wedge D) \vee (\neg D \wedge D) = (\neg a \vee \neg D) \wedge D = \neg(a \wedge D) \wedge D = \neg^D h(a)$. The right-to-left inequality for $\Diamond$ is proved as follows: $h(\Diamond a) = \Diamond a \wedge D \geq \Diamond(a \wedge D) \wedge D = \Diamond h(a) \wedge D = \Diamond^D h(a)$, where the inequality follows from the monotonicity of $\Diamond$. Conversely, $\Diamond^D h(a) = \Diamond(a \wedge D) \wedge D \geq \Diamond a \wedge \Box D \wedge D = \Diamond a \wedge D = h(\Diamond a)$. Here the third step follows from Lemma 4.3. □

Our next goal is to show that the constant interpretation of a nominal is either an atom of $\mathbf{A}_D$ or $\bot$. But first, we need the following result:

Lemma 4.6 *Let $\mathbf{A}$ be a perfect BAO and $a, b \in At\mathbf{A}$. Then $a \leq \Diamond_{i_1}^{-1} \cdots \Diamond_{i_m}^{-1} b$ iff $b \leq \Diamond_{i_m} \cdots \Diamond_{i_1} a$.*

Proof For the left-to-right implication, let a and b be two atoms in $\mathbf{A}$, and assume $a \leq \Diamond_{i_1}^{-1} \cdots \Diamond_{i_m}^{-1} b$. Suppose that $b \nleq \Diamond_{i_m} \cdots \Diamond_{i_1} a$. Then, since b is an atom, $b \leq \neg\Diamond_{i_m} \cdots \Diamond_{i_1} a$, and so $\Diamond_{i_m} \cdots \Diamond_{i_1} a \leq \neg b$. Now, by adjunction, $a \leq \Box_{i_1}^{-1} \cdots \Box_{i_m}^{-1} \neg b$, which means that $\Diamond_{i_1}^{-1} \cdots \Diamond_{i_m}^{-1} b \leq \neg a$. Hence, by our assumption, $a \leq \neg a$, and so $\bot = a \wedge \neg a = a$, which is a contradiction.

The proof of the converse implication is similar. □

Lemma 4.7 *For each $\mathbf{i} \in \mathsf{NOM}$, $s_{\mathbf{i}}^D$ is either $\bot$ or an atom of $\mathbf{A}^\delta$, and hence, of $\mathbf{A}_D$.*

Proof Suppose that $s_{\mathbf{i}}^D \neq \bot$. We show that if $a, b \in At\mathbf{A}^\delta$ such that $a, b \leq s_{\mathbf{i}}^D$, then $a = b$. So let $a, b \in At\mathbf{A}^\delta$ such that $a, b \leq s_{\mathbf{i}}^D$. Then $a, b \leq D$, and so $a \leq (\Diamond^{-1})^{n_1} d$ and $b \leq (\Diamond^{-1})^{n_2} d$ for some natural numbers n_1 and n_2. We thus have that $d \leq \Diamond^{n_1} a$ and $d \leq \Diamond^{n_2} b$ by Lemma 4.6, and so, since $a, b \leq s_{\mathbf{i}}^D$ we find $d \leq \Diamond^{n_1}(a \wedge s_{\mathbf{i}}^D)$ and $d \leq \Diamond^{n_2}(b \wedge s_{\mathbf{i}}^D)$. Hence, $d \leq \Diamond^{n_1}(a \wedge s_{\mathbf{i}} \wedge D) \leq \Diamond^{n_1}(a \wedge s_{\mathbf{i}})$ and $d \leq \Diamond^{n_2}(b \wedge s_{\mathbf{i}} \wedge D) \leq \Diamond^{n_2}(b \wedge s_{\mathbf{i}})$. So, by axiom (Nom), $d \leq \Box^m(\neg s_{\mathbf{i}} \vee a)$ and $d \leq \Box^m(\neg s_{\mathbf{i}} \vee b)$ for all natural numbers m. Since $\Diamond^{-1}$ and $\Box$ are adjoint, $(\Diamond^{-1})^m d \leq \neg s_{\mathbf{i}} \vee a$ and $(\Diamond^{-1})^m d \leq \neg s_{\mathbf{i}} \vee b$, which means that, for all $m \in \mathbb{N}$, it holds that $(\Diamond^{-1})^m d \leq (\neg s_{\mathbf{i}} \vee a) \wedge (\neg s_{\mathbf{i}} \vee b)$. It thus follows that $D \leq (\neg s_{\mathbf{i}} \vee a) \wedge (\neg s_{\mathbf{i}} \vee b) = \neg s_{\mathbf{i}} \vee (a \wedge b)$. Now, if $a \neq b$, $D \leq \neg s_{\mathbf{i}}$, so $s_{\mathbf{i}} \leq \neg D$. Hence, $s_{\mathbf{i}} \wedge D \leq \bot$, i.e., $s_{\mathbf{i}}^D = \bot$, contradicting our assumption that $s_{\mathbf{i}}^D \neq \bot$. □

Corollary 4.8 *For any set Σ of $\mathcal{H}$-formulas, the logic* $\mathbf{H} \oplus \Sigma$ *is sound and strongly complete with respect to the class of all descriptive two-sorted general frames defined by Σ.*

Proof The soundness is obvious. For the completeness, let Γ be a local $\mathbf{H} \oplus \Sigma$-consistent set of formulas. By Theorem 4.2 there is a hybrid algebra $\mathfrak{A} = (\mathbf{A}, X_A)$, an assignment ν on $\mathfrak{A}$ and an element $a \neq \bot$ in A such that $a \leq \widetilde{\nu}(\gamma)$ for all $\gamma \in \Gamma$. Now, consider the two-sorted general ultrafilter frame $\mathfrak{A}_* = (Uf\mathfrak{A}, Q_\Diamond, \widehat{A}, X_A{\uparrow})$ of $\mathfrak{A}$, and define V: $\mathsf{PROP} \to \mathcal{P}(W)$ by $V(p) = \widehat{\nu(p)}$. Clearly, V is admissible. Using structural induction on ψ, we can show that for all $\psi \in \mathcal{H}$ and all ultrafilters $u \in Uf\mathfrak{A}$, it holds that $(\mathfrak{A}_*, V), u \Vdash \psi$ iff $\widetilde{\nu}(\psi) \in u$. As $a \leq \widetilde{\nu}(\gamma)$ for all $\gamma \in \Gamma$, we have $\widetilde{\nu}(\gamma) \in a{\uparrow}$ for all $\gamma \in \Gamma$, and therefore $(\mathfrak{A}_*, V), a{\uparrow} \Vdash \gamma$ for all $\gamma \in \Gamma$. The fact that $\mathfrak{A}_* \Vdash \Sigma$ follows from Theorem 3.11, and by Theorem 3.12, $\mathfrak{A}_*$ is descriptive. □

4.2 Completeness of $\mathbf{H}^+ \oplus \Sigma$

As we will now show, the logic $\mathbf{H}^+ \oplus \Sigma$ is complete with respect to permeated hybrid algebras which validate Σ. We first state and prove the main theorem, and consequently prove the lemmas needed for the main proof.

Theorem 4.9 *For any set Σ of $\mathcal{H}$-formulas, the logic* $\mathbf{H}^+ \oplus \Sigma$ *is sound and strongly complete with respect to the class of all permeated hybrid algebras which validate Σ.*

Proof Suppose Γ is a local $\mathbf{H}^+ \oplus \Sigma$-consistent set of formulas. As before, we will use the orthodox interpretation of $\mathcal{H}$ as the point of departure for the proof. Let NOM' be a denumerably infinite set of nominals disjoint from NOM. We know from Lemma 5.3.9 in ten Cate (2005) that Γ is contained in a maximal local $\mathbf{H}^+ \oplus \Sigma$-consistent set of formulas Γ^+ in the extended language such that

(i) Γ^+ contains at least one nominal, say $\mathbf{i}_0$, and
(ii) for each formula of the form $\Diamond^n(\mathbf{i} \wedge \Diamond\varphi)$ in Γ^+, there is a nominal $\mathbf{j}$ for which the formula $\Diamond^n(\mathbf{i} \wedge \Diamond(\mathbf{j} \wedge \varphi))$ is in Γ^+.

Now, consider the Lindenbaum-Tarski algebra of $\mathbf{H}^+ \oplus \Sigma$ over PROP. For simplicity, denote this algebra by $\mathbf{A}$. In the usual way, we can show that $\mathbf{A} \models \mathbf{H}^+ \oplus \Sigma^\approx$, and hence it is an orthodox interpretation of $\mathcal{H}$. Using standard propositional reasoning we can show that the set $[\Gamma^+] = \{[\gamma] \mid \gamma \in \Gamma^+\}$ is an ultrafilter of $\mathbf{A}$. Hence, for every finite subset $\Gamma' \subseteq \Gamma^+$, it holds that $\bigwedge[\Gamma'] > \bot$ in $\mathbf{A}$.

Next we consider the (orthodox) canonical extension $\mathbf{A}^\delta$ of $\mathbf{A}$. As in the proof of Theorem 4.2, we have $\bigwedge[\Gamma^+] > \bot$ in $\mathbf{A}^\delta$. Since $\mathbf{A}^\delta$ is atomic, there is some atom d in $\mathbf{A}^\delta$ such that $d \leq \bigwedge[\Gamma^+]$. Define, as in the proof of Theorem 4.2, a sequence $d_0, d_1, d_2, \ldots$ of elements of $\mathbf{A}^\delta$, by setting $d_0 = d$ and $d_{n+1} = \Diamond^{-1} d_n$. Set $D = \bigvee_{n \in \mathbb{N}} d_n$. Also define $\mathbf{A}_D$ as in the proof of Theorem 4.2. As before it can be checked that $\mathbf{A}_D$, so defined, is an algebra. Furthermore, using the map h: $A \to A_D$ defined by $h(a) = a \wedge D$, we can show in the same way as in the proof of Lemma 4.5

that $\mathbf{A}_D$ is a homomorphic image of $\mathbf{A}$. Hence, $\mathbf{A}_D \models \mathbf{H}^+ \oplus \Sigma^{\approx}$, and so $\mathbf{A}_D$ is an orthodox interpretation of $\mathcal{H}$. Now, consider the assignment ν_D: $\mathsf{PROP} \to A_D$ given by $\nu_D(p) = h(\nu(p))$, where h is the homomorphism from $\mathbf{A}$ onto $\mathbf{A}_D$, and ν is the natural map taking p to $[p]$. Using the fact that h is a homomorphism, we can show by structural induction that $\widetilde{\nu}_D(\psi) = h(\widetilde{\nu}(\psi))$ for all $\mathcal{H}$-formulas ψ. Since $d \leq D$ and $d \leq \bigwedge[\Gamma^+]$, we have $\widetilde{\nu}_D(\gamma) = h(\widetilde{\nu}(\gamma)) = \widetilde{\nu}(\gamma) \wedge D = [\gamma] \wedge D \geq \bigwedge[\Gamma^+] \wedge D \geq d > \bot$, for all $\gamma \in \Gamma$.

In the same way as in the proof of Lemma 4.7, we can show that for each $\mathbf{i} \in \mathsf{NOM} \cup \mathsf{NOM}'$, $s_{\mathbf{i}}^D$ is either $\bot$ or an atom of $\mathbf{A}_D$. So let $\mathfrak{A}_D = (\mathbf{A}_D^-, X_{A_D})$, where $\mathbf{A}_D^-$ is the reduct of $\mathbf{A}_D$ obtained by omitting the constant interpretations of nominals and $X_{A_D} = \{s_{\mathbf{i}}^D \mid s_{\mathbf{i}}^D > \bot\}$. We claim that $X_{A_D} \neq \varnothing$. To see this, recall that $\mathbf{i}_0 \in \Gamma^+$, so $d \leq \bigwedge[\Gamma^+] \leq [\mathbf{i}_0]$. It follows that $d \leq s_{\mathbf{i}_0}^D$, and so, at least $s_{\mathbf{i}_0}^D > \bot$. In particular, since $s_{\mathbf{i}_0}^D$ and d are both atoms, $s_{\mathbf{i}_0}^D = d$. Now, from the fact that $\mathbf{H}^+ \oplus \Sigma$ is closed under (*Sorted substitution*), we have $\mathfrak{A}_D \models \mathbf{H}^+ \oplus \Sigma^{\approx}$. Furthermore, by Lemma 4.15, $\mathfrak{A}_D$ is permeated. From here we split our reasoning into two cases depending on whether the constant interpretations of all nominals in $\mathbf{A}_D$ are atoms or not:

Case 1: $s_{\mathbf{i}}^D > \bot$ for all $\mathbf{i} \in \mathsf{NOM} \cup \mathsf{NOM}'$. In this case, consider the assignment ν'_D which extends ν_D from PROP to $\mathsf{PROP} \cup \mathsf{NOM} \cup \mathsf{NOM}'$, obtained by simply setting $\nu'_D(\mathbf{i}) = s_{\mathbf{i}}^D$ for each $\mathbf{i} \in \mathsf{NOM} \cup \mathsf{NOM}'$. It is clear that $\widetilde{\nu}'_D(\psi) = \widetilde{\nu}_D(\psi)$ for all $\mathcal{H}$-formulas ψ, and hence, $\widetilde{\nu}'_D(\gamma) \geq d > \bot$, for all $\gamma \in \Gamma$.

Case 2: $s_{\mathbf{i}}^D = \bot$ for some $\mathbf{i} \in \mathsf{NOM} \cup \mathsf{NOM}'$. Here we have that $(\mathfrak{A}_D)_0 \models \mathbf{H}^+ \oplus \Sigma^{\approx}$, and hence, by Proposition 3.2, $\mathfrak{A}_D \times \mathfrak{A}_D \models \mathbf{H}^+ \oplus \Sigma^{\approx}$. By Proposition 3.4, we know that $\mathfrak{A}_D \times \mathfrak{A}_D$ is also permeated. The remaining part of the proof is similar to that of Case 2 in Theorem 4.2. □

Let us now prove the lemmas used in the proof of Theorem 4.9. In what follows, unless stated otherwise, Γ, $\mathbf{A}$, $\mathbf{A}^\delta$, $\mathbf{A}_D$, ν, ν_D and $\mathfrak{A}_D$ will be as in the proof of Theorem 4.9.

Lemma 4.10 *Let* $\mathbf{A}$ *be a perfect BAO,* $a \in A$ *and* $b \in At\mathbf{A}$*. Then we have* $a \wedge \Diamond_{i_1}^{-1} \cdots \Diamond_{i_n}^{-1} b > \bot$ *iff* $b \leq \Diamond_{i_n} \cdots \Diamond_{i_1} a$.

Proof For the left-to-right implication, assume $a \wedge \Diamond_{i_1}^{-1} \cdots \Diamond_{i_n}^{-1} b > \bot$. Then there is a $x \in At\mathbf{A}$ such that $x \leq a \wedge \Diamond_{i_1}^{-1} \cdots \Diamond_{i_n}^{-1} b$, and so $x \leq a$ and $x \leq \Diamond_{i_1}^{-1} \cdots \Diamond_{i_n}^{-1} b$. Hence, by Lemma 4.6, $b \leq \Diamond_{i_n} \cdots \Diamond_{i_1} x$, which means that $b \leq \Diamond_{i_n} \cdots \Diamond_{i_1} a$.

For the converse, suppose $a \wedge \Diamond_{i_1}^{-1} \cdots \Diamond_{i_n}^{-1} b = \bot$. Then $\Diamond_{i_1}^{-1} \cdots \Diamond_{i_n}^{-1} b \leq \neg a$, and so, by adjunction, $b \leq \Box_{1_n} \cdots \Box_{i_1} \neg a = \neg \Diamond_{i_n} \cdots \Diamond_{i_1} a$. Hence, $b \nleq \Diamond_{i_n} \cdots \Diamond_{i_1} a$. □

Lemma 4.11 *Let* $\mathbf{A}$ *be a BAO and* a, b *elements of* $\mathbf{A}$*. Then* $\Box_{i_1} \cdots \Box_{i_m} a \wedge \Diamond_{i_1} \cdots \Diamond_{i_m} b \leq \Diamond_{i_1} \cdots \Diamond_{i_m} (a \wedge b)$.

Proof The proof is straightforward. □

Lemma 4.12 *For any element* a *in* $\mathbf{A}_D$*, if* $d \leq \Diamond^n a$*, then* $d \leq (\Diamond^D)^n a$.

Proof The proof is by induction on n. For $n = 1$, assume $d \leq \Diamond a$. Then $d \wedge D \leq \Diamond a \wedge D$, and so, since $d \leq D$, we have $d \leq \Diamond^D a$. Now, suppose that for all a in

$\mathbf{A}_D$, the claim holds for $n = k$. For $n = k + 1$, assume $d \leq \Diamond^{k+1}a$. Then $d \leq \Diamond^k \Diamond a$. But by Lemma 4.3, $d \leq D \leq \Box^k D$, so using Lemma 4.11 we find that $d \leq \Diamond^k \Diamond a \wedge \Box^k D \leq \Diamond^k(\Diamond a \wedge D) = \Diamond^k \Diamond^D a$. Now, since $\Diamond^D a \in A_D$, we can use the inductive hypothesis to get $d \leq (\Diamond^D)^k \Diamond^D a = (\Diamond^D)^{k+1} a$. □

Lemma 4.13 *For all $\mathcal{H}$-formulas γ, it holds that $\gamma \in \Gamma^+$ iff $d \leq \widetilde{\nu}(\gamma)$ iff $d \leq \widetilde{\nu}_D(\gamma)$.*

Proof Suppose $\gamma \in \Gamma^+$. Then $[\gamma] \in [\Gamma^+]$, and so $\bigwedge[\Gamma^+] \leq [\gamma]$. But $d \leq \bigwedge[\Gamma^+]$, so $d \leq [\gamma] = \widetilde{\nu}(\gamma)$. Conversely, assume $\gamma \notin \Gamma^+$. Since Γ^+ is a maximal consistent set of formulas, we have $\neg\gamma \in \Gamma^+$. Hence, using the left-to-right direction, $d \leq \widetilde{\nu}(\neg\gamma) = \neg\widetilde{\nu}(\gamma)$, and so $d \nleq \widetilde{\nu}(\gamma)$. Now, assume $d \leq \widetilde{\nu}(\gamma)$. Then $d \wedge D \leq \widetilde{\nu}(\gamma) \wedge D$, and so $d \leq \widetilde{\nu}_D(\gamma)$. Conversely, assume $d \leq \widetilde{\nu}_D(\gamma)$. Then $d \leq \widetilde{\nu}(\gamma) \wedge D$, which gives $d \leq \widetilde{\nu}(\gamma)$. □

Lemma 4.14 *Let a be any element in $\mathbf{A}_D$ and n a natural number such that $a \wedge (\Diamond^{-1})^n d > \bot$. Then there is an $s_{\mathbf{j}}^D$ such that $a \wedge s_{\mathbf{j}}^D \wedge (\Diamond^{-1})^n d > \bot$.*

Proof Let a be any element of $\mathbf{A}_D$ such that $a \wedge (\Diamond^{-1})^n d > \bot$. The proof is by induction on n. For $n = 0$, we have $a \wedge d > \bot$. But $d = s_{\mathbf{i}_0}^D$, so we are done. Now, suppose that for every $a \in A_D$, the claim holds for all $n = k$. For $n = k + 1$, assume that $a \wedge (\Diamond^{-1})^{k+1} d > \bot$. Then $d \leq \Diamond^{k+1} a = \Diamond^k \Diamond a$ by Lemma 4.10, and therefore, another application of Lemma 4.10, but in the opposite direction, yields $\Diamond a \wedge (\Diamond^{-1})^k d > \bot$. Hence, since $(\Diamond^{-1})^k d \leq D$, we have $\Diamond a \wedge ((\Diamond^{-1})^k d \wedge D) = (\Diamond a \wedge D) \wedge (\Diamond^{-1})^k d = \Diamond^D a \wedge (\Diamond^{-1})^k d > \bot$. Since $\Diamond^D a$ is an element of $\mathbf{A}_D$, we can use the inductive hypothesis to conclude that there is a nominal $\mathbf{j} \in \mathsf{NOM} \cup \mathsf{NOM}'$ such that $\Diamond^D a \wedge s_{\mathbf{j}}^D \wedge (\Diamond^{-1})^k d > \bot$. But then $d \leq \Diamond^k(\Diamond^D a \wedge s_{\mathbf{j}}^D)$ by Lemma 4.10. Therefore, by Lemma 4.12, $d \leq (\Diamond^D)^k(\Diamond^D a \wedge s_{\mathbf{j}}^D)$. Now, since ν and h are surjective, there is some ψ such that $d \leq (\Diamond^D)^k(\Diamond^D \widetilde{\nu}_D(\psi) \wedge s_{\mathbf{j}}^D) = \widetilde{\nu}_D(\Diamond^k(\Diamond\psi \wedge \mathbf{j}))$. We thus have that $\Diamond^k(\mathbf{j} \wedge \Diamond\psi) \in \Gamma^+$ by Lemma 4.13, which means there is a nominal $\mathbf{k}$ in $\mathsf{NOM} \cup \mathsf{NOM}'$ such that $\Diamond^k(\mathbf{j} \wedge \Diamond(\mathbf{k} \wedge \psi)) \in \Gamma^+$. Hence, by Lemma 4.13, we get $d \leq \widetilde{\nu}_D(\Diamond^k(\mathbf{j} \wedge \Diamond(\mathbf{k} \wedge \psi))) = (\Diamond^D)^k(s_{\mathbf{j}}^D \wedge \Diamond^D(s_{\mathbf{k}}^D \wedge a)) \leq \Diamond^k(s_{\mathbf{j}}^D \wedge \Diamond(s_{\mathbf{k}}^D \wedge a))$. So, by Lemma 4.10, $\bot < (\Diamond^{-1})^k d \wedge s_{\mathbf{j}}^D \wedge \Diamond(s_{\mathbf{k}}^D \wedge a) \leq (\Diamond^{-1})^k d \wedge \Diamond(s_{\mathbf{k}}^D \wedge a)$. But then $d \leq \Diamond^k \Diamond(s_{\mathbf{k}}^D \wedge a) = \Diamond^{k+1}(s_{\mathbf{k}}^D \wedge a)$, and therefore, $(\Diamond^{-1})^{k+1} d \wedge s_{\mathbf{k}}^D \wedge a > \bot$. □

Finally, we are ready to show that $\mathfrak{A}_D$ is permeated.

Lemma 4.15 *$\mathfrak{A}_D$ is permeated.*

Proof For the first condition, let $b \in A_D$ such that $b > \bot$. Then $b \wedge (\Diamond^{-1})^m d > \bot$ for some $m \in \mathbb{N}$. By Lemma 4.14, there is an $s_{\mathbf{j}}^D$ such that $b \wedge s_{\mathbf{j}}^D \wedge (\Diamond^{-1})^m d > \bot$. Then $d \leq \Diamond^m(b \wedge s_{\mathbf{j}}^D)$ by Lemma 4.10. Now, $s_{\mathbf{j}}^D \neq \bot$, for otherwise, $d \leq \Diamond^{m(b)}\bot = \bot$, a contradiction. Hence, $s_{\mathbf{j}}^D$ is an atom by Lemma 4.7, and furthermore, $s_{\mathbf{j}}^D \in X_{A_D}$. We also claim that $s_{\mathbf{j}}^D \leq b$, for if not, $b \wedge s_{\mathbf{j}}^D = \bot$, giving $d \leq \bot$, a contradiction.

To prove the second condition, let $b \in A_D$ and $s_{\mathbf{i}}^D \in X_{A_D}$, and assume that $s_{\mathbf{i}}^D \leq \Diamond^D b$. Then $s_{\mathbf{i}}^D \leq \Diamond b \wedge D$, which means that $s_{\mathbf{i}}^D \leq \Diamond b \wedge (\Diamond^{-1})^m d$ for some $m \in \mathbb{N}$.

Hence, $s_{\mathbf{i}}^D \wedge \Diamond b \wedge (\Diamond^{-1})^m d = s_{\mathbf{i}}^D > \bot$, and so by Lemma 4.10 $d \leq \Diamond^m(s_{\mathbf{i}}^D \wedge \Diamond b)$. We thus have $d \leq (\Diamond^D)^m(s_{\mathbf{i}}^D \wedge \Diamond b)$ by Lemma 4.12, and therefore, since $d \leq D \leq (\Box^D)^m D$, $d \leq (\Diamond^D)^m(s_{\mathbf{i}}^D \wedge \Diamond b) \wedge (\Box^D)^m D$. But since $\mathbf{A}_D$ validates the axioms, we can show in the same way as in Lemma 4.11 that for all $a, b \in A_D$ and all $n \in \mathbb{N}$, $(\Box^D)^n a \wedge (\Diamond^D)^n b \leq (\Diamond)^D)^n(a \wedge b)$, and thus $d \leq (\Diamond^D)^m(s_{\mathbf{i}}^D \wedge \Diamond b \wedge D) = (\Diamond^D)^m(s_{\mathbf{i}}^D \wedge \Diamond^D b)$. Now, since ν and h are both surjective, there is some ψ such that $d \leq (\Diamond^D)^m(s_{\mathbf{i}}^D \wedge \Diamond^D \widetilde{\nu}_D(\psi)) = \widetilde{\nu}_D(\Diamond^m(\mathbf{i} \wedge \Diamond\psi))$. Hence, by Lemma 4.13, $\Diamond^m(\mathbf{i} \wedge \Diamond\psi) \in \Gamma^+$. This means that there is a $\mathbf{j} \in \mathsf{NOM} \cup \mathsf{NOM}'$ such that $\Diamond^m(\mathbf{i} \wedge \Diamond(\mathbf{j} \wedge \psi)) \in \Gamma^+$. Another application of Lemma 4.13, but in the opposite direction, gives $d \leq \widetilde{\nu}_D(\Diamond^m(\mathbf{i} \wedge \Diamond(\mathbf{j} \wedge \psi))) = (\Diamond^D)^m(s_{\mathbf{i}}^D \wedge \Diamond^D(s_{\mathbf{j}}^D \wedge b))$.

Now, first note that $s_{\mathbf{j}}^D \neq \bot$, for else $d = \bot$, which is a contradiction. This means that $s_{\mathbf{j}}^D$ is an atom and in X_{A_D}. Second, $s_{\mathbf{j}}^D \leq b$, for otherwise, $s_{\mathbf{j}}^D \wedge b = \bot$, giving $d = \bot$, again a contradiction. Finally, to see that $s_{\mathbf{i}}^D \leq \Diamond^D s_{\mathbf{j}}^D$, suppose for the sake of a contradiction that it is not. Then $s_{\mathbf{i}}^D \wedge \Diamond^D s_{\mathbf{j}}^D = \bot$, and so we have that $d \leq (\Diamond^D)^m(s_{\mathbf{i}}^D \wedge \Diamond^D(s_{\mathbf{j}}^D \wedge b)) \leq (\Diamond^D)^m(s_{\mathbf{i}}^D \wedge \Diamond^D s_{\mathbf{j}}^D) = (\Diamond^D)^m \bot = \bot$, which is a contradiction. $\Box$

In the same way as in the proof of Corallary 4.8 we can prove the following corollary:

Corollary 4.16 *For any set Σ of $\mathcal{H}$-formulas, the logic* $\mathbf{H}^+ \oplus \Sigma$ *is sound and strongly complete with respect to the class of all strongly descriptive two-sorted general frames defined by* Σ.

5 Algebraic Completeness: Axiomatic Extensions of H(@) and H⁺(@)

We now turn our attention to the language and logics with the satisfaction operator @. In broad strokes, the strategy is similar to that used for axiomatic extensions of **H**. However, the presence of the axioms involving the @-operator ensure that the constant interpretations of nominals in orthodox interpretations cannot be $\bot$, simplifying the proof significantly.

5.1 Completeness of $\mathbf{H}(@) \oplus \Sigma$

We first give the statement of the main result together with its proof, and subsequently prove the lemmas needed.

Theorem 5.1 *For any set Σ of $\mathcal{H}(@)$-formulas, the logic* $\mathbf{H}(@) \oplus \Sigma$ *is sound and strongly complete with respect to the class of all hybrid @-algebras which validate* Σ.

Proof Suppose Γ is a local $\mathbf{H}(@) \oplus \Sigma$-consistent set of formulas. By the Lindenbaum Lemma, Γ is contained in a maximal (locally) consistent set Γ^+ (see Blackburn et al. 2001, Lemma 4.20). As in the proof of Theorem 4.2, we will temporarily treat the nominals as modal constants. Let $\mathbf{A}$ denote the Lindenbaum-Tarski algebra of $\mathbf{H}(@) \oplus \Sigma$ over PROP. In the usual way, we can show that $\mathbf{A}$ validates precisely the theorems of $\mathbf{H}(@) \oplus \Sigma$, and hence $\mathbf{A}$ is an orthodox interpretation of $\mathcal{H}(@)$. A standard propositional argument shows that the set $[\Gamma^+] = \{[\gamma] \mid \gamma \in \Gamma^+\}$ is an ultrafilter of $\mathbf{A}$, and hence, for every finite subset $\Gamma' \subseteq \Gamma^+$, it holds that $\bigwedge[\Gamma'] > \bot$ in $\mathbf{A}$.

Next, consider the orthodox canonical extension $\mathbf{A}^\delta$ of $\mathbf{A}$. Since all axioms of $\mathbf{H}(@)$ are Sahlqvist—fixing the nominal coordinate, @ plays the role of a unary diamond—they are canonical and hence valid in $\mathbf{A}^\delta$. By the compactness of the embedding of $\mathbf{A}$ into $\mathbf{A}^\delta$, $\bigwedge[\Gamma^+] > \bot$ in $\mathbf{A}^\delta$, so since $\mathbf{A}^\delta$ is atomic, there is an atom $d \in \mathbf{A}^\delta$ such that $d \leq \bigwedge[\Gamma^+]$. Now, let $d_0 = d$, and suppose that d_n is already defined. Let $d_{n+1} = \Diamond^{-1} d_n \vee \bigvee_{\mathbf{i} \in \mathsf{NOM}} @_{s_\mathbf{i}}^{-1} d_n$, and set $D = \bigvee_{n \in \mathbb{N}} d_n$. Furthermore, let and $\mathbf{A}_D = (A_D, \wedge^D, \vee^D, \neg^D, \bot^D, \top^D, \Diamond^D, @^D, \{s_\mathbf{j}^D\}_{\mathbf{j} \in \mathsf{NOM}})$, where $A_D = \{a \wedge D \mid a \in A\}$, $\wedge^D$ and $\vee^D$ are the restriction of $\wedge$ and $\vee$ to A_D,

$$\neg^D a = \neg a \wedge D \qquad \Diamond^D a = \Diamond a \wedge D \qquad \top^D = D \qquad \bot^D = \bot \qquad s_\mathbf{j}^D = s_\mathbf{j} \wedge D$$

and, for all $\mathbf{j} \in \mathsf{NOM}$,

$$s_\mathbf{j}^D = s_\mathbf{j} \wedge D \qquad\qquad @^D_{s_\mathbf{j}^D} a = @_{s_\mathbf{j}} a \wedge D.$$

By Lemma 5.4 below, A_D is closed under the above operations, so $\mathbf{A}_D$ is an algebra. Furthermore, from Lemma 5.5 it follows that $\mathbf{A}_D \models \Sigma^{\approx}$ and that $\mathbf{A}_D$ is an orthodox interpretation of $\mathcal{H}(@)$. Now, let $\nu_D\colon \mathsf{PROP} \to A_D$ be the assignment given by $\nu_D(p) = h(\nu(p))$, where h is the homomorphism from $\mathbf{A}$ onto $\mathbf{A}_D$ defined in Lemma 5.5, and ν is the natural map taking p to $[p]$. Using the fact that h is a homomorphism, we can show by structural induction that $\widetilde{\nu}_D(\psi) = h(\widetilde{\nu}(\psi))$ for all $\mathcal{H}(@)$-formulas ψ that use variables from PROP. As $d \leq D$ and $d \leq \bigwedge[\Gamma^+]$, we have $\widetilde{\nu}_D(\gamma) = h(\widetilde{\nu}(\gamma)) = \widetilde{\nu}(\gamma) \wedge D = [\gamma] \wedge D \geq \bigwedge[\Gamma^+] \wedge D \geq d > \bot$, for all $\gamma \in \Gamma$.

Now, if we can find a suitable designated set of atoms in $\mathbf{A}_D$, we can drop the constant interpretations and we would be done. Luckily, unlike for the language $\mathcal{H}$, the @-operator makes things easier for us as it ensures that all $s_\mathbf{j}^D$ are atoms of $\mathbf{A}_D$ (see Lemma 5.10). So let $\mathfrak{A}_D = (\mathbf{A}_D^-, X_{A_D})$, where $\mathbf{A}_D^-$ is the reduct of $\mathbf{A}_D$ obtained by omitting the constant interpretations of nominals and $X_{A_D} = \{s_\mathbf{j}^D \mid \mathbf{j} \in \mathsf{NOM}\}$. Then it follows from the foregoing that $\mathfrak{A}_D$ is a hybrid @-algebra, and since $\mathbf{H}(@) \oplus \Sigma$ is closed under *Sorted substitution*, $\mathfrak{A}_D \models \mathbf{H}(@) \oplus \Sigma^{\approx}$. Finally, consider the assignment ν'_D which extends ν_D from PROP to $\mathsf{PROP} \cup \mathsf{NOM}$ by simply setting $\nu'_D(\mathbf{j}) = s_\mathbf{j}^D$ for each $\mathbf{j} \in \mathsf{NOM}$. Using structural induction, we can prove that $\widetilde{\nu}'_D(\psi) = \widetilde{\nu}_D(\psi)$ for all $\mathcal{H}(@)$-formulas ψ, and hence $\widetilde{\nu}'_D(\gamma) = \widetilde{\nu}_D(\gamma) \geq d > \bot$, for all $\gamma \in \Gamma$. □

We now state and prove the lemmas used in the proof of Theorem 5.1. In what follows, unless stated otherwise, $\mathbf{A}$, $\mathbf{A}^\delta$, $\mathbf{A}_D$, ν, ν_D and $\mathfrak{A}_D$ will be as in the proof of Theorem 5.1.

Lemma 5.2 $D \leq \Box D$.

Proof By the definition of D, $\Box D = \Box \bigvee_{n\in\mathbb{N}} d_n$, and so, by the monotonicity of $\Box$, $\Box D \geq \bigvee_{n\in\mathbb{N}} \Box d_n$. Hence, $\Box D \geq \left(\bigvee_{n\in\mathbb{N}-\{0\}} \Box\left(\Diamond^{-1} d_{n-1} \vee \bigvee_{\mathbf{i}\in\mathsf{NOM}} @_{s_\mathbf{i}}^{-1} d_{n-1}\right)\right) \vee \Box d_0$ by the definition of d_n. From this inequality we obtain $\Box D \geq \bigvee_{n\in\mathbb{N}-\{0\}} \Box\Diamond^{-1} d_{n-1}$. But we know that $\Diamond^{-1}$ and $\Box$ form an adjoint pair, so $\Box D \geq \bigvee_{n\in\mathbb{N}-\{0\}} d_{n-1}$, and hence, by the definition of D, $\Box D \geq D$. □

Lemma 5.3 *For all* $\mathbf{j} \in \mathsf{NOM}$, *we have* $D \leq @_{s_\mathbf{j}} D$.

Proof By the definition of D, $@_{s_\mathbf{j}} D = @_{s_\mathbf{j}} \bigvee_{n\in\mathbb{N}} d_n$, and so, by the monotonicity of $@$, $@_{s_\mathbf{j}} D \geq \bigvee_{n\in\mathbb{N}} @_{s_\mathbf{j}} d_n$. Hence, $@_{s_\mathbf{j}} D \geq \left(\bigvee_{n\in\mathbb{N}-\{0\}} @_{s_\mathbf{j}}(\Diamond^{-1} d_{n-1} \vee \bigvee_{\mathbf{i}\in\mathsf{NOM}} @_{s_\mathbf{i}}^{-1} d_{n-1})\right) \vee @_{s_\mathbf{j}} d_0$, from which we obtain $@_{s_\mathbf{j}} D \geq \bigvee_{n\in\mathbb{N}-\{0\}} @_{s_\mathbf{j}} @_{s_\mathbf{j}}^{-1} d_{n-1}$. But $@_{s_\mathbf{j}}$ and $@_{s_\mathbf{j}}^{-1}$ form an adjoint pair, so $@_{s_\mathbf{j}} D \geq \bigvee_{n\in\mathbb{N}-\{0\}} d_{n-1}$, and therefore, by the definition of D, $@_{s_\mathbf{j}} D \geq D$. □

Using these results, we can prove the following two lemmas:

Lemma 5.4 A_D *is closed under the operations* $\wedge^D$, $\vee^D$, $\neg^D$, $\Diamond^D$ *and* $@^D$.

Proof To see that A_D is closed under $@^D$, let $a \in A_D$. Then there is some $a' \in A$ such that $a = a' \wedge D$. Hence, $@^D_{s_\mathbf{j}^D} a = @_{s_\mathbf{j}}(a' \wedge D) \wedge D \leq @_{s_\mathbf{j}} a' \wedge D$, and conversely, $@^D_{s_\mathbf{j}^D} a = @_{s_\mathbf{j}}(a' \wedge D) \wedge D = @_{s_\mathbf{j}} a' \wedge @_{s_\mathbf{j}} D \wedge D = @_{s_\mathbf{j}} a' \wedge D$, where the last step follows from Lemma 5.3. Hence, $@^D_{s_\mathbf{j}^D} a = @_{s_\mathbf{j}} a' \wedge D$, and so, since A is closed under $@$, we have $@_{s_\mathbf{j}} a' \in A$. We thus have that $@^D_{s_\mathbf{j}} a \in A_D$ by the definition of A_D. The other cases are proved in the same way as in Lemma 4.4 using Lemma 5.2. □

Lemma 5.5 *The map* $h \colon A \to A_D$ *defined by* $h(a) = a \wedge D$ *is a surjective homomorphism from* $\mathbf{A}$ *onto* $\mathbf{A}_D$.

Proof That h is surjective is obvious. In verifying that h is a homomorphism, all cases except those for $\neg$, $\Diamond$, and $@_\mathbf{j}$ are straightforward. The cases for $\neg$ and $\Diamond$ are proved in exactly the same way as in Lemma 4.5. So we need only check $@$. We first prove the left-to-right inequality: $h(@_{s_\mathbf{j}} a) = @_{s_\mathbf{j}} a \wedge D = @_{s_\mathbf{j}} a \wedge D \wedge D \leq @_{s_\mathbf{j}} a \wedge @_{s_\mathbf{j}} D \wedge D = @_{s_\mathbf{j}}(a \wedge D) \wedge D = @^D_{s_\mathbf{j}^D}(a \wedge D) = @^D_{s_\mathbf{j}^D}(h(a))$, where the inequality follows from Lemma 5.3. Conversely, $@^D_{h(s_\mathbf{j})}(h(a)) = @^D_{s_\mathbf{j}^D}(h(a)) = @_{s_\mathbf{j}} h(a) \wedge D = @_{s_\mathbf{j}}(a \wedge D) \wedge D \leq @_{s_\mathbf{j}} a \wedge D = h(@_{s_\mathbf{j}} a)$, where the inequality follows from the fact that $a \wedge D \leq a$ and the monotonicity of $@$. □

Finally, we wish to show that all the $s_\mathbf{i}^D$ are actually atoms of $\mathbf{A}_D$. But first we need the following results:

Lemma 5.6 *Let* $\mathbf{A}$ *be a perfect BAO,* $x \in At\mathbf{A}$ *and* $a \in A$. *If* $x \leq \Diamond_i a$, *then there is an atom* y *in* $At\mathbf{A}$ *such that* $y \leq a$ *and* $x \leq \Diamond_i y$.

Proof Let $x \in At\mathbf{A}$ and $a \in A$ such that $x \leq \Diamond_i a$. Since $\mathbf{A}$ is atomic, we have $x \leq \Diamond_i a = \Diamond_i \bigvee\{y \in At\mathbf{A} \mid y \leq a\} = \bigvee\{\Diamond_i y \in At\mathbf{A} \mid y \leq a\}$. Hence, as x is an atom, there is an atom y such that $x \leq \Diamond_i y$ and $y \leq a$. □

Note that this property also holds for $\Diamond^{-1}$ and $@_{\mathbf{i}}^{-1}$ since they also distribute over arbitrary joins in a perfect BAO.

Lemma 5.7 *Let* $\mathbf{A}$ *be a perfect BAO and* $a \in A$ *and* $x \in At\mathbf{A}$. *If* $x \leq \Diamond_{i_1}^{-1}\Diamond_{i_2}^{-1}a$, *then there is an atom* $y \in At\mathbf{A}$ *such that* $y \leq a$ *and* $y \leq \Diamond_{i_2}\Diamond_{i_1}x$.

Proof Let $x \in At\mathbf{A}$ and $a \in A$ such that $x \leq \Diamond_{i_1}^{-1}\Diamond_{i_2}^{-1}a$. Then there is an atom z such that $z \leq \Diamond_{i_2}^{-1}a$ and $x \leq \Diamond_{i_1}^{-1}z$ by Lemma 5.6 above. Likewise, there is an atom y such that $y \leq a$ and $z \leq \Diamond_{i_2}^{-1}y$. Hence, by Lemma 4.6, $z \leq \Diamond_{i_1}x$ and $y \leq \Diamond_{i_2}z$. From the first inequality it follows that $\Diamond_{i_2}z \leq \Diamond_{i_2}\Diamond_{i_1}x$, and we can thus conclude that $y \leq \Diamond_{i_2}\Diamond_{i_1}x$. □

Lemma 5.8 *Let* $\mathbf{A}$ *be a perfect orthodox interpretation of* $\mathcal{H}(@)$. *Then* $@_{s_{\mathbf{j}}}^{-1}@_{s_{\mathbf{i}}}^{-1}a \leq @_{s_{\mathbf{j}}}^{-1}a$ *for all* $a \in A$ *and* $\mathbf{i}, \mathbf{j} \in \mathsf{NOM}$.

Proof Let $x \in At\mathbf{A}$ and $a \in A$ such that $x \leq @_{s_{\mathbf{j}}}^{-1}@_{s_{\mathbf{i}}}^{-1}a$. It then follows from Lemma 5.7 that there is an atom $y \in At\mathbf{A}$ such that $y \leq a$ and $y \leq @_{s_{\mathbf{i}}}@_{s_{\mathbf{j}}}x$. Hence, as $@_{s_{\mathbf{i}}}@_{s_{\mathbf{j}}}x \leq @_{s_{\mathbf{j}}}x$, we have $y \leq @_{s_{\mathbf{j}}}x$, and so $x \leq @_{s_{\mathbf{j}}}^{-1}y$ by Lemma 4.6. Since $y \leq a$, we get $x \leq @_{s_{\mathbf{j}}}^{-1}y \leq @_{s_{\mathbf{j}}}^{-1}a$. It thus follows that $@_{s_{\mathbf{j}}}^{-1}@_{s_{\mathbf{i}}}^{-1}a \leq @_{s_{\mathbf{j}}}^{-1}a$. □

Lemma 5.9 *Let* $\mathbf{A}$ *be a perfect orthodox interpretation of* $\mathcal{H}(@)$. *Then* $@_{s_{\mathbf{j}}}^{-1}\Diamond^{-1}a \leq @_{s_{\mathbf{j}}}^{-1}a$ *for all* $a \in A$ *and* $\mathbf{i}, \mathbf{j} \in \mathsf{NOM}$.

Proof Let $a \in A$ and $x \in At\mathbf{A}$ such that $x \leq @_{s_{\mathbf{j}}}^{-1}\Diamond^{-1}a$. Then there is an atom $y \in At\mathbf{A}$ such that $y \leq a$ and $y \leq \Diamond @_{s_{\mathbf{j}}}x$ by Lemma 5.7. But $\Diamond @_{s_{\mathbf{j}}}x \leq @_{s_{\mathbf{j}}}x$, so $y \leq @_{s_{\mathbf{j}}}x$. This implies that $x \leq @_{s_{\mathbf{j}}}^{-1}y$ by Lemma 4.6. Hence, as $y \leq a$, we obtain $x \leq @_{s_{\mathbf{j}}}^{-1}y \leq @_{s_{\mathbf{j}}}^{-1}a$, and thus, $@_{s_{\mathbf{j}}}^{-1}\Diamond^{-1}a \leq @_{s_{\mathbf{j}}}^{-1}a$. □

Lemma 5.10 *For each* $\mathbf{i} \in \mathsf{NOM}$, $s_{\mathbf{i}}^D$ *an atom of* $\mathbf{A}^\delta$, *and hence of* $\mathbf{A}_D$.

Proof First, $s_{\mathbf{i}}^D \neq \bot$, for otherwise, since $\mathbf{A}_D$ is an orthodox interpretation of $\mathcal{H}(@)$, $D = @_{s_{\mathbf{i}}^D}^D s_{\mathbf{i}}^D = @_{s_{\mathbf{i}}^D}^D \bot = \bot$, which is not possible. Now, let $a, b \in At\mathbf{A}^\delta$ such that $a \leq s_{\mathbf{i}}^D$ and $b \leq s_{\mathbf{i}}^D$. The proof will be complete if we can show that $a = b$. In order to do this, we show that $d \leq @_{s_{\mathbf{i}}}a$ and $d \leq @_{s_{\mathbf{i}}}b$, from which it then follows that $d \leq @_{s_{\mathbf{i}}}(a \wedge b)$ and hence, since $@_{s_{\mathbf{i}}}\bot = \bot$, it cannot be the case that $a \neq b$.

We only consider the case for a, the one for b is identical. Since $a \in At\mathbf{A}^\delta$, there is a finite string of $\Diamond^{-1}$'s and $@_{\mathbf{j}}^{-1}$'s, denote it by Q, such that $a \leq Qd$. Applying Lemmas 5.8 and 5.9 to this inequality, we find that either $a \leq (\Diamond^{-1})^{m_1}d$, or $a \leq @_{s_{\mathbf{j}_1}}^{-1}d$ or $a \leq (\Diamond^{-1})^{m_1}@_{s_{\mathbf{j}_1}}^{-1}d$.

In the case when $a \leq (\Diamond^{-1})^{m_1} d$, we have $d \leq \Diamond^{m_1} a = \Diamond^{m_1}(s_\mathbf{i} \wedge a) \leq \Diamond^{m_1} @_{s_\mathbf{i}} a \leq @_{s_\mathbf{i}} a$, where the first inequality follows from Lemma 4.6, the equality from the fact that $a \leq s_\mathbf{i}$, while the second and third inequalities follow from the facts that $s_\mathbf{i} \wedge a \leq @_{s_\mathbf{i}} a$ and $\Diamond @_{s_\mathbf{i}} a \leq @_{s_\mathbf{i}} a$ and $\Diamond @_{s_\mathbf{i}} b \leq @_{s_\mathbf{i}} b$ in any orthodox interpretation of $\mathcal{H}(A)$.

In the case when $a \leq @^{-1}_{s_{\mathbf{j}_1}} d$, observe that $d \leq @_{s_{\mathbf{j}_1}} a = @_{s_{\mathbf{j}_1}}(s_\mathbf{i} \wedge a) \leq @_{s_{\mathbf{j}_1}} @_{s_\mathbf{i}} a \leq @_{s_\mathbf{i}} a$. Here the first inequality follows by Lemma 4.6 (fixing the nominal coordinate, @ is completely join-preserving, hence a diamond), the equality is again justified by the fact that $a \leq s_\mathbf{i}$, while the second and third inequalities are direct applications of the definition of an orthodox interpretation of $\mathcal{H}(@)$.

Lastly, consider the case when $a \leq (\Diamond^{-1})^{m_1} @^{-1}_{s_{\mathbf{j}_1}} d$. Then, using reasoning similar to the previous two cases, we find that $d \leq @_{s_{\mathbf{j}_1}} \Diamond^{m_1} a \leq @_{s_{\mathbf{j}_1}} \Diamond^{m_1}(s_\mathbf{i} \wedge a) \leq @_{s_{\mathbf{j}_1}} \Diamond^{m_1} @_{s_\mathbf{i}} a \leq @_{s_{\mathbf{j}_1}} @_{s_\mathbf{i}} a \leq @_{s_\mathbf{i}} a$. □

The proof of the following corollary is similar to that of Corollary 4.8:

Corollary 5.11 *For any set* Σ *of* $\mathcal{H}(@)$*-formulas, the logic* $\mathbf{H}(@) \oplus \Sigma$ *is sound and strongly complete with respect to the class of all descriptive two-sorted general frames defined by* Σ.

5.2 Algebraic Completeness of $\mathbf{H}^+(@) \oplus \Sigma$

The completeness result in this section in proven with a construction similar to that used in the previous subsection. As before, we first formulate and prove the main result, and afterwards give the lemmas needed in its proof.

Theorem 5.12 *For any set* Σ *of* $\mathcal{H}(@)$*-formulas, the logic* $\mathbf{H}^+(@) \oplus \Sigma$ *is sound and strongly complete with respect to the class of all permeated hybrid @-algebras which validate* Σ.

Proof Suppose Γ is a local $\mathbf{H}^+(@) \oplus \Sigma$-consistent set of formulas. Let NOM' be a denumerably infinite set of nominals disjoint from NOM. We know from Blackburn et al. (2001, Lemma 7.25) that Γ is contained in a maximal local $\mathbf{H}^+(@) \oplus \Sigma$-consistent set of formulas Γ^+ in the language extended with NOM' such that

(i) Γ^+ contains at least one nominal, say $\mathbf{i}_0$, and
(ii) for each formula of the form $@_\mathbf{i}\Diamond\varphi \in \Gamma^+$, there exists a nominal $\mathbf{j}$ such that $@_\mathbf{i}\Diamond\mathbf{j} \in \Gamma^+$ and $@_\mathbf{j}\varphi \in \Gamma^+$.

Now, consider the Lindenbaum-Tarski algebra of $\mathbf{H}^+(@) \oplus \Sigma$ over PROP. Denote it by $\mathbf{A}$. In the usual way, $\mathbf{A} \models \mathbf{H}^+(@) \oplus \Sigma^{\approx}$, and hence $\mathbf{A}$ is an orthodox interpretation of $\mathcal{H}(@)$. Standard propositional reasoning shows that $[\Gamma^+] = \{[\gamma] \mid \gamma \in \Gamma^+\}$ is an ultrafilter of $\mathbf{A}$, and thus $\bigwedge[\Gamma'] > \bot$ in $\mathbf{A}$ for every finite subset $\Gamma' \subseteq \Gamma^+$.

Next, consider the orthodox canonical extension $\mathbf{A}^\delta$ of $\mathbf{A}$. Then $\bigwedge[\Gamma^+] > \bot$ in $\mathbf{A}^\delta$, for otherwise the compactness of the embedding of $\mathbf{A}$ into $\mathbf{A}^\delta$ would yield a finite subset $\Gamma' \subseteq \Gamma^+$ such that $\bigwedge[\Gamma'] \leq \bot$ in $\mathbf{A}$.

Let d in $At\mathbf{A}^\delta$ such that $d \leq \bigwedge[\Gamma^+]$ and denote it by d_0.

Now, let $d_{n+1} = \Diamond^{-1} d_n \vee \bigvee_{\mathbf{i} \in \mathsf{NOM} \cup \mathsf{NOM}'} @_{s_\mathbf{i}}^{-1} d_n$, and $D = \bigvee_{n \in \mathbb{N}} d_n$. Define

$$\mathbf{A}_D = (A_D, \wedge^D, \vee^D, \neg^D, \bot^D, \top^D, \Diamond^D, @^D, \{s_\mathbf{j}^D\}_{\mathbf{j} \in \mathsf{NOM} \cup \mathsf{NOM}'})$$

such that $A_D = \{a \wedge D \mid a \in A\}$, $\wedge^D$ and $\vee^D$ are the restriction of $\wedge$ and $\vee$ to A_D,

$$\neg^D a = \neg a \wedge D \qquad \Diamond^D a = \Diamond a \wedge D \qquad \top^D = D \qquad \bot^D = \bot \qquad s_\mathbf{j}^D = s_\mathbf{j} \wedge D$$

and, for all $\mathbf{j} \in \mathsf{NOM}$,

$$s_\mathbf{j}^D = s_\mathbf{j} \wedge D \qquad\qquad @^D_{s_\mathbf{j}^D} a = @_{s_\mathbf{j}} a \wedge D.$$

In the same way as in Lemma 5.4, we can then show that $\mathbf{A}_D$ is an algebra, and furthermore, in exactly the same way as in Lemma 5.5, we can show that the map h: $A \to A_D$ defined by $h(a) = a \wedge D$ is a surjective homomorphism from $\mathbf{A}$ onto $\mathbf{A}_D$, and hence, it follows that $\mathbf{A}_D \models \mathbf{H}(@) \oplus \Sigma^{\approx}$ and that $\mathbf{A}_d$ is an orthodox interpretation of $\mathcal{H}(@)$. Consider the assignment ν_D defined by $\nu_D(p) = h(\nu(p))$, where ν is the natural map taking p to $[p]$. Using the fact that h is a homomorphism, we can show by structural induction that $\widetilde{\nu}_D(\psi) = h(\widetilde{\nu}(\psi))$ for all $\mathcal{H}(@)$-formulas ψ that use variables from PROP. Hence, since $d \leq D$ and $d \leq \bigwedge[\Gamma^+]$, we have $\widetilde{\nu}_D(\gamma) = h(\widetilde{\nu}(\gamma)) = \widetilde{\nu}(\gamma) \wedge D = [\gamma] \wedge D \geq \bigwedge[\Gamma^+] \wedge D \geq d > \bot$, for all $\gamma \in \Gamma$.

In the same way as in the proof of Lemma 5.10, we can show that for each $\mathbf{i} \in \mathsf{NOM} \cup \mathsf{NOM}'$, $s_\mathbf{i}^D$ is an atom of $\mathbf{A}_D$. So let $\mathfrak{A}_D = (\mathbf{A}_D^-, X_{A_D})$, where $\mathbf{A}_D^-$ is the reduct of $\mathbf{A}_D$ obtained by omitting the constant interpretations of nominals, and $X_{A_D} = \{s_\mathbf{i}^D \mid \mathbf{i} \in \mathsf{NOM} \cup \mathsf{NOM}'\}$. The set X_{A_D} is thus clearly non-empty. In particular, $s_{\mathbf{i}_0}^D \in X_{A_D}$. Moreover, $s_{\mathbf{i}_0}^D = d$. To see this, recall that $\mathbf{i}_0 \in \Gamma^+$, so $d \leq \bigwedge[\Gamma^+] \leq [\mathbf{i}_0]$. It follows that $d \leq s_{\mathbf{i}_0}^D$, and so, since $s_{\mathbf{i}_0}^D$ and d are both atoms, $s_{\mathbf{i}_0}^D = d$.

Since $\mathbf{H}^+(@) \oplus \Sigma$ is closed under (*Sorted substitution*), it follows from the foregoing that $\mathfrak{A}_D \models \mathbf{H}(@)^+ \oplus \Sigma^{\approx}$. Now, consider the assignment ν_D' which extends ν_D from PROP to $\mathsf{PROP} \cup \mathsf{NOM} \cup \mathsf{NOM}'$, obtained by simply setting $\nu_D'(\mathbf{i}) = s_\mathbf{i}^D$ for each $\mathbf{i} \in \mathsf{NOM} \cup \mathsf{NOM}'$. It is clear that $\widetilde{\nu}_D'(\psi) = \widetilde{\nu}_D(\psi)$ for all $\mathcal{H}(@)$-formulas ψ, and hence that $\widetilde{\nu}_D'(\gamma) = \widetilde{\nu}_D(\gamma) \geq d > \bot$, for all $\gamma \in \Gamma$. Finally, by Lemma 5.15, $\mathfrak{A}_D$ is permeated. □

In what follows Γ, $\mathbf{A}$, $\mathbf{A}^\delta$, $\mathbf{A}_D$, ν, ν_D and $\mathfrak{A}_D$ will be as in the proof of the above theorem. The following lemma is needed to prove that $\mathfrak{A}_D$ is permeated.

The proof of the following lemma is the same as that of Lemma 4.13:

Lemma 5.13 *For all $\mathcal{H}(@)$-formulas γ, it holds that $\gamma \in \Gamma^+$ iff $d \leq \widetilde{\nu}(\gamma)$ iff $d \leq \widetilde{\nu}_D(\gamma)$.*

Lemma 5.14 *Let a be an element of $\mathbf{A}_D$. Then, for all $n \in \mathbb{N}$, $d \leq @^D_{s^D_{\mathbf{i}}}(\Diamond^D)^n a$ implies there is a nominal $\mathbf{j} \in \mathsf{NOM} \cup \mathsf{NOM}'$ such that $s^D_{\mathbf{j}} \leq a$.*

Proof The proof is by induction on n. For $n = 0$, suppose that $d \leq @^D_{s^D_{\mathbf{i}}}a$. This implies that $s^D_{\mathbf{i}} \leq a$, for if not, $d \leq @^D_{s^D_{\mathbf{i}}}a = \bot$, which is a contradiction. Hence, $\mathbf{i}$ can be taken as the desired nominal. Now, suppose that for every $a \in A_D$, the claim holds for $n = k$. For $n = k + 1$, assume $d \leq @^D_{s^D_{\mathbf{i}}}(\Diamond^D)^{k+1}a = @^D_{s^D_{\mathbf{i}}}(\Diamond^D)^k\Diamond^D a$. By the inductive hypothesis, we have a nominal $\mathbf{k} \in \mathsf{NOM} \cup \mathsf{NOM}'$ such that $s^D_{\mathbf{k}} \leq \Diamond^D a$. This means that that $D = @^D_{s^D_{\mathbf{k}}}s^D_{\mathbf{k}} \leq @^D_{s^D_{\mathbf{k}}}\Diamond^D a$, and so $d \leq @^D_{s^D_{\mathbf{k}}}\Diamond^D a$. Now, since both ν and h are surjective, there is a ψ such that $d \leq @^D_{s^D_{\mathbf{k}}}\Diamond^D\widetilde{\nu}_D(\psi) = \widetilde{\nu}_D(@_{\mathbf{k}}\Diamond\psi)$. In the same way as in Lemma 4.13, we can show that for every $\mathcal{H}(@)$-formula γ, $\gamma \in \Gamma^+$ iff $d \leq \widetilde{\nu}_D(\gamma)$. But then $@_{\mathbf{k}}\Diamond\psi \in \Gamma^+$, which means there is a nominal $\mathbf{j} \in \mathsf{NOM} \cup \mathsf{NOM}'$ such that $@_{\mathbf{k}}\Diamond\mathbf{j} \in \Gamma^+$ and $@_{\mathbf{j}}\psi \in \Gamma^+$. Hence, $d \leq \widetilde{\nu}_D(@_{\mathbf{j}}\psi) = @^D_{s^D_{\mathbf{j}}}\widetilde{\nu}_D(\psi) = @^D_{s^D_{\mathbf{j}}}a$, and so $D = @^D_{s^D_{\mathbf{i}_0}}d \leq @^D_{s^D_{\mathbf{i}_0}}@^D_{s^D_{\mathbf{j}}}a \leq @^D_{s^D_{\mathbf{j}}}a$, where the equality follows from the fact that $s^D_{\mathbf{i}_0} = d$ and the last inequality from the axiom (*Agree*). Therefore, $@^D_{s^D_{\mathbf{j}}}a = D$, which implies that $s^D_{\mathbf{j}} \leq a$. □

Lemma 5.15 *$\mathfrak{A}_D$ is permeated.*

Proof To prove the first condition, let $b \in \mathbf{A}_D$ such that $b > \bot$. Then we have $b \wedge \bigvee_{n\in\mathbb{N}} d_n \vee \bigvee_{\mathbf{i}\in\mathsf{NOM}\cup\mathsf{NOM}'} @^{-1}_{\mathbf{i}}d_n > \bot$, which means we have the following cases by Lemmas 5.8 and 5.9:

Case 1: $b \wedge (\Diamond^{-1})^{n_0}d > \bot$, for some $n_0 \in \mathbb{N}$. We thus know from Lemma 4.10 that $d \leq \Diamond^{n_0}b$, which implies $d \leq (\Diamond^D)^{n_0}b$ (the proof of this is similar to that of Lemma 4.12). Now, recall that $d \leq \bigwedge[\Gamma^+] \leq [\mathbf{i}_0]$ and $d \leq D$, so $d \leq [\mathbf{i}_0] \wedge D = s^D_{\mathbf{i}_0}$ and hence, as both d and $s^D_{\mathbf{i}_0}$ are atoms, $d = s^D_{\mathbf{i}_0}$. Hence, $d = s^D_{\mathbf{i}_0} \wedge (\Diamond^D)^{n_0}b$, which means that $d \leq @^D_{s_{\mathbf{i}_0}}(\Diamond^D)^{n_0}b$ by (*introduction*). It thus follows from Lemma 5.14 that there is a nominal $\mathbf{j} \in \mathsf{NOM} \cup \mathsf{NOM}'$ such that $s^D_{\mathbf{j}} \in X_{A_D}$ and $s^D_{\mathbf{j}} \leq b$.

Case 2: $b \wedge @^{-1}_{s_{\mathbf{j}}}d > \bot$, for some $\mathbf{j} \in \mathsf{NOM} \cup \mathsf{NOM}'$. Then $d \leq @_{s_{\mathbf{j}}}b$ by Lemma 4.10, and so, applying Lemma 5.14, we obtain a nominal $\mathbf{k} \in \mathsf{NOM} \cup \mathsf{NOM}'$ such that $s^D_{\mathbf{k}} \in X_{A_D}$ and $s^D_{\mathbf{k}} \leq b$.

Case 3: $b \wedge (\Diamond^{-1})^{n_0}@^{-1}_{s_{\mathbf{j}}}d > \bot$, for some $n_0 \in \mathbb{N}$ and some $\mathbf{j} \in \mathsf{NOM} \cup \mathsf{NOM}'$. As in the previous cases $d \leq @_{s_{\mathbf{j}}}\Diamond^{n_0}b$ by Lemma 4.10. Hence, by Lemma 5.14, there is a nominal $\mathbf{k} \in \mathsf{NOM} \cup \mathsf{NOM}'$ such that $s^D_{\mathbf{k}} \in X_{A_D}$ and $s^D_{\mathbf{k}} \leq b$.

For the second condition, let $s^D_{\mathbf{i}} \in X_{A_D}$ and $b \in A_D$ such that $s^D_{\mathbf{i}} \leq \Diamond^D b$. Then we have $D = @^D_{s^D_{\mathbf{i}}}\Diamond^D b$. By the surjectivity of ν and h there is a formula ψ such that $b = \widetilde{\nu}_D(\psi)$, and hence $d \leq D = @^D_{s^D_{\mathbf{i}}}\Diamond^D b = @^D_{s^D_{\mathbf{i}}}\Diamond^D\widetilde{\nu}_D(\psi) = \widetilde{\nu}_D(@_{\mathbf{i}}\Diamond\psi)$. This means that $@_{\mathbf{i}}\Diamond\psi \in \Gamma^+$ (see Lemma 5.13), and therefore there is a nominal $\mathbf{j} \in \mathsf{NOM} \cup \mathsf{NOM}'$ such that $@_{\mathbf{i}}\Diamond\mathbf{j} \in \Gamma^+$ and $@_{\mathbf{j}}\psi \in \Gamma^+$. Hence, by Lemma 5.13, $d \leq \widetilde{\nu}_D(@_{\mathbf{i}}\Diamond\mathbf{j}) = @^D_{s^D_{\mathbf{i}}}\Diamond^D s^D_{\mathbf{j}}$ and $d \leq \widetilde{\nu}_D(@_{\mathbf{j}}\psi) = @^D_{s^D_{\mathbf{j}}}b$. It follows from Proposition 3.7 that $s^D_{\mathbf{j}} \leq b$ and $s^D_{\mathbf{i}} \leq \Diamond^D s^D_{\mathbf{j}}$, as desired. □

Corollary 5.16 *For any set* Σ *of* $\mathcal{H}(@)$*-formulas, the logic* $\mathbf{H}^{+}(@) \oplus \Sigma$ *is sound and strongly complete with respect to the class of all strongly descriptive two-sorted general frames defined by* Σ.

6 Global Completeness

We now turn our attention to the question of algebraic completeness for the global consequence relations associated with hybrid logics.

Definition 6.1 Let C be a class of hybrid (@-)algebras. A formula φ is a *global consequence over* C of a set of formulas Γ (notation: $\Gamma \models_{\mathsf{C}} \varphi$) if for each hybrid (@-)algebra $\mathfrak{A} \in \mathsf{C}$ and assignment v on $\mathfrak{A}$, if $\widetilde{v}(\gamma) = 1$ for all $\gamma \in \Gamma$, then $\widetilde{v}(\varphi) = 1$. A hybrid logic Λ is *strongly globally complete* with respect to a class C of hybrid (@-)algebras if $\Gamma \models_{\mathsf{C}} \varphi$ implies $\Gamma \vdash_{\Lambda} \varphi$ for all Γ and φ. Weak global completeness is the case when the set $\Gamma = \varnothing$.

We note that $\mathbf{H}$ is not globally complete with respect to the class of descriptive two-sorted general frames. Indeed, observe that $\mathbf{j}$ is a global consequence of $\mathbf{i}$ over such frames, as $\mathbf{i}$ is globally satisfiable only on one-point descriptive two-sorted general frames.

Now, consider the model $\mathfrak{M} = (W, R, \{S_{\mathsf{i}}\}_{\mathsf{i} \in \mathsf{NOM}}, V)$, where $W = \{u, v\}$, $R = \varnothing$, $S_{\mathsf{i}} = W$ and $S_{\mathsf{j}} = \{v\}$ for all nominals $\mathbf{j} \neq \mathbf{i}$. This model is based on a frame for $\mathbf{H}$ if we interpret the nominals as modal constants. Indeed, in such a frame, all sorted-substitution instances of (*Nom*) are valid if, in each point-generated subframe, the interpretation of every nominal is at most a singleton—this is the case here. Moreover, on this model $\mathfrak{M} \Vdash \mathbf{i}$ but $\mathfrak{M} \nVdash \mathbf{j}$. Since global derivability in $\mathbf{H}$ is sound with respect to the class of all such models,[3] it follows that $\mathbf{i} \nvdash^{g}_{\mathbf{H}} \mathbf{j}$. The same holds for $\mathbf{H}^{+}$.

In contrast to these negative results for $\mathbf{H}$ and $\mathbf{H}^{+}$, we are able to prove the following general global completeness results for axiomatic extensions of $\mathbf{H}(@)$ and $\mathbf{H}^{+}(@)$.

Theorem 6.2 *For any set* Σ *of* $\mathcal{H}(@)$*-formulas, the logic* $\mathbf{H}(@) \oplus \Sigma$ *is sound and globally strongly complete with respect to the class of all hybrid @-algebras which validate* Σ.

Proof As soundness is straightforward, we will only prove strong completeness. Suppose $\Gamma \nvdash^{g}_{\Gamma} \varphi$. Define a sequence $\Gamma_0, \Gamma_1, \ldots$ of sets of formulas, beginning with $\Gamma_0 = \Gamma$. Suppose Γ_n is already defined, then let $\Gamma_{n+1} = \{\Box\gamma \mid \gamma \in \Gamma_n\} \cup \{@_{\mathbf{j}}\gamma \mid \mathbf{j} \in \mathsf{NOM}, \gamma \in \Gamma_n\}$. The set $\Phi = \{\neg\varphi\} \cup \bigcup_{n \geq 0} \Gamma_n$ is locally consistent, for otherwise

[3] I.e. models based on frames where nominals are treated as modal constants interpreted as non-empty sets which have at most singleton intersections with each point-generated submodel. It is easy to check that in such models the $\mathbf{H}$ axioms are valid. For the sake of the rules, it is enough to note that modus ponens and necessitation preserve global truth on models; that (every sorted-substitution instance of) the premise of (*NameLite*) is false; and that sorted substitution only applies to axioms.

we would have $\Gamma \vdash^g_\Gamma \varphi$. Extend Φ to a maximal (locally) consistent set Φ^+. Now proceed exactly as in the proof of Theorem 5.1, substituting Γ^+ with Φ^+, to define elements d and D and to construct the algebra $\mathbf{A}_D$ and assignment ν_D on $\mathbf{A}_D$. Now $d \leq D$ and $d \leq \bigwedge[\Phi^+]$, we have $d \leq \bigwedge[\Phi^+] \wedge D \leq [Q\gamma] \wedge D = Q[\gamma] \wedge D \leq Q[\gamma] \wedge QD = Q([\gamma] \wedge D) = Q\widetilde{\nu}_D(\gamma)$, for all strings Q of $\Box$'s and $@_{s_j}$'s, and all $\gamma \in \Gamma$. Here the third inequality follows from Lemmas 5.2 and 5.3. But we know $\Diamond^{-1} \dashv \Box$ and that $@_{s_j}^{-1} \dashv @_{s_j}$, so $Q'd \leq \widetilde{\nu}_D(\gamma)$ for all strings Q' of $\Diamond^{-1}$'s and $@_{s_j}^{-1}$'s, and all $\gamma \in \Gamma$. This implies that $\bigvee_{n \in \mathbb{N}} d_n \leq \widetilde{\nu}_D(\gamma)$ for all $\gamma \in \Gamma$. Therefore, by the definition of D, $D \leq \widetilde{\nu}_D(\gamma)$, i.e., $\widetilde{\nu}_D(\gamma) = \top_{\mathbf{A}_D}$ for all $\gamma \in \Gamma$. Also note that $\widetilde{\nu}_D(\neg\varphi) \geq d > \bot$. We now again proceed further exactly as in the proof of Theorem 5.1 to turn $\mathbf{A}_D$ into a hybrid @-algebra $\mathfrak{A}_D$ and ν_D into an assignment ν'_D on $\mathfrak{A}_D$ such that $\widetilde{\nu}'_D(\gamma) = \top_{\mathbf{A}_D}$ for all $\gamma \in \Gamma$ and $\widetilde{\nu}'_D(\neg\varphi) > \bot$. $\Box$

By the duality outlined in Sect. 3.3 we have this corollary:

Corollary 6.3 *For any set Σ of $\mathcal{H}(@)$-formulas, the logic $\mathbf{H}(@) \oplus \Sigma$ is sound and globally strongly complete with respect to the class of all descriptive two-sorted general frames defined by Σ.*

For logics enriched with the non-orthodox rules we also have global completeness with respect to classes of permeated hybrid @-algebras. The proof is similar that of Theorem 6.2, except that the set Φ must be extended to a maximal consistent set Φ^+ also satisfying conditions (i) and (ii) in the proof of Theorem 5.12, so that we can appeal to Lemma 5.15 to prove that the constructed hybrid @-algebra is permeated.

Theorem 6.4 *For any set Σ of $\mathcal{H}(@)$-formulas, the logic $\mathbf{H}^+(@) \oplus \Sigma$ is sound and globally strongly complete with respect to the class of all permeated hybrid @-algebras which validate Σ.*

Corollary 6.5 *For any set Σ of $\mathcal{H}(@)$-formulas, the logic $\mathbf{H}^+(@) \oplus \Sigma$ is sound and globally strongly complete with respect to the class of all strongly descriptive two-sorted general frames defined by Σ.*

7 Conclusion

We introduced the notions of hybrid algebras and hybrid @-algebras together with their permeated subclasses. We considered some basic truth preserving operations for these structures and explored their duality with two-sorted general frames. We proved general, strong, local and global completeness results for axiomatic extensions of the basic hybrid logics with respect to classes of hybrid algebras.

The availability of this semantics, so close to the familiar algebraic semantics for modal logic, has proven extremely useful to the authors in their study of hybrid logic. It has enabled us to adapt existing algebraic techniques for modal logic to obtain new result for hybrid logics, including Sahlqvist-type theorems (Conradie and Robinson 2017) and finite model properties (Conradie and Robinson 2019; Robinson 2015).

In this paper we have considered only two hybrid languages, and have not mentioned languages containing e.g. the universal modality of the 'down-arrow' binder. In is in fact easy to extend the results obtained to languages with the universal modality, using constructions and ideas similar to those we have employed here. This is worked out in full detail in Robinson (2015). As already mentioned, the 'down-arrow' binder seems to require a very different treatment (see Litak 2006).

Having established an algebraic semantics for a logic, one naturally wants to know how strong the connection between the logic and the algebras is. In particular, is the logic algebraized by (classes of) the algebras, in the sense of Blok and Pigozzi (1989)? Answering this for hybrid logics and hybrid algebras calls for a proper notion hybrid algebraizability. Initial explorations in this direction indicate that results on hybrid algebraizability might mirror those for modal logic, at least as far as the basic logics are concerned. We will pursue this further in future work.

We have considered completeness of standard Hilbert-style systems, but the literature also contains natural deduction, Gentzen and tableaux systems for hybrid logic (see e.g. Seligman 1997; Blackburn 2000a; Bolander and Braüner 2006; Braüner 2010). The completeness of the latter systems with respect to hybrid algebras remains to be investigated.

References

Areces, C., and B. ten Cate. 2006. Hybrid logics. In *Handbook of modal logics*, ed. P. Blackburn, J. van Benthem, and F. Wolter, pp. 821–868. Elsevier.

Bolander, T., and T. Braüner. 2006. Tableau-based decision procedures for hybrid logic. *Journal of Logic and Computation* 16 (6): 737–763.

Blackburn, P., M. de Rijke, and Y. Venema. 2001. Modal logic, volume 53 of *Cambridge tracts in theoretical computer science*. Cambridge University Press.

Blackburn, P. 2000a. Internalizing labelled deduction. *Journal of Logic and Computation* 10 (1): 137–168.

Blackburn, P. 2000b. Representation, reasoning, and relational structures: a hybrid logic manifesto. *Logic Journal of the IGPL* 8 (3): 339–625.

Blok, W.J., and D. Pigozzi. 1989. *Algebraizable logics*, No. 396. American Mathematical Soc.

Braüner, T. 2010. *Hybrid logic and its proof-theory*, vol. 37. Springer.

Blackburn, P., and M. Tzakova. 1999. Hybrid languages and temporal logic. *Logic Journal of IGPL* 7 (1): 27–54.

Conradie, W., and C. Robinson. 2017. On Sahlqvist theory for hybrid logics. *Journal of Logic and Computation* 27 (3): 867–900.

Conradie, W., and C. Robinson. 2019. Analogues of Bull's theorem for hybrid logic. *Logic Journal of the IGPL* 27 (3): 281–313.

Davey, B., and H. Priestley. 2002. *Introduction to lattices and order*. Cambridge University Press.

Gargov, G., and V. Goranko. 1993. Modal logic with names. *Journal of Philosophical Logic* 22: 607–636.

Goldblatt, R.I. 1976. Metamathematics of modal logic I. *Reports on Mathematical Logic* 6: 41–78.

Goldblatt, R. 1989. Varieties of complex algebras. *Annals of Pure and Applied Logic* 44 (3): 173–242.

Gargov, G., S. Passy, and T. Tinchev. 1987. Modal environment for Boolean speculations. In *Mathematical logic and its applications*, pp. 253–263. Springer.

Litak, T. 2006. Algebraization of hybrid logic with binders. In *Relations and Kleene algebra in computer science*, vol. 4136, ed. Renate A. Schmidt, 281–295., Lecture notes in computer science Berlin, Heidelberg: Springer.

Prior, A.N. 1957a. *Past, present and future*. Oxford University Press.

Prior, A.N. 1957b. *Time and modality*. Oxford University Press.

Robinson, C. 2015. *Algebraic methods for hybrid logics*. PhD thesis, University of Johannesburg. https://ujdigispace.uj.ac.za/handle/10210/13869.

Seligman, J. 1997. The logic of correct description. In *Advances in intensional logic*, pp. 107–135. Springer.

ten Cate, B. 2005. *Model theory for extended modal languages*. PhD thesis, Institute for Logic, Language and Computation, Universiteit van Amsterdam.

Venema, Y. 2006. Algebras and coalgebras. In *Handbook of modal logic*, ed. P. Blackburn, J. van Benthem, and F. Wolter, pp. 331–426. Elsevier.

The manufacturer's authorised representative in the EU is Springer Nature Customer Service Centre GmbH, Europaplatz 3, 69115 Heidelberg, Germany. If you have any concerns regarding our products, please contact ProductSafety@springernature.com

Printed and bound by CPI Group (UK) Ltd, Croydon, CR0 4YY
25/09/2026
02222898-0020